Enterprise Systems Engineering

Advances in the Theory and Practice

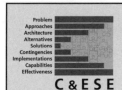

COMPLEX AND ENTERPRISE SYSTEMS ENGINEERING

Series Editors: Paul R. Garvey and Brian E. White
The MITRE Corporation
www.enterprise-systems-engineering.com

Architecture and Principles of Systems Engineering
Charles Dickerson and Dimitri N. Mavris
ISBN: 978-1-4200-7253-2

Designing Complex Systems: Foundations of Design in the Functional Domain
Erik W. Aslaksen
ISBN: 978-1-4200-8753-6

Enterprise Systems Engineering: Advances in the Theory and Practice
George Rebovich, Jr. and Brian E. White
ISBN: 978-1-4200-7329-4

Model-Oriented Systems Engineering Science: A Unifying Framework for Traditional and Complex Systems
Duane W. Hybertson
ISBN: 978-1-4200-7251-8

FORTHCOMING

Complex Enterprise Systems Engineering for Operational Excellence
Kenneth C. Hoffman and Kirkor Bozdogan
ISBN: 978-1-4200-8256-2
Publication Date: November 2010

Engineering Mega-Systems: The Challenge of Systems Engineering in the Information Age
Renee Stevens
ISBN: 978-1-4200-7666-0
Publication Date: June 2010

Leadership in Decentralized Organizations
Beverly G. McCarter and Brian E. White
ISBN: 978-1-4200-7417-8
Publication Date: October 2010

Systems Engineering Economics
Ricardo Valerdi
ISBN: 978-1-4398-2577-8
Publication Date: December 2011

RELATED BOOKS

Analytical Methods for Risk Management: A Systems Engineering Perspective
Paul R. Garvey
ISBN: 978-1-58488-637-2

Probability Methods for Cost Uncertainty Analysis: A Systems Engineering Perspective
Paul R. Garvey
ISBN: 978-0-8247-8966-4

AUERBACH PUBLICATIONS

www.auerbach-publications.com
To Order Call: 1-800-272-7737 • Fax: 1-800-374-3401
E-mail: orders@crcpress.com

Enterprise Systems Engineering

Advances in the Theory and Practice

Edited by
George Rebovich, Jr.
Brian E. White

CRC Press is an imprint of the
Taylor & Francis Group, an **informa** business
AN AUERBACH BOOK

Complex and Enterprise Systems Engineering Series

CRC Press
Taylor & Francis Group
6000 Broken Sound Parkway NW, Suite 300
Boca Raton, FL 33487-2742

© 2011 by Taylor and Francis Group, LLC
CRC Press is an imprint of Taylor & Francis Group, an Informa business

No claim to original U.S. Government works

Printed in the United States of America on acid-free paper
10 9 8 7 6 5 4 3 2 1

International Standard Book Number: 978-1-4200-7329-4 (Hardback)

This book contains information obtained from authentic and highly regarded sources. Reasonable efforts have been made to publish reliable data and information, but the author and publisher cannot assume responsibility for the validity of all materials or the consequences of their use. The authors and publishers have attempted to trace the copyright holders of all material reproduced in this publication and apologize to copyright holders if permission to publish in this form has not been obtained. If any copyright material has not been acknowledged please write and let us know so we may rectify in any future reprint.

Except as permitted under U.S. Copyright Law, no part of this book may be reprinted, reproduced, transmitted, or utilized in any form by any electronic, mechanical, or other means, now known or hereafter invented, including photocopying, microfilming, and recording, or in any information storage or retrieval system, without written permission from the publishers.

For permission to photocopy or use material electronically from this work, please access www.copyright.com (http://www.copyright.com/) or contact the Copyright Clearance Center, Inc. (CCC), 222 Rosewood Drive, Danvers, MA 01923, 978-750-8400. CCC is a not-for-profit organization that provides licenses and registration for a variety of users. For organizations that have been granted a photocopy license by the CCC, a separate system of payment has been arranged.

Trademark Notice: Product or corporate names may be trademarks or registered trademarks, and are used only for identification and explanation without intent to infringe.

Library of Congress Cataloging-in-Publication Data

Enterprise systems engineering : advances in the theory and practice / editors, George Rebovich, Jr. and Brian E. White.
    p. cm. -- (Complex and enterprise systems engineering)
"A CRC title."
Includes bibliographical references and index.
ISBN 978-1-4200-7329-4 (alk. paper)
    1. Systems engineering. 2. Support services (Management) I. Rebovich, George. II. White, Brian E. III. Title. IV. Series.

TA168.E58 2010
658.4'038011--dc22
                                                                                        2010005490

Visit the Taylor & Francis Web site at
http://www.taylorandfrancis.com

and the CRC Press Web site at
http://www.crcpress.com

Contents

Foreword vii
Preface xi
Acknowledgments xiii
Editors xv
Contributors xvii

1   Introduction 1
    JOSEPH K. DEROSA

2   Systems Thinking for the Enterprise 31
    GEORGE REBOVICH JR.

3   Pilots and Case Studies 63
    KIMBERLY A. CRIDER

4   Capabilities-Based Engineering Analysis 95
    STEVEN J. ANDERSON and MICHAEL J. WEBB

5   Enterprise Opportunity and Risk 161
    BRIAN E. WHITE

6   Architectures for Enterprise Systems Engineering 181
    CARLOS TROCHE, JOSEPH K. DEROSA, J. HOWARD BIGELOW JR.,
    TERENCE J. BLEVINS, JULIO C. FONSECA, MICHAEL J. WEBB,
    MICHAEL R. COIRIN, HERMAN L. KARHOFF, JAY M. VITTORI,
    BHUPENDER (SAM) SINGH, RICHARD W. FOX, JEFFERY S. COOK,
    CINDY A. PLAINTE, MARK D. BAEHRE, and MICHAEL R. McFARREN

7   Enterprise Analysis and Assessment 205
    JOHN J. ROBERTS

8   Enterprise Management 231
    ROBERT S. SWARZ

9 Agile Functionality for Decision Superiority 293
KEVIN A. CABANA, LINDSLEY G. BOINEY, LEWIS A. LOREN,
CHRISTOPHER D. BERUBE, ROBERT J. LESCH, LINSEY B. O'BRIEN,
CRAIG A. BONACETO, and HARCHARANJIT SINGH

10 Enterprise Activities: Evolving toward an Enterprise........................395
STEPHEN C. ELGASS, L. SHAYN HAWTHORNE, CHRIS A.
KAPRIELIAN, PAUL S. KIM, ANDREW K. MILLER, LAURA R. RICCI,
MAURA A. SLATTERY, J. REID SLAUGHTER, PETER A. SMYTON,
and RICHARD E. STUEBE

Index 439

Foreword

In September 2004, the Center for Air Force Command and Control Systems of The MITRE Corporation commissioned a Focus Group to study what we call Enterprise Systems Engineering (ESE). It was becoming increasingly clear—and of crisis proportions—that the traditional systems engineering processes and tools were inadequate to face the scale and complexity of the systems our government clients needed. In our role to manage not-for-profit federally funded research and development centers chartered by the federal government to act in the public interest, we considered this a critical problem that need to be addressed. A large amount of money was being spent in the development of these complex systems, but they seemed to be habitually late-to-market and exceeded the budget. Worse still, many were obsolete before they were deployed or the program was cancelled before delivering anything at all. Clearly, the public interest was not adequately met. Many concerted attempts by the government, by contractors, and by us to better implement systems engineering as we knew it seemed not to improve the situation. Undoubtedly, a new approach was needed. I was fortunate enough to cochair that Focus Group. This activity led many of us, especially the authors of this book, on a fascinating journey into a new and exciting area of study that would significantly expand the scope and understanding of systems engineering.

We came to realize that something had changed in the way people worked together, both within the general population and the systems engineering discipline. Seldom did isolated groups work on local problems to build stove-piped solutions. Seldom were systems developed or used in a social, political, economic, or technical vacuum. It seemed that everyone and everything were interconnected and interdependent. Whether it was caused by the cultural shift brought on by the Internet and World Wide Web or a genuine increase in the complexity of the missions, something was qualitatively different.

Changes in the sociocultural and technical landscape had repercussions on the practice of systems engineering. Requirements changed faster than they could be met; risks shifted in a never-ending dynamic of actions and reactions within the network of stakeholders; configurations changed with rapidly changing technology cycles; and integrated testing was faced with trying to match the scale and complexity of the operational enterprise. Processes that managed requirements, risks, configurations, and tests, four of the "Horsemen" of traditional systems engineering, were confounded by their complex environment. Our journey started with a keen awareness of the problem, but with few solutions.

The Focus Group was not the only group within MITRE to begin this journey, nor was MITRE the only organization to recognize the need. Even before the Focus Group was launched, some seminal work was done by Doug Norman of MITRE in collaboration with Yaneer Bar Yam of the New England Complex Systems Institute (NECSI). In March 2004, the Massachusetts Institute of Technology's (MIT's) Engineering Systems Division hosted a "call to arms" symposium for engineering leaders to focus on complex systems engineering. In November 2004, we launched an MIT–MITRE research collaboration. Complex systems engineering programs began or were on the drawing boards at institutions such as the University of California at San Diego, Johns Hopkins University, Stevens Institute of Technology, University of Vermont, and the Software Engineering Institute at Carnegie Mellon University. The Defense Science Technology Organization within the Australian government began a vigorous research program in complex systems and eventually applied many of the Bar Yam–Norman techniques to counter-insurgency operations in Afghanistan. A new era in systems engineering had been born.

The Focus Group's first step was to define terms. What is an *enterprise*? What are its *boundaries*? What are the *scales* of the enterprise and the corresponding systems engineering? How do you define *requirements*? These definitions turned out to be more difficult than we had imagined. However one drew the boundary of the enterprise, there were influences coming in from outside that boundary. Systems engineering had to be done coherently at the multiple and seemingly incommensurate levels of the enterprise: the individual systems, groups of systems performing a cooperative mission, and the enterprise as a whole. Requirements changed routinely. We established a baseline taxonomy to begin the work: the enterprise boundary was defined to include all those elements we could either control or influence; we partitioned the systems engineering into three engineering tiers—individual systems, system-of-systems, and enterprise; and the requirements were defined in terms of enterprise outcome spaces without any obvious allocation down to systems or subsystems. There was no established theory to guide our choices.

Next we turned to the practice. After all, these complex systems had been extensively studied for a number of years, and experienced systems engineers were coping with the complexity in their own ways. We consulted our practicing systems engineers who were involved in these complex systems. This led to the following five methods that seemed to have some success in dealing with complex systems and turned them into what we called Enterprise Systems Engineering (ESE) processes.

1. Capabilities-based engineering analysis that focused on the high-level requirements the whole enterprise must satisfy
2. Enterprise architecture that brought a coherent view of the whole enterprise

3. Enterprise analysis and assessment that evaluated the effectiveness of the enterprise rather than the ability of system components to meet specifications
4. Strategic technical planning that laid out a minimum set of standards or patterns that every system in the enterprise would follow
5. Technology planning that looked for cues in the environment that would indicate the emerging technologies that could be implemented next into the enterprise

We fitted these new processes together with the existing traditional systems engineering processes into a new framework. We considered the traditional processes in the usual way from requirements to manufacturing, integrated testing, and deployment. We embedded that as an inner loop within an outer loop of inputs and constraints supplied by the ESE processes. Thus for a single program, the requirements, risks, and so on were fixed through the development cycle, unless and until something in the ESE processes caused them to change. When that happened, the change was made and again the individual system development proceeded as if it were fixed. This seemed to match with reality—our development tools assumed that things were known and fixed, and our experience indicated that things changed abruptly from the outside a number of times throughout the life of a program. In the second year of the Focus Group, we launched several pilot programs to test the effectiveness of those ESE processes. Both the processes and the pilots are reported in the chapters of this book.

In spite of some success in creating an enterprise approach where there had previously been none, there was something missing. We had no theory to explain our approaches and we had no approaches leveraged from a theory. Without a theory, the intellectual content was not there and, more importantly, no one would be eager to adopt any new techniques without some semblance of a firm theoretical foundation. The next step in the process closed the theory-and-practice loop, and led us to believe that we had indeed begun a new phase of systems engineering.

There were two main sources of theoretical underpinnings for this new brand of ESE. The first was in the realm of complex systems. In January 2004, MITRE Technical Report on Complex Systems Engineering published by Norman and Kuras reflected the groundbreaking work that Norman had carried out with NECSI. It recognized the correspondence between information-intensive networks in man-made enterprises and complex ecosystems in the natural world. It recommended an approach called the systems engineering "regimen" based on evolutionary theory and complexity science. More importantly, it opened a floodgate of theoretical knowledge from complexity science, as exemplified by the authors such as Stuart Kauffman, John Holland, Robert Axelrod, Stephen Wolfram, Duncan Watts, and a number of researchers from the Santa Fe Institute. Complexity addressed networks of interrelated nodes just like those of our clients. The networks were generally open systems with porous boundaries, much like those we encountered in our information systems. The natural systems were dynamic and unpredictable in a way similar to what we were experiencing and trying desperately to model. For the first time, we had a mathematical basis for ESE. We had found part of the body of knowledge that was needed to engineer the enterprise.

The second source of theoretical knowledge came from the management literature exemplified by authors such as Russell Ackoff, Jamshid Gharajedaghi, Peter Senge, John Sterman, and Herbert Simon, to name a few. They focused on the complex interactions between people in purposeful endeavors. Their systems were learning and adaptive, and their organizations displayed ordered and explainable social structure. The social and business environments they studied were identical to the socioeconomic environments with which ESE had to cope.

Armed with this new knowledge and the benefit of several ongoing pilots in ESE, those involved in the Focus Group initiative wrote dozens of papers and made numerous presentations at symposia sponsored by professional societies such as the Institute of Electrical and Electronics Engineers, the International Council on Systems Engineering (INCOSE), and the International Symposium on Complex Systems. The "Regimen" paper was eventually published by Norman and Kuras as a chapter in a book edited by Bar Yam and his team from NECSI, and another version was presented by Kuras and White at the INCOSE 2005 Symposium. During this period, the Systems Engineering Division of the Center for Air Force Command and Control conducted a set of monthly seminars to share the information internally and took to writing a number of interrelated reports on the various aspects of ESE. This book is the result of that effort.

Thus this book is as much about history as it is about systems engineering. It is a snapshot of the early thinking in ESE—a snapshot taken before the memory of its perspective corrupted by time, as surely it will be. It was taken at one place—the MITRE Corporation, where systems engineering is the essence of its culture. The editors have compiled it not as a pedagogical exposition of ESE, but more as an anthology of related works focused on the burgeoning field. The definitive story of how to systems engineer an enterprise is not yet written. Many across the world will eventually write on it. Therefore this book is also about the future. It points an arrow in the direction of systems engineering evolution. We all stand on the threshold of a new era in engineering—one that is as much a part of social change as it is about technological change.

Joseph K. DeRosa, PhD
Director of Systems Engineering
Advanced Systems Engineering Center
The MITRE Corporation
Bedford, Massachusetts

Preface

This book is intended to guide systems engineering practitioners who are expanding their horizons from the engineering of systems to the engineering of enterprises. The focus is on engineering the enterprise within a public sector setting. The chapters range from the complex characteristics and behaviors of enterprises to the challenges they pose for engineering and technology. They examine the impacts of enterprise processes and leading-edge technologies on the evolution of an enterprise. The book should be of interest to systems engineers, systems architects, and engineering managers working in the defense, aerospace, telecommunications, and IT sectors.

George Rebovich, Jr.
Dr. Brian E. White
The MITRE Corporation
Bedford, Massachusetts

Acknowledgments

The editors thank Dr. Joseph K. DeRosa, former Director of The MITRE Corporation's Air Force Center's Systems Engineering Directorate, for directing the effort to "write the book" on enterprise systems engineering (ESE).

Chapter 4: The authors of this chapter have drawn extensively from their MITRE colleagues for inputs, comments, and reviews; special thanks to prominent contributors: Howard Bigelow, Michael Bloom, Kevin Cabana, Michael Coirin, Joe DeRosa, Joe Duquette, Julio Fonseca, Marie Francesca, Elaine Goyette, Elizabeth Harding-Laramee, Chuck Marley, Keith McCaughin, John Roberts, Tony Romano, Peter Smyton, Robert Swarz, Shelby Sullivan, Carlos Troche, Jay Vittori, Brian White, and Don Willette.

Chapter 5: The author thanks Prof. Oli de Weck of MIT and an anonymous INCOSE reviewer for their thorough and helpful reviews of the draft paper submitted to, and later revised and published in, the *2006 INCOSE Symposium* (White, 2006). Several MITRE colleagues provided useful suggestions to improve the content and presentation of this material. Special thanks to the principal reviewer, Paul Garvey. Other major reviewers included John Brtis, Mike Kuras, and Michelle Steinbach. Further inputs were received from Norm Bram, Anne Cady, Joe DeRosa, Pam Engert, Dan Lambert, Michael Moore, and John Roberts. Margaret Hill and Natalie Doyle-Hennin provided expert editing support.

Chapter 6: The chapter could not have been completed without the earnest contributions from the other contributors as well as all the previous ones for their inspiration, suggestions, and conviction. A special acknowledgment to Doug Maldonado and Kathleen McDermott—the former for his editorial suggestions and the latter for making this chapter a concrete reality.

Reference

White, B. E. "Enterprise Opportunity and Risk." *INCOSE Symposium*. Orlando, FL, 9–13 July, 2006.

Editors

George Rebovich, Jr. is director of the Systems Engineering Practice Office (SEPO) at The MITRE Corporation. He holds a corporate position and reports to MITRE's corporate chief engineer. He has held a variety of systems engineering positions at MITRE including leading technical departments and groups, directing sponsor projects and technology projects, and serving as the chief system engineer for a C4I System program. Rebovich holds a BS from the U.S. Military Academy, an MS (in mathematics) from Rensselaer Polytechnic Institute, a Certificate of Administration and Management from Harvard University, and is a graduate of the U.S. Army Command and General Staff College. Before joining MITRE, Rebovich served in the U.S. Army, including tours on duty in the United States, Europe, and as an artillery forward observer with the 101st Airborne Division in Vietnam. He was a former assistant professor of mathematics at the U.S. Military Academy.

Brian E. White was the director of the Systems Engineering Process Office (SEPO) at The MITRE Corporation. He holds a corporate position and reports to MITRE's corporate chief engineer. Dr. White's current professional interests are focused on applying complexity theory and complex systems principles to improving the practice of systems engineering in situations where the interactions of people dominate the technological issues. As reflected in Dr. White's conference papers and internal presentations since 2003, he has contributed ideas for defining and improving the engineering of complex systems. Dr. White is founding coeditor of the Complex and Enterprise Systems Engineering (ESE) book series established in 2007 with Taylor & Francis (T&F). In addition to coediting this book and authoring Chapter 5, he coauthored another series book entitled *Leadership in Decentralized Organizations*. He is also the coauthor of a chapter called Emergence of SoS, Socio-Cognitive Aspects for another T&F book (not of this series) titled *Systems of Systems Engineering: Principles and Applications*. From 2005 to 2008, Dr. White was the International Council on Systems Engineering (INCOSE) assistant director for Systems Science and the founding chair of the Systems Science Enabler Group (SSEG). He received a PhD and an MS in computer sciences from the University of Wisconsin, and SM and SB degrees in electrical engineering from MIT. He served as an Air Force Intelligence Officer, and for 8 years was at MIT's Lincoln Laboratory. Dr. White spent 5 years as a principal engineering manager at Signatron, Inc. In his 26 years at The MITRE Corporation, he has held a variety of senior technical staff and project/resource management positions. During his professional career, Dr. White has published more than 100 significant technical papers and reports.

Contributors

Steven J. Anderson
The MITRE Corporation
Langley Air Force Base,
 Virginia

Mark D. Baehre
The MITRE Corporation
McLean, Virginia

Christopher D. Berube
The MITRE Corporation
Bedford, Massachusetts

J. Howard Bigelow Jr.
The MITRE Corporation
Bedford, Massachusetts

Terence J. Blevins
The MITRE Corporation
McLean, Virginia

Lindsley G. Boiney
The MITRE Corporation
Bedford, Massachusetts

Craig A. Bonaceto
The MITRE Corporation
Bedford, Massachusetts

Kevin A. Cabana
The MITRE Corporation
Bedford, Massachusetts

Michael R. Coirin
The MITRE Corporation
Langley Air Force Base,
 Virginia

Jeffrey S. Cook
The MITRE Corporation
Langley Air Force Base,
 Virginia

Kimberly A. Crider
The MITRE Corporation
Hanscom Air Force Base,
 Massachusetts

Joseph K. DeRosa
The MITRE Corporation
Bedford, Massachusetts

Stephen C. Elgass
The MITRE Corporation
Colorado Springs, Colorado

Julio C. Fonseca
The MITRE Corporation
Langley Air Force Base,
 Virginia

Richard W. Fox
The MITRE Corporation
Norfolk Naval Base,
 Virginia

L. Shayn Hawthorne
The MITRE Corporation
Colorado Springs, Colorado

Chris A. Kaprielian
The MITRE Corporation
Bedford, Massachusetts

Herman L. Karhoff
The MITRE Corporation
Dayton, Ohio

Paul S. Kim
The MITRE Corporation
Bedford, Massachusetts

Robert J. Lesch
The MITRE Corporation
Bedford, Massachusetts

Lewis A. Loren
The MITRE Corporation
Bedford, Massachusetts

Michael R. McFarren
The MITRE Corporation
McLean, Virginia

Andrew K. Miller
The MITRE Corporation
Colorado Springs, Colorado

Linsey B. O'Brien
The MITRE Corporation
Bedford, Massachusetts

Cindy A. Plainte
The MITRE Corporation
Bedford, Massachusetts

George Rebovich Jr.
The MITRE Corporation
Bedford, Massachusetts

Laura R. Ricci
The MITRE Corporation
Bedford, Massachusetts

John J. Roberts
The MITRE Corporation
Bedford, Massachusetts

Bhupender (Sam) Singh
The MITRE Corporation
Bedford, Massachusetts

Harcharanjit Singh
The MITRE Corporation
Bedford, Massachusetts

Maura A. Slattery
The MITRE Corporation
Bedford, Massachusetts

J. Reid Slaughter
The MITRE Corporation
Colorado Springs, Colorado

Peter A. Smyton
The MITRE Corporation
Bedford, Massachusetts

Richard E. Stuebe
The MITRE Corporation
Colorado Springs, Colorado

Robert S. Swarz
The MITRE Corporation
Bedford, Massachusetts

Carlos Troche
The MITRE Corporation
Langley Air Force Base,
    Virginia

Jay M. Vittori
The MITRE Corporation
Langley Air Force Base,
 Virginia

Michael J. Webb
Old Dominion University
Norfolk, Virginia

Brian E. White
The MITRE Corporation
Bedford, Massachusetts

Chapter 1

Introduction

Joseph K. DeRosa

Contents

1.1 Rationale, Scope, and Context 3
    1.1.1 Rationale: A Tsunami of Change 3
    1.1.2 Scope: The Enterprise Is the System 4
    1.1.3 Context: All That's Difficult Is Not Complex 5
1.2 ESE Is Different—Operating between Order and Randomness 5
1.3 Essential Elements of ESE 7
    1.3.1 Development through Adaptation 9
    1.3.2 Strategic Technical Planning 11
        1.3.2.1 Real-World Experience: The Efficiency–Innovation Trade Space 14
        1.3.2.2 Loose Couplers in Network Science 15
        1.3.2.3 Choosing Loose Couplers—Power Laws from Complexity Science 15
        1.3.2.4 Small Worlds in Network Science 16
        1.3.2.5 Self-Organizational Constructs from Complexity Science 16
        1.3.2.6 Enterprise Resiliency and Change—Phase Change in Networks 17
    1.3.3 Enterprise Governance 17
        1.3.3.1 Collective Knowledge Framework 19
        1.3.3.2 Individuals and the Enterprise—Game Theory and Governance of the Commons 19
        1.3.3.3 Shape Enterprise Success Criteria 20
        1.3.3.4 Incentivize Enterprise Behavior 21

2 ■ Enterprise Systems Engineering

    1.3.4 Enterprise SE Processes 21
1.4 Previewing the Chapters 23
    1.4.1 Overview of the Authoring Process 23
    1.4.2 Systems Thinking for the Enterprise 23
    1.4.3 Pilots and Case Studies 24
    1.4.4 Capabilities-Based Engineering Analysis 24
    1.4.5 Enterprise Opportunity and Risk 24
    1.4.6 Architecture for ESE 25
    1.4.7 Enterprise Analysis and Assessment 25
    1.4.8 Enterprise Management 26
    1.4.9 Agile Functionality for Decision Superiority 26
    1.4.10 Enterprise Activities (Evolving toward an Enterprise) 27
1.5 Preplanning the Future of ESE 27
References 27

This book is about the future of systems engineering (SE). Specifically, it focuses on the theory and practice of what we call enterprise systems engineering (ESE). Aspects of SE that were traditionally insignificant or not at all known are becoming significant for systems that must operate in complex enterprises. Both the generally accepted theory and practice are going through significant changes. Although technology acts as an accelerator, change is not so much driven by technology as by what emerged from that technology: the interconnectedness and interdependence of today's enterprises. We are observing problems and failures with the application of traditional SE methods to the development of complex systems. As we accumulate more such experience, ESE takes on all the earmarks of Kuhn's scientific revolution [1]. The International Council on Systems Engineering (INCOSE) has noted that the increasing complexity in both problems and the systems required to deal with them was already stretching the validity of present SE processes [2].

    One can trace the theory and practice of what we will refer to as traditional systems engineering (TSE) back to the 1946 Army Air Forces' Project RAND (Research ANd Development) and the Air Force's 1953 Semi-Automatic Ground Environment (SAGE) air defense system. Present-day SE has been shaped by many of the early SE contributors [e.g., Bell Laboratories, IBM, MIT Lincoln Laboratory, and its Federally Funded Research and Development Center (FFRDC)* spin-off, the MITRE Corporation]. Tools and techniques such as system dynamics [3], operations research, computer modeling, and simulation were developed or applied to solve problems and design systems. TSE is essentially a reductionist process, proceeding logically from requirements to delivery of a system. There are a number

---

* FFRDCs are independent, not-for-profit entities chartered to act in the public interest. MITRE was established in 1958 and currently operates three FFRDCs: for the Department of Defense, the Federal Aviation Administration, and the Internal Revenue Service/Veterans Administration.

of fundamental textbooks [4] on TSE and many engineering schools that offer graduate degrees in SE. Much of the TSE discipline and practice is captured in the INCOSE SE Handbook [5].

At this point in time, however, there is no agreed-upon starting point or logical curriculum for the theory and practice of ESE. Our current body of knowledge consists of the legacy of theory and practice embodied in TSE as well as snippets from the pioneers and zealots of systems thinking, and the practitioners and inventors of our modern-day systems.

This chapter highlights some of the author's observations regarding the theory and practice, and attempts to accomplish three things:

1. Set the rationale, scope, and context of ESE
2. Introduce some important theories and practices
3. Conjecture on the future possible world of SE

Subsequent chapters present more detail of some of these concepts and practices, and those authors make their own conjectures on the theory and practice.

1.1 Rationale, Scope, and Context

When we use the term "enterprise" to modify SE, it is to emphasize that our scope is beyond the bounds of a single system or component. Of course, SE has always taken an expansive view, applying multiple disciplines in its methods and considering the environment in which a system is immersed. However, a qualitative change in both the theory and practice is under way. Systems engineers are being asked to shape larger networks of interdependent people, processes, and technologies into capabilities of unprecedented complexity [6].

1.1.1 Rationale: A Tsunami of Change

The Internet and the World Wide Web (WWW) have changed the way we work forever. A wave of social and cultural change is upon us. The convergence of rapidly advancing computer and communications technologies have shaped a new social, cultural, and economic fiber. Cell phones proliferate and have become much more than communications devices. We as a people are thinking, acting, and working differently. As systems engineers, we are being asked to solve problems that result from this giant coevolving network and, at the same time, responsibly create enterprises of social, cultural, and economic value. We are moving at an accelerated pace from building systems that solve problems in specialized fields to building global enterprises that solve problems in cooperative ventures. The compartmented knowledge of the few is being outpaced by the collective knowledge, tools, and inventiveness of the many. These changes, of necessity, affect the way we engineer systems.

We have mastered the techniques of TSE that built the great dams, that circled our globe with telecommunications satellites, and that put a man on the moon. However, the complexity and scope of today's information-intensive and interdependent systems create the need for new tools, methods, and understanding. It is not that TSE will fall by the wayside as did the horse and buggy when the automobile was introduced, but rather it will become one ingredient in the SE of more complex enterprises. Whatever the future holds for SE, we can be sure that the problems we are being asked to solve are fundamentally different—because of the interconnectedness and complexity of the world in which those problems exist. Change is upon us. From the popular children's game of Hide and Seek, "here we come, ready or not!"

1.1.2 Scope: The Enterprise Is the System

In this book, we define an enterprise to be the purposeful combination of people who make decisions, their workflow processes, and the technical systems they use to carry out their decision making and workflow. Business needs are satisfied and operational goals are met through this complex web of interactions distributed across geography and time [7]. Likewise, SE itself is distributed in a complex web across geography and time. No longer is design confined to one shop or is enterprise management confined to one organization. Systems engineers must interact with other systems engineers. Engineering processes must integrate with other engineering processes. Technology implementations must be compatible with other technology implementations. The SE and acquisition enterprise builds the operational enterprise. Enterprises are built by enterprises.

In the mid-1980s, John Gage of Sun Microsystems made the prophetic observation that the network is the computer [8]. This implied that functionality was migrating from end-user devices to the network. A good example is that our phone messages migrated from a system called an answering machine to a service in the network called voice mail. Entertainment, business, and defense systems, as well as myriad others, are on a similar path. A corollary to the axiom "the network is the computer" is that when an individual computer becomes a node in the network, its behavior is necessarily shaped by the network characteristics. Thinking has shifted from node to network.

Likewise, those little boxes with fixed boundaries we called systems are being replaced by a cloud of capabilities obtained from the enterprise. That cloud is implemented with distributed technical infrastructure by multiple organizations. Although TSE methods tell us how to build an individual system or capability, these methods must now be shaped by ESE considerations. We must expand and complement our TSE tool set through a mind-set that encompasses the combination of people, processes, and technology that make up the enterprise—the scope of our SE necessarily expands to the enterprise. In the new SE paradigm, the enterprise is the system.

1.1.3 Context: All That's Difficult Is Not Complex

To illustrate the nature of the complex context in which modern SE is practiced, we postulate a grand engineering challenge with three linked activities:

1. Design and build an (specified) information system that needs hundreds of operators and over 10 million lines of unprecedented software code to perform its functions
2. Construct several three-story steel buildings with a total floor space of 170,500 square feet to house the bulk of the people and computers in this system
3. Hollow out the inside of a granite mountain as the site in which these buildings are constructed

As far-fetched as this work seems, it was a real-world engineering challenge. Starting with the groundbreaking work in May 1961 [9] to the delivery of the final Integrated Capability increment in 1998, this is precisely what the U.S. Air Force did to construct the North American Aerospace Defense Command (NORAD) Combat Operations Center. This Center protects the North American continent by detecting unfriendly missiles, aircraft, and space objects. The mountain protected the Center from a nuclear blast. Phase 1 consisted of construction to remove over half a million cubic yards of rock from Cheyenne Mountain in Colorado Springs and shore up the chambers for building construction and operations. The costs were $12M (in 1961 dollars), and it was completed on schedule in 1 year with another 2 years of various modifications. The final increment of the software (called the Cheyenne Mountain Upgrade Program) delivered over 10 million lines of code and became fully operational on October 14, 1998 [10]—about 10 years late and 1 billion dollars over the budget!

Some would say that digging a hole in a granite mountain is more difficult than writing a software. Although writing a software is not backbreaking work, it is complex. Complexity arises from interdependencies of piece parts. The number of possible interdependencies in software is astronomical compared to those in rock formations. Indeed, digging out the interior of a mountain is difficult, but building a huge system consisting of people, processes, and technology is difficult and complex. "Complexity" trumps "difficulty" in SE.

1.2 ESE Is Different—Operating between Order and Randomness

Early in the history of systems thinking, Warren Weaver [11] made a subtle but far-reaching observation regarding the regions in which problems exist. He noted that problems involving the behavior patterns of organized groups of persons and a wide range of similar problems in the biological, medical, psychological, economical, and political sciences lie in a region that is other than the regions of simplicity or randomness. This single observation led to the partitioning of much of science and mathematics into what Weinberg [12] later defined in terms of three regions:

1. Region I he called *Organized Simplicity*. Here, there are systems, usually with small populations having a great deal of structure, explained by simple equations (e.g., Newton's laws and Kirchhoff's laws)
2. Region II he called *Unorganized Complexity*. Here systems are sufficiently random to be statistically regular (e.g., noise in communications systems and gases in thermodynamic systems)
3. Region III he called *Organized Complexity*. Here there is no simple or highly redundant structure, no statistical regularity, no equilibrium, no simple linear chain of cause and effect. Here the problems lie within the realm of complexity science

The TSE has developed well-established theories and accepted practices both for simple systems operating in Region I and for statistically regular systems operating in Region II. For (organized) complex systems operating in Region III there is neither a well-established theory nor accepted practices.

Within Region I, the theory of linear systems with its assumption of superposition and ensuing causality gives a firm mathematical basis to TSE as well as a formula for its practice. Through ordinary differential equations (which focus on the components of the system) or equivalently through functional analysis (which focuses on the response of a system to an impulse stimulus), a systems engineer can predict the output of a system to any input. It is all linked through an elegant theory involving Fourier (Harmonic) Analysis and eigenfunctions of the differential equations.* A tight little Region I theory: system responses tied to system stimulus through harmonic analysis. A well-defined Region I practice: break a problem down into parts, design a subsystem for each part separately, and integrate the separate subsystems together.

Within Region II, random process theory has been successfully applied to the problem of finding signals embedded in noise. During the war years of the midtwentieth century, visionaries like Norbert Wiener [13] in the United States and Kolmogorov [14] in the Soviet Union turned the problem of dealing with noise in communications and radar systems into a generalized harmonic analysis similar to that of Region I. Shannon Theory [15] and the Information Age soon followed. Still

---

* Any input can be expressed as a sum of sinusoidal (harmonic) functions through Fourier analysis, and the output can be computed by adding up the outputs due to each sinusoid separately (multiplied by a constant). The sinusoids are eigenfunctions of the differential equations and the constants are the Fourier transforms of the system's impulse response (and eigenvalues of the differential equations).

at the heart of SE for problems that fell in Region I and Region II was a theory based on the linearity assumption and a practice based on reductionist/constructionist thinking (i.e., splitting the problem into parts and solving each part separately). There was no discussion of any problems in a region between signal and noise, little was said of the Weaver's "other" region, and no hint of Weinberg's Region III.

Within Region III, the region of organized complexity, there is no obvious theory on which to base the practice of (enterprise) SE. Problems in the biological, sociological, and behavioral sciences inhabit this region. Problems in cellular automata, learning, and artificial intelligence inhabit it. When dealing with the combination of people, process, and technology that make up our enterprises, we are working in Region III. Enterprise systems engineers, as well as program managers (PMs) and the customers they collectively serve, are faced with difficult and unanswered questions when operating between order and randomness

- When, where, and how do we promote standard processes and technologies to bring order to an enterprise?
- How do we organize teams that are both effective and efficient in a complex environment?
- What is the "right" combination of innovation and integration for the enterprise at hand?
- How do we bring about an enterprise to be both adaptive and resilient, that is, how do we introduce transformative change and avoid network catastrophe?

These are difficult questions. ESE holds the promise of answering them. Our current intuition tells us that the solutions will incorporate a percolating mix of organized and unorganized components drawn from complexity science and evolutionary methods. ESE is different from the SE that has come before—it operates on problems in the middle, organized complexity region between simple order and randomness.

The next section outlines some of the fundamentals drawn from multiple disciplines that I believe are key to developing ESE theory and practice.

1.3 Essential Elements of ESE

The roots of ESE theory were the result of deliberate cultivation by some of the pioneers of systems thinking such as Ashby [16] and Wiener [17] in the study of Cybernetics, von Bertalanffy [18], and Weinberg [12] in the study of General Systems Theory, and Ackoff [19] and Simon [20] in the business literature. In that same era, the field of complexity science was gaining ground from the well springs of biology, chemistry, economics, and evolutionary theory. It came to a nexus in 1984 with the establishment of the Santa Fe Institute (SFI). Gell-Mann [21],

Kauffman [22], and Holland [23] are among the many thought leaders associated with SFI.

The roots of ESE practice were the result of millions of seeds sewn across the engineering landscape by experimentalists, tinkerers, and hackers. They practiced a brand of SE that was exploratory and experimental rather than preplanned and execution-oriented. Intel Corporation cofounder Gordon E. Moore postulated a law for electronics-based information systems that tracked their consistent improvement by a factor of two every 24 months [24]. (A monthly compounded bank account growing at that rate would be earning 35% annual interest.) This phenomenal growth has spawned an explosion in innovation and evolutionary dynamics in the development of enterprises such as the Internet and WWW. Out of the huge variety of new systems capabilities offered, selective pressures of the marketplace culled out the most fit to compete in the next evolutionary spiral. We are experiencing in a very real way what the evolutionary philosopher Daniel Dennet called "a design without a designer" [25].

The elements of the theory and practice of ESE lie somewhere in the milieu of systems thinking, complexity science, and the evolution of the Internet and WWW. At this stage in the development of ESE, the practice leads the theory. However, as the two coevolve, each will inform the other in a never-ending virtuous cycle. Although it is not clear at this point in time which concepts and related practices will ultimately prove most useful, I have made a first attempt to cull out those elements of ESE that seem to be gaining favor in the practice and that have some basis in theory. These are given in Table 1.1.

Table 1.1 Key Elements of the Theory and Practice of ESE

Essential Elements of ESE	Descriptions
Development through adaptation	Deals with uncertainty and conflict in the enterprise through adaptation (adaptive stance): variety, selection, exploration, and experimentation.
Strategic technical planning	Brings a balance of integration and innovation to the enterprise. Leverages the practice of layered architectures with loose couplers and theory of order and chaos in networks.
Enterprise governance	Sets enterprise fitness as measure of system success. Leverages game theory, ecology, and practices of "satisficing" and governing the commons.
ESE processes	Brings new ways of working horizontally across the enterprise. Leverages business literature and practice.

How these essential elements are being used in practice and how they relate to theory are discussed below. Although these, as well as other elements are beginning to cut paths through the ESE landscape, the final map is not yet drawn.

1.3.1 Development through Adaptation

The first essential element of ESE comes from evolutionary development. Evolutionary development is a proven way to deal with uncertainty and conflict in complex systems. Evolutionary development implies that subsequent generations of an enterprise are modified descendants of preceding generations. Adaptation occurs when the modifications create enterprise capabilities that succeed and endure in their environment, and that adaptation redefines the starting point for the next generation (version) of the enterprise. If the rate of adaptation keeps pace with the rate of change of the environment, the development of the new capability is said to be agile.

For example, the familiar Web Browser has gone through a number of adaptations in its evolutionary development from the first instantiation in 1991 by Tim Berners-Lee [26], through the overwhelming popularity of Netscape in the mid-to-late 1990s, to the current hegemony of Internet Explorer. In that adaptation process, uncertainty in requirements was resolved, conflict over whether it would be based on proprietary or open standards was resolved, and competition for market share was resolved.

The challenge we face in ESE is to find the essential ingredients of adaptation and apply them in an appropriate and repeatable manner. Fortunately, adaptation and evolutionary development are not new. We can gain insights from the fields of Evolutionary Biology, Computer Science, and Social Sciences (e.g., economics and sociology). Evolutionary Biology shows us the remarkable ability of (Darwinian) evolution to seek out solutions in complex environments [27,28]. Computer Science formulates adaptation into an engineering process [29], while social science has successfully viewed fields such as economics [30] and urban development [31] as adaptive and evolutionary processes. Although there are differences between these fields and ESE, there is a consistent pattern observed in adaptation

1. An innovation process presents a *variety* of differentiated options to the enterprise
2. *Fitness criteria* determine which will survive, flourish, and perish
3. A *selection* process culls out, amplifies, and integrates successful options into the enterprise

In the software development world, two landmark papers in 1986—by Barry Boehm [32] and Frederick Brooks [33]—crystallized a spiral approach to software development that makes use of iterative and incremental prototypes with user feedback accepting, rejecting, or redefining the requirements at each iteration. The incremental prototypes present a limited amount of variety to the process, while the user feedback supplies the fitness criteria and the selection. A number of similar "Agile Methods" have appeared since those papers. Recent work published by Stevens et al. [34] advocates venture capital techniques in which many small bets and staged commitments provide variety and selection in the development of complex enterprises.

We have defined an enterprise to be a network of interdependent people, processes, and supporting technologies. The reality of such situations is that there may be multiple technologies involved in delivering capabilities through dissimilar processes with no single authority in charge. The interdependencies generate uncertainties in both the requirements and the effects of system design activities. The different perspectives of all the stakeholders across the enterprise generate real and imagined conflicts regarding the purpose of the enterprise and the proper direction of its development. Therefore, the requirements are not always known or knowable. Even if they are known at the level of overall desired capabilities, the many interacting stakeholders, processes, and technologies in the system and its environment create a level of complexity that makes the allocation of those requirements down to stable subsystem requirements nearly impossible. Simon [35] defined an ill-formed problem as one that had to be reformulated at each move toward a solution, for example, in the game of chess, after each move of the opponent. Similarly, in a changing technology base and in purposive environments, requirements have to be reformulated each time the environment adjusts to the current development path.

Adaptive development exploits the power in an evolutionary approach to find enterprise designs in a design space that is too large and complex to prespecify. It stands in marked contrast to the traditional waterfall or V-Models, which develop capability through top-down, requirements-driven approaches. The enterprise could be a multinational corporation wishing to consolidate the information technology (IT) infrastructure of its subsidiaries to bring coherence to its planning and budgeting processes, or it could be a government organization charged with delivering joint command and control (C2) capability with separate Army, Naval, and Air Force components. Such enterprises are known in the literature as complex adaptive systems—they are more like ecosystems than well-defined and controlled systems. An adaptive stance is needed in their development.

To summarize, development through adaptation refers to the process of constructing a variety of innovative and differentiated options to the enterprise, having them selected out by some enterprise-level fitness criteria and integrating the successful ones into the next generation of the enterprise. The innovation is opportunistic at each stage of development. The enterprise develops not through conforming to a plan but rather through coevolution as its various parts adapt to change. The design of each system or component now contains its context in the enterprise as a design variable. The development satisfies and suffices ("satisfices") the many stakeholders rather than optimizes to meet their needs, that is, solutions are found that are "good enough."

There is much work to be done to turn this method into a well-defined and pervasive tool of ESE. Moreover, to gain support, it must be demonstrated that any increase in cost to present a variety of options to the enterprise is offset by the opportunity benefit of not having failed programs. Lastly, it is our contention that for classes of systems for which the complexity overwhelms our ability to plan and direct the outcomes, we have no choice but to use adaptation for development.

1.3.2 Strategic Technical Planning

The second essential element in the practice of ESE is strategic technical planning (STP)—the *a priori* laying down of a small number of key enterprise rules that govern the interaction between participants in the enterprise. The STP provides the basic contract to which the participants of the enterprise agree before joining the enterprise. The STP serves to organize and integrate disparate activities across the enterprise. Frequently expressed through an Architectural Pattern, the rules are understandable, unambiguous, and accepted by stakeholders. A successful STP is a simple STP.

Early in the development of a network-centric enterprise for the U.S. Air Force, MITRE put together an STP that encouraged both innovation and integration across the enterprise [36]. It was approved by the Commander of the Air Force's Electronics Systems Center (AF/ESC) for their systems developments in 2002 and updated in 2005 [37]. The pattern in the STP is a common one for IT-intensive enterprises, and it has two main attributes:

1. *A Layered Architecture* in which the elements of the enterprise are logically (and sometimes physically) composed of a few well-defined and universal layers.
2. *Loose Coupling* in which the interactions between entities in adjacent layers are governed by a set of protocols and expected behaviors rather than by intimate knowledge of each other [38,39].

Architectural Patterns had their beginnings in the groundbreaking work by Christopher Alexander [40] around 1977 in which the construction of towns and buildings was based on patterns of behavior and on previously successful structural patterns. While the architectural plan consisted only of classic patterns tested in the real world and reviewed by multiple architects over time, they allowed the architects to adapt them to particular situations. Alexander's work in codifying architectural patterns was brought into software engineering practice through Design Patterns [41], which developed patterns for common problems in the design of object-oriented software systems. Patterns for the enterprise systems have also been cataloged in other works [42,43].

Layered architectures for computer communications networks were formalized by the International Organization for Standardization (ISO) in the 1979 publication

12 ■ *Enterprise Systems Engineering*

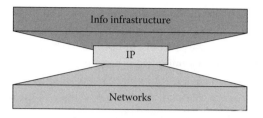

Figure 1.1 Loose coupling between layers.

of the Open System Interconnect (OSI) Seven Layer Reference Model [44]. In October 1989, the Internet Engineering Task Force (IETF) published its Internet (TCP/IP) Model specification [45] for the four layers that defined the Internet architecture. The relentless push of Moore's Law and the dot-com boom–bust of the 1990s left behind a relatively stable *n*-tiered architecture [46] for information systems. The Web 2.0 and "mashup"* technologies that dominate the current internet technology conversation are made possible through this stable *n*-tiered architecture.

Integration across the enterprise is achieved through the loose couplers. A new implementation of an entity or a new entity can be inserted in a layer without affecting any other layers as long as the loose coupler is utilized. Innovation is enabled by the lack of dependencies between layers.†

Figure 1.1 illustrates how the Internet protocol (IP) is used to form such an STP pattern. The IP is used as a loose coupler between a myriad of application data types and a myriad of communications networks. Thus, even though the data in the Information Infrastructure layer may be associated with many and varied applications (e.g., e-mail, payroll, targeting, e-commerce, etc.), and even though the networks in the networks layer may be many and varied (e.g., Public Switched Telephone Network, Asynchronous Transfer Mode networks, Ethernet, etc.) as long as the entities in both layers recognize the IP protocol, any application's data can be carried over to any media and any media can carry data from any application. Moreover, new or improved media or applications can come to the network as long as they "speak" IP. The loose coupler integrated the network and independent layers enabled innovation within each layer.

Other loose couplers share the same pattern—disparate enterprise entities interacting by using agreed-upon protocols and by getting predictable behaviors. Powerful loose couplers exist, not only within information systems, but also within the business world. For example,

- *Hypertext transfer protocol* [47] *(HTTP)* is a loose coupler between disparate services and clients. The services can expose different data and different

---

* A mashup is a Web application that combines data and/or services from multiple sources.
† The concept of loose couplers benefited from many private conversations with Dr. Seth Landsman of MITRE.

capabilities and exist on different networks. The clients requesting data from any of the services can be desktop or cell phone Web browsers, or noninteractive machine-to-machine entities. Both the services and clients interact predictably via the HTTP protocol. Since each party only knows of the other through the loose coupler protocol, a service or client may be added or replaced with no apparent impact to others in the enterprise.

- *Domain name system (DNS)* [48] loosely couples a domain name (e.g., google.com), used by application layer clients, to a service endpoint (i.e., IP address), used by the information infrastructure layer capabilities. The translation from a domain name is seamless and invisible to the user and client. Thus, the mapping of a domain name to a service endpoint can be changed without affecting the client, and new mappings between domain name and endpoint can come online without affecting existing mappings.
- *A purchase requisition (PR)* in business loosely couples personnel who need things to those who buy things. Typically the protocol includes the item needed, the preferred supplier, the need date, and an approval signature, and the expected behavior is that purchasing department will order the item and have it delivered. Once the PR process is in place, the company can venture into new products and new vendors. In fact, as long as the PR protocol is followed, the purchasing department need not be in the same company. Outsourcing is enabled by loose coupling.

The STP concept adopted by AF/ESC has matured over the years since 2002. The latest thinking for an STP in C2 enterprises was presented in a recent IT workshop that MITRE runs for military General Officers. It was derived from a paper by the Workshop leader R. Byrne [36]. Figure 1.2 shows the C2 elements organized in four layers, one each for sensors, networks, data, and applications. It also shows several loose couplers labeled GeoRSS, IP, CoT, and Mashups.

The bottom layer has many of the diverse sensors in the world. Innovation in sensor technology simply adds more sensors in this layer. Above it is a loose coupler interface to the networks called GeoRSS, an emerging WWW standard that allows all these diverse sensors to easily be accessed by anything outside of the sensor layer. Likewise, there are many different networks, data (or information) formats, and applications in the C2 enterprise. IP is the pervasive Internet loose coupler that allows diverse communications networks to interplay. CoT is an increasingly used XML schema to which all existing data sources (and sinks) can translate to exchange basic "what, where, when" information [49]. Mashups enable diverse applications to share their services in a standardized way and thus allow new combinations of services to be composed as needed. For example, when a Google Map is mashed up with housing prices, the combined product is composed rapidly between two loosely coupled services that were designed independently of each other. This simple C2 STP balances innovation (diverse solutions) inside each layer with integration (through loose couplers) between layers.

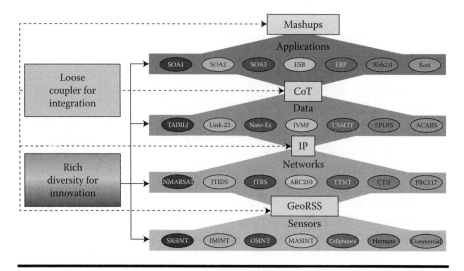

Figure 1.2 A sample command and control of STP architecture.

In general, the STP defines the fundamental design patterns (e.g., layers and the loose couplers) to which all designs and agreements in the enterprise must adhere. Enterprise architects use the STP to communicate and coordinate across the enterprise. The design patterns will be slightly adapted to particular situations much as they are in buildings, cities, and object-oriented design. The STP brings a balance of integration and innovation to the enterprise. Real-world experience as well as theory shed light on this integration–innovation balance and its trade space. Some of this is briefly discussed below.

1.3.2.1 Real-World Experience: The Efficiency–Innovation Trade Space

We are being asked to engineer enterprises so that they operate efficiently and allow adaptability. At the same time they must be resilient against catastrophic failure without being so rigid as to be trapped in stasis. ESE must shape the evolution of an enterprise to strike a balance between efficiency to keep costs down and innovation to allow adaptability for the long term.

This efficiency–innovation trade was observed in the performance of world auto makers during the 1970s. The U.S. auto makers had invested heavily in efficient, well-integrated operations. They were unable to adapt their production enterprise to the new requirements of changing fuel economy and emission standards. Their Japanese competitors were tuned for a mix of both efficiency and innovation. They more quickly adapted and ultimately increased their market share.

Today, the U.S. Department of Defense acquisition enterprise has an extremely ordered process for development of new capabilities and is struggling to tune itself to the pace of change in IT systems. However, taking a lesson from the Asian auto manufacturers, it should be possible to build an acquisition enterprise with both increased efficiency and increased innovation. The challenge of ESE is to bring the right mix of integration and innovation to the enterprise as a whole through the systems (services, components) we engineer. The STP appears to be one way of doing that. To better understand the mechanisms and to invent new mechanisms, we turn to network science and complexity science for insights.

1.3.2.2 Loose Couplers in Network Science

Understanding how actual and theoretical networks operate is fundamental to ESE. A wide variety of networks from social networks to computer-communications networks to biological networks exhibit features described by theoretical models that interpolate between perfectly ordered and perfectly random graphs [50]. A lattice of identically related repeating parts is a perfectly ordered (redundant) network—quite predictable, of low complexity and with little variation. The perfectly random networks studied by the Hungarian mathematicians Erdos and Renyi [51] are at the other end of the spectrum of models—disordered and varying unpredictably.* Enterprise networks display features unlike either of these two extremes, (e.g., they exhibit high clustering, small distances between any randomly chosen nodes, and power-law distributions of the number of links per node).

Empirical data from the WWW, with its hundreds of million nodes, reveal that any Web site is hyperlinked to any other Web site through less than 20 "clicks" (on the average) and the clustering of nodes is orders of magnitude greater than if it were random [52]. Similar results are obtained for the router network of the Internet. In network science, a mechanism called preferential attachment leads to a network structure in which a small number of nodes (called hubs) have many connections and a large number of nodes have few. These hubs bring organization to the network just as airport hubs bring organization to the air transportation network. The "amazon.com" hub couples people with many and varied tastes to share their opinions through book reviews just as IP acts as a hub between systems with many and varied local communications networks. The hubs from network science correspond to the loose couplers of the STP.

1.3.2.3 Choosing Loose Couplers—Power Laws from Complexity Science

The numbers of connections per node in many real networks (from biological to the Internet and WWW) oftentimes exhibit a power-law probability distribution, that is, they are scale-free. Power laws abound in complex systems. This puts

---

* As noted previously, behavior can be unpredictable, while statistics of behavior can be predictable.

engineering analysis in a fundamentally different regime of mathematics than the law of large numbers and Gaussian (Normal) distributions. We find ourselves working with Pareto 80-20 rules rather than Gaussian means and variances. This gives a heuristic indicator of how to choose the loose couplers in an enterprise: look for those protocols and behaviors that may account for only 20% of the design but account for 80% of the usage.

1.3.2.4 Small Worlds in Network Science

A network can create what is known as a small world, in which the number of hops it takes to get from one node to another is a relatively small number compared to the number of nodes in the network. The idea that we are connected to any other human being on the planet through at most six acquaintances was popularized by the movie "Six Degrees of Separation" and studied by network scientists [53]. If we had a simple network with k links per node, then you could reach $N = k^d$ nodes in d steps. Thus d scales as log N/log k. In a network with 1 billion nodes and 10 connections per node, it would take only nine steps to reach any node from any other. Reach is one aspect of the power of networks and is certainly the hallmark of the WWW.

1.3.2.5 Self-Organizational Constructs from Complexity Science

The process by which the structure and dynamics of an enterprise evolve is a mixture of design and self-organization. We saw that loose couplers and network clusters can be viewed as organizational constructs—strategic convergence points between the layers in an STP or hubs in a network exhibiting preferential attachment. However, the way agreed-upon patterns emerge is through usage, akin to the way cow paths shape the landscape as well as follow the landscape they shape. Management hierarchies are common in business firms and cross-corporate teams frequently self-organize to get a job done. Darwin's theory of evolution rests both on the gradual accumulation of useful variations and on what appear to be organized substrates in the metabolic network. Clearly there are self-organizational constructs shaping the structure and dynamics of real-world networks.

Steven Johnson articulated the power of self-organization in the ways ant colonies, brains, cities, and software evolve [31]. He observed how the vast connectivity of the Internet and WWW has unleashed that power. However, the heavy lifting in the theory of self-organization has been done by complexity scientists—many associated with SFI. Among them, Stuart Kauffman stands out as a leader.

Kauffman studied the behavior of (Boolean) networks [27] with N nodes in which the behavior of each node is a Boolean function of inputs from K other nodes. He has related this NK model to real-world metabolic as well as engineered networks. He found that natural selection is not powerful enough to have produced the order we find in the universe. Thus there needed to be a spontaneous self-organizing force at work as well. He showed that massively disordered systems can spontaneously crystallize a very high degree of order. He also showed that selection tends to favor complex systems capable of adaptation (poised at the boundary between order and chaos). He found that as certain parameters were varied, the networks exhibited three broad regions of behavior: an ordered regime filled with large frozen components, a chaotic regime with no frozen components, and a middle (complex) regime where the frozen components are just beginning to percolate and the unfrozen components are just ceasing to percolate. Thus he articulated the theoretical basis for Weaver's observations and laid the groundwork for future research in ESE.

1.3.2.6 Enterprise Resiliency and Change—Phase Change in Networks

The other key phenomenon of networks that does not generally take place in linear causal chains is that of phase changes. As some network parameter is varied past a certain critical point, the network behavior changes abruptly. This is observed in physics as temperature is continually increased—for example, water first experiences a phase change from ice to liquid and then from liquid to steam. A classic paper [54] from mathematics indicates that this is a property of the network (not of the physics), and a pivotal paper [6] from biology articulates the nonlinear effects in its declaration that "More is Different."

This has implications for ESE. As the number of layers or the number of loose couplers is varied, the enterprise can exhibit a phase change from relatively orderly to relatively chaotic behavior. A better understanding of this phenomenon will help us tune our enterprises to trade efficiency for adaptability, to be resilient against catastrophic failures, and/or to effect enterprise transformation. Phase change in networks is a topic that needs empirical data and research to become a tool of ESE.

In summary, the STP serves as an organizational construct for an enterprise. It is a propagating form of organization where it creates more of the enterprise in the form of itself and therefore amplifies its impact. It has been effectively used in the evolution of the Internet and WWW and has roots in both network and complexity science.

1.3.3 Enterprise Governance

Governance shapes the collective actions of stakeholders in the enterprise. In the previous section, we talked about the STP shaping the technical landscape in which an enterprise is developed. Enterprise governance generalizes that concept to include shaping the political, operational, economical, and technical (POET) landscape. It is common in TSE to view the nontechnical factors as outside the purview of the systems engineer. However, in complex systems the technical solution is so intertwined with the nontechnical variables that enterprise systems engineers must consider nontechnical context as a design variable. Napster solved the technical problem of sharing music over the Internet, but Apple solved the enterprise problem of delivering iTunes over the Internet [55]. Apple's governance delivered a POET solution.

There are two common approaches to the governance of an enterprise: central control by a hierarchical authority and reliance on market forces to solve enterprise problems. Both approaches have a certain unassailable logic based on their assumptions, but neither alone has been unassailable in its success. There are numerous cases of large, centrally-controlled projects that failed under the crush of a complex environment, and there are numerous cases of deregulated (free market) developments that delivered degraded products or services at higher prices. Complexity overwhelms the capacity of any hierarchical authority to understand, let alone control development of an enterprise. Further, markets do not necessarily act fairly or efficiently. Nor do markets necessarily drive toward the desired outcome.

The Internet and WWW have been governed by a framework that is a hybrid of hierarchical authority and free market. The Internet Architecture Board (IAB), IETF, Internet Corporation for Assigned Names and Numbers (ICANN), and the WWW Consortium (W3C) establish rules and procedures to enforce enterprise policies and resolve disputes that involve conflict. However, the way these authorities arrive at their rules is by utilization of grass roots consortia that represent the knowledge in the enterprise marketplace. Internet and WWW governance is a synergistic mix of legitimate authority and collective knowledge of the many.

Much like the way the members of a sports team compete for positions and salary within the team, but at the same time they cooperate to win a game, with the Internet, there is a greater good that results from cooperation. Similarly, in the development of an enterprise, there is a greater good to be realized from cooperation of the participants. The enterprise governance framework must be properly set up to realize that greater good. At this point in time, there is no widely accepted theory or practice for enterprise governance. However, from our experience with competitive–cooperative games and with our knowledge of complex systems, we can postulate several objectives of effective enterprise governance

1. Create a *framework* that supports inclusion of collective knowledge and effort
2. Balance the interests of *individuals and the enterprise* (player and team)
3. Shape enterprise *success criteria* (rules of the game)
4. *Incentivize* enterprise behavior (rewards)

We discuss each of these in the following sections and identify some fields of study that inform them.

1.3.3.1 Collective Knowledge Framework

The development of an enterprise requires the collective action of a number of stakeholders: owners or sponsors, suppliers, workers, operators, and developers. The knowledge required to determine what to produce, when to produce viability in the POET environment, and so on, is distributed across those stakeholders. The framework needs to support the mining, capture, and sharing (proliferation) of that knowledge. Think of the stakeholders as existing in an ecosystem, each competing and cooperating to achieve their ends while keeping the ecosystem in balance. Lean thinking emphasizes transparency (where data are widely shared with stakeholders), and concurrent engineering embodies team values of cooperation, trust, and sharing. The ecosystem analogy highlights the concept of "satisficing"—a satisfy and suffice strategy that is adequate for the set of stakeholders without necessarily being optimal for any subset.

A clear case of where the trade between individual and enterprise paid off is in the transition from developing "stovepipe" communications systems to the Internet. Our challenge in ESE is to gain a clear understanding of governance frameworks that promote enterprise outcomes and those that do not. We need to learn how to answer some difficult questions. What was there about the Internet framework that made the transition to IP so much more pronounced than the transition to object-oriented design? What is the right framework for the emerging service-oriented architecture developments?

1.3.3.2 Individuals and the Enterprise—Game Theory and Governance of the Commons

At the heart of the enterprise governance problem is the conflict between what is best for the enterprise and what is perceived to be best for any individual agent or system in the enterprise. (In a healthy ecosystem, some rabbits get eaten by some foxes.) Governance should strike a balance between encouraging cooperation and competition to keep the enterprise (ecosystem) healthy and flourishing.

The Prisoners Dilemma problem [28] of game theory and social science represents the conflict between optimizing at the local level and cooperating for a greater good. If a prisoner betrays his partner in crime, he gets a lighter sentence. If he attempts to cooperate with his partner by remaining silent, one of the following two things could happen. If his partner also remains silent, he gets away with the crime or if his partner betrays him, he gets a heavier sentence. Thus, an individual who cooperates with his partner in their criminal enterprise may reap a greater gain or risk a greater penalty, whereas if he does not cooperate, he is assured of a lesser penalty. The study of this problem can have far-reaching consequences for ESE.

PMs of systems that fit into a larger enterprise have experienced a similar tension. Some have cooperated for the greater good and have been penalized (in terms of program cost, schedule, and performance) by partners who did not live up to their end of the bargain. The question is: how does a governance body create a development environment that encourages cooperation for the common good? For the answer, we may have to look in other fields where collective action is required to devise sociotechnical solutions that serve the common good.

Governance of the Commons is a field of study in which a common-pool resource (the "commons") is managed with the participants cooperating in the appropriation and provisioning of that resource such that it is not degraded or ruined (Tragedy of the Commons). Ostrom [56] has examined a number of common-pool resource situations involving grazing lands, water rights, and fishing grounds. Successful governance frameworks have a common pattern

1. The participants with the technical knowledge (i.e., the sheep and cow herders, water suppliers, fishermen) make the agreements on how to share the resource
2. They look to the governing authority for enforcement (in the form of licensing, sanctions, and fines)
3. They also perform the policing of the agreements (again because they possess the technical knowledge)

There are similarities to governance of the information-intensive enterprise we have been discussing. Much of the infrastructure in an enterprise is a common-pool resource. The information regarding the political climate, operations, economics, and technology is scattered across many stakeholders. Even the capacities of the regulators, contractors, financiers, and engineers are limited. It is likely that principles from governance of the commons could be applied to enterprise governance.

1.3.3.3 Shape Enterprise Success Criteria

In an evolutionary environment, development proceeds according to some fitness criteria. Fitness for natural systems is limited by biology and chemistry of the organisms and their environment. However, engineered systems can favor development in any one of a myriad of directions defined by the governance scheme. Behavior of complex systems can be counterintuitive since autonomous agents in the enterprise can act or manipulate the world on their own behalf. Therefore, the fitness criteria must coevolve with the enterprise.

Some key work done by Dorner [57] on complex decision making indicated that those governance authorities who continually reevaluate courses of action perform better in a complex enterprise than those that use a plan-and-execute strategy. Because human cognition is limited and fallible, the most desirable courses of action are not necessarily known or knowable for either the individual or the enterprise. Therefore, governance must experiment, collect feedback and, if necessary, change course with its enterprise success criteria.

The behavior of an enterprise will be determined by the way systems developers in the enterprise construct and weave together component capabilities. This will in turn be determined by how the governance procedures define success. The levels of agility, flexibility, resilience, interoperability, and so on, will be set forward in the success criteria. To use the sports analogy again, how the game is scored determines how the game is played.

1.3.3.4 Incentivize Enterprise Behavior

In the commercial marketplace, profit incentivizes behavior. If there is some wider social goal to be realized, governments will bias the ability to achieve profit to support that goal, for example, tax credits for fewer emissions. In the development of an enterprise those same principles apply. If PMs are promoted for achieving cost, schedule, and performance goals without regard for how the developed systems contribute to the enterprise goals, then enterprise goals will not be reached. Figure 1.3 indicates this with a simple picture. If the incentives match curve A, the PM will benefit but the enterprise will not. We must devise and execute reward schemes that match curve B, where the individual PM benefits when the enterprise benefits.

In summary, enterprise governance applies to the full set of POET factors in an enterprise. It capitalizes on the knowledge of the many and resolves conflict between the enterprise and the individual systems in the enterprise. It sets enterprise fitness as a measure of system success, and it leverages game theory, ecology, and practice of "satisficing."

1.3.4 Enterprise SE Processes

Much of the early thinking in processes that support enterprise development occurred first in the business literature. Early work by Ackoff and Emory [19] set the stage for understanding complex organizational dynamics and purposeful systems. Peter Senge [58] established a number of elements in the building of learning

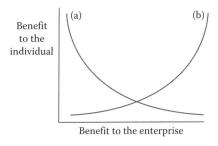

Figure 1.3 Opposing incentives. (a) Favors selfish behavior, (b) favors enterprise behavior.

(adaptive) organizations. John Sterman [59] modeled learning organizations using the tools of Systems Dynamics. A recent book by Gharajedaghi [60] summarizes much of the previous work. Recent work by Ancona and Bresman [61] reveal how to create teams to solve enterprise problems and how to use experimentation and exploration to do so. The paths these authors followed were not vertical within a single development organization. They brought new ways of working horizontally (or more realistically networked) across an enterprise. The focus was decidedly business management; however, the principles applied equally well to ESE.

At the 2006 INCOSE Symposium [62] and subsequently at the 2006 International Conference on Software and Systems Engineering Applications [63], we introduced a model for ESE derived from first principles in systems thinking, complexity science, and evolutionary development. It heavily leveraged the principles espoused by the business authors mentioned above and the complexity authors discussed previously. It consisted of an inner loop that encompassed the traditional ANSI/EIA 632 SE processes [64] and was shaped by a set of newly-defined ESE processes. It also included an outer loop accounting for the socioeconomic business factors of the enterprise environment (see Figure 1.4).

We hypothesized that the new ESE processes were fundamental and set out to verify the hypothesis through a number of pilots and case studies. As a result of that effort we added Stakeholder Analysis as fundamental to the suite of ESE processes. Thus, the current suite of six ESE processes consists of the following:

1. Capabilities-based engineering analysis (CBEA)—focuses on the high-level requirements the whole enterprise must satisfy

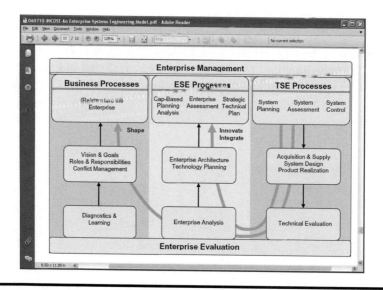

Figure 1.4 Enterprise systems engineering processes.

2. Enterprise architecture—brings a coherent view of the whole enterprise
3. Enterprise analysis and assessment—evaluates the effectiveness of the enterprise rather than the ability of system components to meet specifications
4. STP—lays out a minimum set of standards or patterns that every system in the enterprise must follow
5. Technology planning—looks for cues in the environment that would indicate the emerging technologies that could next be implemented into the enterprise
6. Stakeholder analysis—examines the equities of those involved in the business and engineering processes of Figure 1.4, stewarding their flow of resources into and out of the enterprise

The TSE processes are discussed in the INCOSE SE Handbook [5] and the five original ESE processes are discussed in two early papers [62,63]. More detailed discussions of Stakeholder Analysis from an enterprise perspective are given in two companion papers [65,66]. These six ESE processes drive much of the subject matter presented in subsequent chapters of this book, and the validation of these ESE processes through the pilots and case studies is discussed in Chapter 3.

1.4 Previewing the Chapters
1.4.1 Overview of the Authoring Process

Each of the chapters was composed by different MITRE Technical Staff in the Systems Engineering Division of the Center for Air Force C2 Systems. They were done after we had completed some preliminary work as a group defining the ESE processes. The groups briefed each other on the works-in-progress as they went along and completed the work in about a year. As a result there are consistent veins of thought that run through them, but also different points of view from the authors.

1.4.2 Systems Thinking for the Enterprise

In Chapter 2, titled "Systems Thinking for the Enterprise," George Rebovich Jr. gives an overview of the ESE terrain. He deftly synthesizes the work of several "systems thinking" visionaries from across the multidisciplinary fields of business, technology, and complexity science—and he adds to it his own experience in the practice of SE.

He characterizes an enterprise as a network of people, processes, technology (as well as other parts) immersed in and interacting with a complex environment that constitutes a higher level enterprise. Following the work of Gharajedaghi [60] and Ackoff [67], he proposes an operational definition of the boundary of an enterprise and espouses the viewpoint that proper SE always looks to the larger enterprise.

However, the key synthesis he makes is to join the idea of the complementarity of differentiation and integration to the evolutionary processes of variety, shaping interactions, and selection. Thus, he melds the ideas from the business literature of Gharajedaghi with those of the complexity scientists, Axelrod and Cohen. He illustrates the application to ESE with a number of enlightening examples and stories. This one chapter establishes the meaning of system thinking as a mode of thought applied to ESE, and poses questions for the systems engineer to consider in harnessing the complexity of the enterprise.

1.4.3 Pilots and Case Studies

In Chapter 3, titled "Pilots and Case Studies," Kimberly Crider moves from theory to practice. She captures the lessons learned from a number of independent pilot activities in ESE, and she analyzes successes and failures, from an ESE viewpoint, through a retrospective look at several past projects, that is, through case studies. The lenses through which the pilots and case studies were viewed are the core ESE processes discussed in Section 1.3.4 (which were based on the systems thinking principles presented by Rebovich Jr. in Chapter 2). She presents a brief description of each ESE process, a set of overall findings and details of lessons learned, and findings for each individual project studied. To our knowledge, this is the first attempt at analyzing the practice of ESE against a proposed theory of how to do it. No doubt, both the theory and practice will both evolve, and this was an important first step in that journey.

1.4.4 Capabilities-Based Engineering Analysis

Chapter 4 on Capabilities-Based Engineering Analysis by Steven Anderson and Mike Webb focuses on the process by which desired capabilities are identified at the enterprise scale and coupled into evolutionary plans at the scale of systems or portfolios of systems. CBEA is one of the core ESE processes. The focus on capabilities provides the guiding principles for enterprise evolution. They give two real-world examples of the capabilities planning approach from the Joint Military Services and U.S. Air Force. CBEA supports the decisionmakers in allocating resources and planning programs that are fit-for-purpose from an enterprise perspective.

1.4.5 Enterprise Opportunity and Risk

In Chapter 5, Brian White takes on the important SE topic of risk, which drives one of the most important processes in TSE. His major thesis is that we need to both expand and complement the risk concept to emphasize opportunity, particularly in an enterprise. Further, he proposes uncertainty management as a more general and more appropriate process in ESE than TSE risk management. Uncertainty management combines opportunity and risk management and recognizes that some events are either as yet unknown or cannot be classified as opportunities or risks until more is known. He sets up the opportunity management process similar to the familiar risk management process in terms of probability of occurrence and impact. However, he injects a subtle but important element into the discussion—that of our inability to totally envision the outcome space in a complex system.

White also reviews what is known as the Regimen for Complex SE [68,69].* Strictly speaking, nonlinear dynamic systems are deterministic. However, because of their complexity, they are essentially unpredictable. By that we mean a small change in one parameter can have a huge effect on the system as a whole—the so-called butterfly effect discovered by Edward Lorenz [70]. Thus, in the development of complex systems, we encounter what seasoned systems engineers call unknown unknowns—outcomes that simply cannot be envisioned *a priori*. The Regimen sets up the theoretical basis for dealing with them and leads to an adaptive and evolutionary strategy. Therefore, dealing with these unknown unknowns becomes a planning process that is executed not before the fact, but rather, as the opportunities and risks emerge.

1.4.6 Architecture for ESE

In Chapter 6, Carlos Troche and a number of architectural professionals, who have worked for many years in the U.S. Department of Defense (DoD) enterprise, give the benefit of their combined perspectives on building and organizing information-intensive enterprises. Their interesting musings on the value and purpose of enterprise architecture as an organizing construct are backed up by a presentation of the practical experience and methods used within the DoD. Enterprise Architecture is one of the core ESE processes, and they present extensive case studies from their experience base. As we have defined an enterprise, it is a complex adaptive system. Of fundamental importance to the enterprise systems engineer is the finding or inducing of order in such a system. Ever since Christopher Alexander [71] laid the foundation for how to broadly think about the buildings, streets, and cities that organize the complex world in which we live and work, architecture has been a tool for visualizing the whole, parts, and their interactions. Nowhere in the ESE processes is the complexity and order of the enterprise more apparent than in its architecture.

1.4.7 Enterprise Analysis and Assessment

In Chapter 7, John Roberts clearly articulates the need for enterprise-level analysis and assessment (EA&A). He focuses on military command and control systems

---

* White has since augmented and updated the Regimen; the methodology is renamed Complex Adaptive Systems Engineering (CASE), cf. Ref. [72].

using a service-oriented architecture and a global infrastructure. The principles apply equally well to ecommerce and eGovernment applications riding on today's Internet and WWW. After establishing the need for a new suite of analysis and assessment tools at the enterprise level, he lays out seven specific recommendations for the composition of those tools. He discusses the roles of stakeholders and government in the process and ends with some practical considerations for implementation. EA&A is one of the core ESE processes, and in many respects, it is a way of keeping score on the enterprise evolution. It is well known that the way you keep score in a game determines how the game is played. EA&A is a key process for shaping the direction of evolution in an enterprise.

1.4.8 Enterprise Management

ESE is a *combined* management and technical process. As is characteristic in complex systems, the interaction of management and SE creates both needs and possibilities that neither could be addressed separately [73]. Chapter 8, by Robert Swarz, addresses the important problem of management processes that should be undertaken to perform the full ESE function. Although it does not focus on the traditional engineering or technical processes, it does discuss how some need to be augmented in ESE. The management processes include strategic planning (i.e., developing a plan, promulgating it, and managing it) as well as a number of organizational and governance issues. Material in this chapter has been liberally taken from Stephen Haines' book on the Systems Thinking Approach® to Strategic Planning and Management (CRC Press, 2000—used with the permission of the author) (www.HainesCentre.com). TSE processes that require extensions and variations in order to use them at the enterprise level include risk and configuration management. New issues associated with technology forecasting, planning, and transition are also discussed.

1.4.9 Agile Functionality for Decision Superiority

Chapter 9 by Kevin Cabana et al. is forward looking. It recognizes the key role that information plays in future warfare and the need for agility in an enterprise to maintain information advantage. The principles are equally applicable to any information-intensive, competitive environment in which it is critical to get the right information to the right people at the right time. The authors establish a framework based on information generation, management, and exploration as canonical functions and use the tenets of complex systems to ensure adaptation and agility in the capabilities these functions combine to enable. Many of the topics discussed are derived from current research in information systems, and much of the discussion includes the social aspects of the enterprise—the people part of people, processes, and technology.

1.4.10 Enterprise Activities (Evolving toward an Enterprise)

Written by a team of experienced systems engineers, Chapter 10 takes a fresh look at ESE activities needed to evolve an enterprise. Their "enterprise activities" focus on DoD Command, Control, Communications, Computers, Intelligence, Surveillance and Reconnaissance (C4ISR) systems, and they break into the three main categories of strategic planning, mission thread analysis, and technical framework development. They present a number of examples from industry and the DoD to illustrate their approach. What is important to note is that their intuition and experience independently led them to use much of the same holistic thinking discussed in Chapter 2 and many of the same processes discussed in the other chapters. What is most useful is that they quickly move from theory to practice giving a number of practical examples of what to do and how to do. In particular, they bring in the important enterprise characteristics of adaptability, flexibility, and scalability, and they introduce the use of a particular STP that uses the ISO stack organizing principles discussed in Section 1.3.2.

1.5 Preplanning the Future of ESE

The practice of ESE is proceeding at a feverish pace. The interconnected nature of our modern world, the collaborative ways in which we now work, and the influence of the Internet and WWW compel us to develop complex enterprises. A theory based on network science, complexity science, and social science is beginning to emerge, although at the present time there is no consensus on what it contains. We can expect that in the years to come, empirical evidence will accumulate on what works and what does not. That will be followed by refinement of the theory to explain what is observed and to predict what will happen in new enterprise developments. We can expect that the universities will begin teaching a consistent version of ESE-101 in its graduate programs and then in its undergraduate programs. In the spirit of this book, some parts of those predictions will gain support and some will fall by the wayside. New predictions will be reformulated, combined with other predictions, and the real future will evolve into something entirely different. Thus is the nature of complex systems.

References

1. T. S. Kuhn, *The Structure of Scientific Revolutions*. Chicago: University of Chicago Press, 1962.
2. H. E. Crisp, International Council on Systems Engineering. "Systems Engineering Vision 2020." INCOSE-TP-2004-004-02 2007 INCOSE Model Based Workshop, Albuquerque, January 2007.

3. http://en.wikipedia.org/wiki/System_dynamics.
4. A. Kossiak and W. N. Sweet, *Systems Engineering Principles and Practice*. New York: John Wiley and Sons, 2002.
5. C. Haskins, ed., *Systems Engineering Handbook*, vol. 3.1. INCOSE-TP-2003-002-03.1, INCOSE, August 2007.
6. P. W. Anderson, "More is Different." *Science*, 177, 4047 (1972): 393–396.
7. T. L. Friedman, *The World is Flat: A Brief History of the Twenty-First Century*. New York: Farrar, Strauss and Giroux, 2005.
8. K. Murphy, "John Gage, Sun's Voice to the Outside World." Developer.com, November 1998 http://www.developer.com/tech/article.php/606241.
9. M. H. Howes, "Methods and costs of constructing the underground facility of North American Air Defense Command at Cheyenne Mountain, El Paso County, Colo." U.S. Dept of Interior, Bureau of Mines Information Circular IC 8294, 1966.
10. P. E. Coyle, III, "Cheyenne Mountain upgrade." *Missile Defense and Related Programs Director of Operational Test and Evaluation FY98 Annual Report*. Department of Defense, February 1999.
11. W. Weaver, "Science and complexity." *American Scientist*, 36, (2008): 536, 1948.
12. G. M. Weinberg, *An Introduction to General Systems Thinking*. New York: Dorset House, 1975.
13. N. Wiener, *Extrapolation, Interpolation and Smoothing of Stationary Time Series with Engineering Applications*. Cambridge: MIT Press, 1930.
14. V. M. Tikhomorov, ed., *Selected works of A.N. Kolmogorov*, the Netherlands: Dordrecht, Springer, 1992.
15. C. E. Shannon, "A mathematical theory of communication." *Bell System Technical Journal*, 27, (1948): 379–423, 623–656.
16. W. R. Ashby, *Introduction to Cybernetics*. London: Chapman & Hall, 1956.
17. N. Wiener, *Cybernetics: or Control and Communication in the Animal and the Machine*. Cambridge: MIT Press, 1948.
18. L. von Bertalanffy, *General Systems Theory*. New York: George Braziller, 1969.
19. R. L. Ackoff and F. E. Emery, *On Purposeful Systems*. London: Tavistock Publications; Aldine, 1972.
20. J. March and H. Simon, *Organizations*. New York: Wiley and Sons, 1958.
21. M. Gell-Mann, *The Quark and the Jaguar: Adventures in the Simple and the Complex*. New York: Holt & Co., 1994.
22. S. Kauffman, *Investigations*. New York: Oxford University Press, 2000.
23. J. H. Holland, *Adaptation in Natural and Artificial Systems*. Cambridge: MIT Press, 1992.
24. "Excerpts from a Conversation with Gordon Moore: Moore's Law," Intel Corporation. 2005. ftp://download.intel.com/museum/Moores_Law/Video-Transcripts/Excepts_A_Conversation_ with_Gordon_Moore.pdf.
25. D. C. Dennet, *Darwin's Dangerous Idea*. New York: Simon & Schuster, 1996.
26. T. Berners-Lee, "The World Wide Web." A very short personal history. http://www.w3.org/People/Berners-Lee/ShortHistory.html.
27. S. Kauffman, *The Origins of Order*. New York: Oxford University Press, 1993.
28. R. Axelrod and M.D. Cohen, *Harnessing Complexity*. New York: Basic Books, 2001.
29. J. H. Holland, *Hidden Order: How Adaptation builds Complexity*. New York: Basic Books, 1995.
30. E. D. Beinhocker, *The Origin of Wealth*. Boston: Harvard Business School Press, 2006.

31. S. Johnson, *Emergence: The Connected Lives of Ants, Brains, Cities and Software.* New York: Simon & Schuster, 2001.
32. B. Boehm, "A spiral model of software development and enhancement." *ACM SIGSOFT Software Eng. Notes,* 11, 4 (August 1986).
33. F. Brooks, "No silver bullet: Essence and accidents of software engineering." *IEEE Computer,* (1987): 10–19.
34. R. Stevens and R. M. King, "Venture capital and IT acquisition: Managing uncertainty." *IEEE International Systems Conference.* Montreal, 2008.
35. H. A. Simon, "Structure of ill-structured problems." *Artificial Intelligence,* 4 (1973): 181–201.
36. R. J. Byrne, *The Yin and Yang of Complex Open Systems (Balancing Integration with Innovation),* Bedford, MA: MITRE Corporation, Technical Paper # 09—0097, February 2009, http://www.mitre.org/work/tech_papers/tech_papers_09/09_0097/.
37. R. J. Byrne, *ESC Strategic Technical Plan (STP) v2.0* http://info.mitre.org/infoservices/selfserve/public_release_docs/2005/05-0518.pdf.
38. D. Kaye, *Loosely Coupled, the Missing Pieces of Web Services.* Marin County, CA: RDS Press, 2003, p. 31.
39. R. J. Byrne, Practicing Enterprise Systems Engineering. Position Paper for *Symposium on Complex Systems.* January 2007. http://cs.calstatela.edu/wiki/images/6/6b/Byrne.doc accessed January 2009.
40. C. Alexander, S. Ishikawa, M. Silverstein, M. Jacobson, I. Ingrid Fiksdahl-King, and S. Angel, *A Pattern Language: Towns, Buildings, Construction.* Center for Environmental Structure, New York: Oxford University Press, 1977.
41. E. Gamma, R. Helm, R. Johnson, and J. Vlissides, *Design Patterns Elements of Reusable Object-Oriented Software.* Reading, Massachusetts: Addison-Wesley Professional, 1994.
42. G. Hohpe and B. Woolf, *Enterprise Integration Patterns.* Reading, Massachusetts: Addison-Wesley Professional, 2003.
43. M. Fowler, *Patterns of Enterprise Application Architecture.* Reading, Massachusetts: Addison-Wesley Professional, 2002.
44. ISO/TC97/SC16, "Reference Model for Open Systems Interconnection." doc N227, June 1979.
45. R. Braden, ed., Requirements for Internet Hosts—Communication Layers. IETF, RFC: 1122, October 1989.
46. W. W. Eckerson, "Three tier client/server architecture: Achieving scalability, performance, and efficiency in client server applications." *Open Information Systems* 10, 1, 3 (1995): 20.
47. R. Fielding et al., "Hypertext Transfer Protocol—HTTP 1.1." *IETF* June 1999, 64. http://tools.ietf.org/html/rfc2616.
48. J. Postel and J. Reynolds, "Domain requirements." *IETF,* RFC 920, 1984, http://tools.ietf.org/html/rfc920.
49. R. W. Miller and D. G. Winkowski, Loose couplers as information design strategy. *IEEE MILCOM 2007,* October 2007.
50. D. J. Watts, *Small Worlds: The Dynamics of Networks between Order and Randomness.* Princeton: Princeton University Press, 1999.
51. M. Newman, A. L. Barabasi, and D. J. Watts, *The Structure and Dynamics of Networks. Princeton Studies in Complexity.* Princeton, New Jersey: Princeton University Press, 2006.
52. R. Albert and A. L. Barabasi, "Statistical mechanics of complex networks." *Reviews in Modern Physics,* 74 (2002): 47–97.

53. A. L. Barabasi, *Linked: How Everything is Connected to Everything Else and What It Means for Business, Science, and Everyday Life.* New York: Plume, 2003.
54. E. Friedgut and G. Kalai, "Every monotone graph has a sharp threshold." *Proceedings of the American Mathematical Society*, 124, 10 (1996): 2993–3002.
55. O. Brafman and R. A. Beckstrom, *The Starfish and the Spider, the Unstoppable Power of Leaderless Organizations.* New York: Penguin Group, 2006, p. 193.
56. E. Ostrom, *Governance of the Commons: The Evolution of Institutions for Collective Action.* New York: Cambridge University Press, 1990.
57. D. Dietrich, *The Logic of Failure: Recognizing and Avoiding Error in Complex Situation.* New York: Perseus Books, 1989.
58. P. Senge, *The Fifth Discipline.* New York: Double Day Publishing, 1990.
59. J. D. W. Morecroft and J. D. Sterman, eds, *Modeling for Learning Organizations.* Portland, Oregon: Productivity Press, 1994.
60. J. Gharajedaghi, *Systems Thinking.* Boston: Elsevier, 1999.
61. D. Ancona and H. Bresman, *X-teams: How to Build Teams That Lead, Innovate and Succeed.* Jackson: Perseus Distribution Services, 2007.
62. J. K. DeRosa, G. Rebovich, and R. Swarz, "An enterprise systems engineering model." *Proceedings of 16th International Symposium, International Conference on Systems Engineering (INCOSE)*, Orlando, July 2006.
63. R. Swarz and J. K. DeRosa, "A framework for enterprise systems engineering processes." *Proceedings of the International Conference on Software and Systems Engineering, CNAM*, Paris, December 2006.
64. ANSI/EIA, *Processes for Engineering a System.* ANSI/EIA-632-1999 http://webstore.ansi.org/RecordDetail.aspx?sku=ANSI%2FEIA-632-1999.
65. L. K. McCaughin and J. K. DeRosa, "Stakeholder analysis to shape the enterprise." *Proceedings of the 6th International Conference on Complex Systems, New England Complex Systems Institute*, Boston, June 2006, http://necsi.org/events/iccs6/viewpaper.php?id=390.
66. L. K. McCaughin and J. K. DeRosa, "Balancing the balanced scorecard for complex systems." *The MITRE Corporation, International Conference on Software & Systems Engineering Applications*, Paris, December 2007.
67. R. Ackoff, *Ackoff's Best: His Classic Writings on Management.* New York: John Wiley & Sons, 1999.
68. D. O. Norman and M. L. Kuras, "Engineering complex systems, Chapter α." *The MITRE Corporation*, January 2004. http://www.mitre.org/work/tech_papers/tech_papers_04/norman_engineering/index.html.
69. M. L. Kuras and B. E. White, "Engineering enterprises using complex-system engineering." *INCOSE 2005 Symposium*, Rochester, July 2005.
70. E. Lorenz, Deterministic "Nonperiodic Flow." *Journal of the Atmospheric Sciences*, 20 (1963): 130–141.
71. A. Christopher et al. *A Pattern Language: Towns, Buildings, Construction.* Center for Environmental Structure, New York: Oxford University Press, 1977.
72. B. E. White, "Complex adaptive systems engineering." *8th Understanding Complex Systems Symposium.* University of Illinois at Urbana-Champaign, May 12–15, 2008; for paper, presentation, and audio clip, refer to: http://www.howhy.com/ucs2008/schedule.html.
73. L. K. McCaughin and J. K. DeRosa, "Combined systems engineering and management in the evolution of complex adaptive systems." *IEEE System Conference*, Honolulu, April 2007.

Chapter 2

Systems Thinking for the Enterprise

George Rebovich Jr.

Contents

2.1 Introduction 32
2.2 The Enterprise 34
2.3 Moving from Classical Systems Engineering to Enterprise Engineering.....38
2.4 Modes of Thought..39
    2.4.1 Complementarity..41
    2.4.2 Interdependence..43
2.5 A Framework for Harnessing Complexity 44
    2.5.1 A Systems View of Development....................................... 44
    2.5.2 A Systems View of Development—Examples..................... 46
    2.5.3 A General Development Framework for Complex Systems............47
    2.5.4 Variation..49
        2.5.4.1 The Exploration–Exploitation Trade.....................50
    2.5.5 Interaction 51
    2.5.6 Selection 53
2.6 Application to an Enterprise...55
    2.6.1 Purposeful Questions 55
        2.6.1.1 Systems Thinking Questions 55
        2.6.1.2 Harnessing Complexity Questions 56
    2.6.2 Examples 57
        2.6.2.1 Portfolio Management...57
        2.6.2.2 Evolving Enterprise Processes and Practices.....................59

2.7 Postscript 60
Acronyms 61
References 61

2.1 Introduction

Governments, large multinational organizations, and indeed all of society are in the midst of a major transformation driven by and deriving its character largely from advances in information technology (IT).

The rate of technical change in information processing, storage, and communications bandwidth is enormous. Expansions in other technologies (e.g., netted-sensors) have been stimulated and shaped by these changes. The information revolution is intensifying interactions among people, businesses, systems, organizations, nations, and processes that were previously separated in distance or time. Surprisingly, future events in this information-abundant world are harder to predict and control with the result that our world is becoming increasingly complex [1]. This is illustrated in Figure 2.1 in which our increasing World Wide Web (WWW) interconnectedness makes us all coproducers of outcomes in airline flight availability. The example illustrated in Figure 2.1 is for air travel pricing and reservations. Our increasing interconnectedness and access through the Internet and the WWW (left side of the figure) puts us all in competition for finite resources such as nonstop connections and seats whose availability and price can change with astonishing speed (right side screen shots). Individuals, as consumers, and commercial airlines, as providers, are coproducers of a complex outcome that maps passengers and prices to flights and seats.

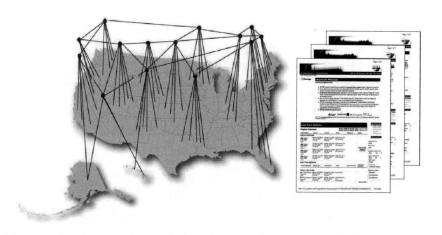

Figure 2.1 Our world is becoming increasingly complex. (Adapted from Rebovich, G. Jr. *Proceedings of 2006 IEEE International Conference on Systems*, April 2006.)

This new complexity* is not a consequence of the interdependencies that arise when large numbers of systems are networked together to achieve some collaborative advantage. When the networked systems are individually adapting to both technology and mission changes, then the environment for any given system or individual becomes essentially unpredictable. The combination of large-scale interdependencies and unpredictability creates an environment that is fundamentally different from that of the system or system of systems (SoS) level. As a result, systems engineering success expands to encompass not only success of an individual system or SoS, but also the network of constantly changing systems. Examples in which this new complexity is evident include the United States Federal Aviation Administration's National Airspace System, the Defense Department's Global Information Grid, the Internal Revenue Service's Tax Systems, and the Department of Homeland Security's Secure Border Initiative's SBInet [3].

In any age, the view we hold of our world is disrupted and threatened by new technologies of the era. But ultimately we synthesize them into a new and more powerful perspective. For example, astronomy threatened the view of a geocentric universe but it ultimately led to a more powerful view of the cosmos and our place in it. And so it is in our day that IT has both revealed the complexity in our social systems and accelerated it [4].

How do we form a new and more powerful perspective for dealing with this complexity? What is needed is a new way of thinking—a systems thinking† that captures the fundamental relationships of information to complexity so that engineers of every kind of enterprise can obtain the benefits and elude the obstacles of this change. This new systems thinking is rooted in several distinct fields but one of them—evolutionary biology—is a tap root. The use of evolutionary concepts to deal with complexity has grown in recent years because biology has changed from a passive, *descriptive* science to an active, *constructive* one.‡ Knowledge of genetics is used to manipulate and modify chromosomes to avoid or cure human diseases by creating new drugs or, in some cases, by altering the genetic makeup of individual humans. We clone life forms and not just simple organisms but some profoundly complex ones. These advances and others like them have fundamentally changed our view of and relationship to biology and its modes of thought.

---

* Complexity theory forms a framework for understanding difficult problems and applying multiple disciplines to their solution.
† *Systems thinking* is the ability and practice of examining the whole system rather than trying to fix isolated problems [5]. It requires a *systems perspective*, which is the act of taking into account all the behaviors of a system as a whole in the context of its environment. This focuses on the interactions and relationship between the system and its environment (Y. Bar Yam, NECSI, personal communication). More fundamentally, *thinking* refers to how one organizes and uses knowledge to cope with situations: it mediates the perception of a problem and the production of a response (*World Book Encyclopedia*).
‡ Kenneth R. Miller, Professor of Biology, Brown University.

As a consequence we find ourselves poised at the leading edge of a new systems thinking which is emerging at the intersection of IT and evolutionary biology.* As will be seen, part of this new systems thinking requires a replacement of the notion that specific engineering outcomes or goals can always be assured with one that seeks to shape, improve, or increase the value of engineering outcomes through thoughtful interventions in the ever-increasing numbers of circumstances in which we are not fully in control.

The purpose of this chapter is to introduce and discuss new and emerging modes of thought that are increasingly being recognized as essential to successful systems engineering within an enterprise. The emphasis is on those that are substantially different in kind or degree from those in classical system engineering. Classical systems thinking and processes will always be relevant to the degree that the problem is or can be constrained to a system or SoS problem. But they do not equip the systems engineer to deal with all that is going on in an enterprise. This new way of thinking is meant to supplement traditional systems thinking, not supplant it.

The presentation approach in this chapter discusses the principles of a concept followed by an everyday example of it. Examples are drawn from diverse sources and situations including organizational, social, technical, political, military, educational, governmental, financial, medical, and even board games.

2.2 The Enterprise

By *enterprise* we mean an entity comprised of interdependent resources (e.g., people, processes, organizations, technology, and funding) that interact with each other (e.g., coordinate functions, share information, and allocate funding) and their environment to achieve goals, as depicted in Figure 2.2.†

Historically, our focus has been on the technologies which have enabled the development of the piece parts—systems and subsystems—contained in the enterprise. Modern systems thinkers such as Ackoff and Gharajedaghi are increasingly taking a holistic view of an enterprise as [6,7]

- A multiminded, sociocultural entity
- Comprised of a voluntary association of members who can choose their goals and means

---

* Some make the case that ecology is more to the point than evolutionary biology. An ecology evolves itself through replacement of its elements and is in a dynamic equilibrium which is akin to the development of a system of IT systems (from a discussion with D. Norman, The MITRE Corporation).
† This definition is similar in its essentials with that of *Enterprise* in Ref. [8].

Systems Thinking for the Enterprise ■ 35

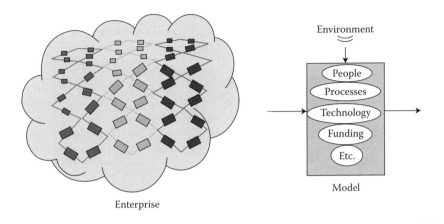

Figure 2.2 Enterprise model.

- An entity whose members share values embedded in a (largely common) culture
- Having the attributes of a purposeful entity*
- An entity whose performance improves through alignment of purposes across its multiple levels

There has been a steady progression toward this enterprise model in the U.S. government organizations that procure and use IT-intensive systems. Witness the reorganizations and restructurings over the past decade that have eliminated various forms of stovepiping as well as organizational initiatives to explicitly coordinate and synchronize goals from the highest levels of an organization down to individual performers.

There is a nested nature to most government enterprises.† At every level, except the very top and bottom, an enterprise itself is part of a larger enterprise and contains subenterprises, each with its own people, processes, technologies, funding, and other resources. As depicted in Figure 2.3, the family of Airborne Early Warning and Control (AEW&C) systems is an enterprise which is nested (or

---

* A *purposeful entity* can achieve the same outcome in different ways (in the same environment) and can achieve different outcomes (in the same or different environments).
† The property of nestedness derives from the goal oriented, hierarchical nature of most government organizations. These kinds of enterprises are the focus of this report. This property is not seen in or to be expected of all enterprises. For example, many chain hotel enterprises are, in fact, associations of independent properties that operate as agents of and on behalf of the hotel enterprise in providing lodging and related services. The hotel corporation has little or no control of and assumes limited responsibility for the agent properties. In these arrangements either party can readily terminate the affiliation between the property and the hotel enterprise.

36 ■ *Enterprise Systems Engineering*

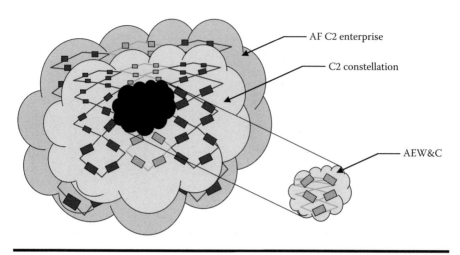

Figure 2.3 Nested nature of enterprises.

contained) in the Command and Control (C2) Constellation enterprise is contained in the Air Force C2 Enterprise.

Alignment of purposes across the levels of the enterprise can improve overall enterprise performance: the subenterprises contribute to the goals of the containing enterprise. This view has important implications for how systems engineers must think about their activities within an enterprise setting: it puts a premium on synthesis.*

For example, at the AEW&C system program level, the view must be that an AEW&C system builds an air picture that serves the higher goal of achieving situation awareness (SA) within the C2 Constellation. This requires the AEW&C systems engineer to ask (and answer) how the AEW&C piece parts developed serve SA in the C2 Constellation *in addition* to how they serve the AFW&C system specification. At the next level of the enterprise,† the view must be that the C2 Constellation develops integrated capabilities to serve the higher goal of providing net-centric C2 for the Air Force C2 Enterprise. The implication is that the systems engineer must address how the C2 Constellation piece parts serve the Air Force C2 Enterprise in addition to how they serve the C2 Constellation. At the highest level,‡ in this example, the view must be that the Air Force C2 Enterprise develops Air Force Net-Centric capabilities to serve the higher goal of providing net-centric C2 for the Joint/Coalition C2 Enterprise. The implication is that the

---

* *Synthesis* is the ability to identify the whole of which a system is a part, explain the behavior or properties of the whole, and disaggregate the whole to identify the role or function of the system in the whole [7].
† *Wing* level in the U.S. Air Force.
‡ Center *Command* level.

systems engineer must address how the Air Force C2 Enterprise piece parts serve joint and coalition net-centric C2 in addition to how they serve the Air Force C2.

This discussion leads to an operational definition of enterprise viewed from the perspective of an individual (system engineer or other participant) or team in the enterprise. It aims to answer the question, "what is my (our) enterprise?"

The enterprise is the set of interdependent elements (systems and resources) that a participating actor or actors either control* or influence.† The remainder of the elements constitutes the enterprise environment [7]. This is depicted in Figure 2.4.

Note that this definition of enterprise and its boundary are virtual constructs that depend on the makeup, authority, and roles of the participating actors in a community of interest. For example, the system engineering team of a system managed by a program office may have virtual control of most engineering decisions being made on the system's day-to-day development activities. If the system is required to be compliant with standards developed by the U.S. Defense Information Systems Agency, the program team may have representation on the standards team, but that representation is one voice of many and so the standard is a program element or variable that the program team can influence but not control. The U.S. Federal Acquisition Regulation requirements, which apply to virtually all U.S. government acquisitions, are elements or variables that apply to our example program, but since they are beyond the control of the team, they are part of the program's environment.

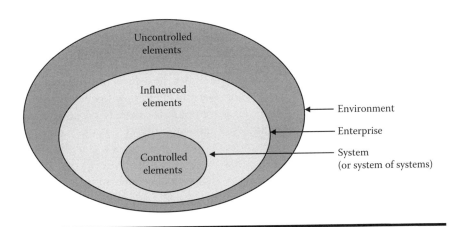

Figure 2.4 Operational definition of enterprise.

---

* To *control* means that an action we can choose to take is necessary and sufficient to produce an outcome.
† To *influence* means that an action we can choose to take is necessary but not sufficient to produce an outcome; our action is a co-producer of an outcome.

The implication and view being advocated here is that all actors or teams in an enterprise setting should know their enterprise and be aware of which enterprise elements or variables they control and which they influence. Environmental elements or factors cannot be controlled or influenced. But the individual or project team may very well need to be aware of and understand implications of environmental factors.

We will return to this controlled–influenced–uncontrolled framework later when the discussion focuses on approaches to shaping, improving, or increasing the value of outcomes for the ever-increasing number of enterprise elements that can be influenced but not controlled.

2.3 Moving from Classical Systems Engineering to Enterprise Engineering

Systems engineering can be defined as a process that integrates multiple disciplines to define and transform requirements into a system while involving environmental, economical, political, and social aspects. Classical systems engineering is a term used to describe engineering of subsystems, systems, and some forms of SoS.

Classical systems engineering is a sequential, iterative development process whose goal is to produce products, many of which are of unprecedented technical complication and sophistication. The INCOSE (ANSI/EIA 632) systems engineering model is a widely recognized representation of classical systems engineering.

An implicit assumption of this classical model is that all relevant factors are largely under the control of or can be well understood and accounted for by the engineering organization, the system engineer, or the program manager. The culture and processes in such an organization normally reflect this assumption.

There are fundamental differences in an enterprise. While some factors may continue to be well understood by or remain under the control of the system engineer or program manager, others are not. Because enterprises frequently embrace diverse agencies, sponsors, and operational communities, there is increased emphasis on working across and bridging organizational cultures, agendas, and socio-political-economic differences. Enterprises exhibit attributes of a complex system. As a result there arise questions of how to develop enterprise processes that approximate natural evolution and, in some cases, how to deliberately mimic, encourage, facilitate, and channel them in constructive directions. Adjectives that are used to describe system engineering and other processes in an enterprise expand to include: *evolutionary, emergent, adaptive, self-organizing, competitive,* and *cooperative.*

For many, even most systems engineers the reality of their existence is that they inhabit and must perform well in two different system engineering environments at the same time: a local, program office environment in which the culture, norms, expectations, and rewards may be more influenced by a classical systems engineering culture and an enterprise engineering culture in which the focus is on enabling the broader enterprise, in some cases even at the expense of the smaller system.

Every system engineer must bring enterprise perspectives, behaviors, skills, and competencies* to their activities whether they are executed at the subsystem, system, SoS, or enterprise scale. In that sense, every system engineer operating in or working on an enterprise must consider themselves as an enterprise engineer, who is required to perform systems thinking for the enterprise.†

2.4 Modes of Thought

Analysis is the ability to decompose an entity into deterministic components, explain each component separately, and aggregate the component behaviors to explain the whole [9].‡

Analysis results in knowledge of an entity: it reveals internal structure. If the entity is a system then *analysis* answers the question, "how does the system work?" For example, to know how an automobile works, you analyze it: that is, you take it apart and determine what each part does. This is essential to important activities such as repairing automobiles or diagnosing and repairing problems of other, more complicated systems.

---

* A *behavior* is a specific action. A *skill* is an ability to perform a complex collection of actions with ease, precision, and adaptability to changing conditions. *Knowledge* is a body of understood information possessed by an individual which is in accord with an established fact. A *competency* is a cluster of job-related behaviors, skills and knowledge that affects a major part of one's role or responsibility; correlates with success; can be measured against accepted standards (that relate to success); and can be improved via training and development. (Adapted from Parry, S. B., *Training Journal*, 33 (7), 48–56, 1996.)

† Thus, an *Enterprise Engineer* is a system engineer who brings an enterprise perspective to his or her activities. The Enterprise Engineer may work on problems at any scale of the enterprise: subsystem, system, SoS, or Enterprise. The individual who does not (or is not required to) bring an enterprise perspective to his or her activities is termed a subsystem engineer; SoS engineer, enterprise system engineer, etc. (from a discussion with R. Swarz, The MITRE Corporation).

‡ This is a classical definition of systems analysis and a very simple one at that. New modes of analysis are emerging. The forces at work in this development are many. Part of it comes from the evolution to net-centric systems in which one can neither state, *a priori*, the extent of a system nor decompose it into deterministic parts which perform specific or guaranteed roles or functions. Yet the need to perform analysis continues, unabated. Additionally, the view of systems and enterprises as sociocultural entities suggests that analysis should include social or human dimensions, not just technical ones. This implies that systems analysis may have important similarities with (and much to learn from) fields as diverse as intelligence analysis, medical analysis, stock market analysis, and weather forecasting. Finally, the role of conceptual framework, limits in cognition, and biases in human decision making are increasingly being recognized as forces which are shaping how we view and do systems analysis: it is not just about the data, it is also about how we, as humans, perceive the data. It is not the intent of this chapter to address these new and emerging aspects of systems analysis, which would require a discussion every bit as long as this one.

Analysis, as one of our earliest and most natural modes of thought, is so fundamental to human thinking that it has almost become synonymous with it. But *analysis* is not the only mode of thought required to answer important questions about a system.

Synthesis is the mode of thought that results in the understanding of an entity: that is, an appreciation of the role or function an entity plays in the larger system of which it is a part. Just as *analysis* answers the "what is it?" question about an entity, *synthesis* answers the "why is it what it is?" question.

Synthesis is the ability to identify the whole of which a system is a part, explain the behavior or properties of the whole and disaggregate the whole to identify the role or function of the system in the whole [7].

Systems thinking requires knowledge *and* understanding—both analysis and synthesis—represented in the same view. That is precisely the point of the AEW&C/C2 Constellation example. The ability to combine analytic and synthetic perspectives in one view is an enabler of alignment of purposes, which is so important to successful engineering in nested or hierarchical enterprises. It allows the systems engineer to ask purposeful questions and trace the implications of potential answers across the enterprise. Would a change in performance at the subsystem level result in a change at the enterprise level? If so, how, and is it important? How would a new enterprise-level need to be met?

There is another reason why *synthesis* is such an important *systems thinking* concept. It expands the solution space of a problem in a way that can make intractable problems solvable. The following example illustrates how.

Example

In the early twentieth century, the City of Chicago ran a large trolley car system. The General Manager (GM) of the trolley system had a goal to improve its performance and efficiency. Trolley cars were run by two operators: a driver and a conductor. The driver was responsible for maintaining the trolley's schedule: that is, arriving and departing the route stops on time. The conductor was responsible for collecting fares and assuring that passengers paid the correct fare and departed at the right stop (fares were based on geographical zones). Both driver and conductor were paid fixed hourly salaries.

The GM reasoned that if the operators' salary structure were changed to a lower hourly base rate plus a performance-indexed rate this would motivate operator performance improvement which would translate into improved overall trolley system performance and efficiency. This was done. The performance-indexed part of the drivers' salary was tied to on-time schedule performance and that of the conductor to assuring fare equity.

But this change immediately caused a conflict. The driver was motivated to depart a stop quickly (to stay on schedule) while the conductor was motivated to more carefully (and, hence, more slowly) issue tickets, collect fares, and check that departing passengers had not gone beyond the zone for which they paid (inspectors, masquerading as ordinary passengers, evaluated driver and conductor performance). The relationship between drivers and conductors became antagonistic with the result that overall trolley system performance actually decreased. The problem, of course, was that the new performance system had created an inherent conflict between the operators.

Many solutions were tried. They all focused on improving the efficiency of one or the other of the trolley operators' job as a mechanism for deconfliction. For example, different internal configurations of the trolley were tried in an attempt to reduce the time it took to execute conductor responsibilities. Nothing worked and the conflict continued.

A consultant was finally brought in. The problem was explained to him in great detail, including all the options that had been considered for improving operator efficiency. The first question the consultant asked was, "how many trolleys are in the system?" The GM thought this was a curious and not particularly relevant question given that the problem was a conflict between two operators on a trolley and he answered, "1250." The consultant wrote that down and asked, "how many trolley stops are in the system?" The GM answered, "850" and went on to express his disappointment in this line of obviously irrelevant questioning. The consultant wrote down 850, looked up at the GM and said, "I have a solution to your problem. There are more conductors than stops. Take the conductors off the trolleys, position them at the stops where they can collect fares from customers lining up at the stop before the trolley arrives and check fare zones from passengers after they exit the trolley. In this way you will completely separate the contentious timelines of the two operators." [7]

The consultant had used a counterintuitive principle to solve the trolley problem. When faced with an apparently intractable problem the usual response is to make it smaller, contain, or constrain it. The consultant in this example was going down a line that appeared to make the problem space larger in seemingly irrelevant ways. But, in fact, he was making the *solution* space larger by viewing the problem in the context of its containing whole. This is a form of *synthesis*.

This example also touches on and leads into the next topic, *complementarity*.

2.4.1 Complementarity

Much of traditional systems thinking has been dominated by a view that opposing tendencies are to be treated as two sides of a zero-sum game: a gain for one requires a loss for the other.

The principle of *complementarity* or *multidimensionality* holds that seemingly opposing tendencies not only coexist, overlap, and interact, but they may also form a complementary relationship. More formally, *complementarity* is the ability to see complementary relations in opposing tendencies and to create feasible wholes with seemingly unfeasible parts [6].

This principle changes the categorization of opposing tendencies from an *or* to an *and* relationship. They can then be viewed as distinct, interdependent dimensions that can interact and be integrated into a new way of thinking. *Complementarity* expands the way of thinking about opposing tendencies from choices between win, lose, or compromise to one in which both (or all) opposing tendencies can win.

Example

Westville (fictitious name) is a suburb of a large New England city. The community has had a long history of quality in its educational system. The reputation is justified. By any measure—results in academic contests, SAT scores, or placement in the most competitive universities—Westville consistently outperformed its more affluent neighboring towns and even most nearby prestigious private schools.

The quality education Westville schools provide was a consequence of several factors: strong parental involvement; the percentage of local taxes that are devoted to the schools; and the effectiveness of school system administrators in applying monies to programs of excellence.

Life in Westville revolved around the children and schools so much so that after their children graduated "empty nesters" in large numbers would migrate out of town in search of communities better aligned with their new interests. They were replaced by younger families eager to take advantage of Westville's quality schools.

But something changed in the early 1990s. Adults whose children had graduated from the school system increasingly viewed Westville as their retirement community of choice. These active seniors began looking to the town for services and facilities that served their needs.

At the same time, education costs began rising faster than the economy. Families with school-aged children advocated additional taxes to maintain the quality of education while retirees, particularly those on fixed incomes, became apprehensive of their ability to afford living in "their" town. Over time, a conflict emerged and grew as a consequence of the shifting demographics of the town. Families with school-aged children and empty nesters had different priorities. Town meetings became heated.

Town members framed the discussion as a contention concerning the amount and distribution of financial resources. The solution took the form of an unstable compromise between opposing factions representing quality schools and senior livability. The school administration would find and implement incremental efficiencies year after year until it became clear to a majority of the town that additional monies were needed for schools at which point the town voted an increase in taxes, to the great consternation of the seniors. Seniors got a few programs aimed at them but they were always judged "too little, too late." Both sides were unhappy most of the time.

A small number of citizens formed a group that aimed at asking the question, "what would a solution look like that maintained school quality and improved senior livability?" The question suggested an interdependence between what had been viewed as two opposing tendencies and opened up new solution possibilities beyond win, lose, or compromise.

In discussing this question it became clear that many empty nesters and retirees,

- Came from the knowledge industry
- Understood the value of education
- Wanted to give something back to their community
- Had a "pipe dream" of contributing to education in some way

An idea emerged out of the group for a program in which empty nesters and retirees could become teacher assistants or even teachers (if certified or willing to undergo certification). Individuals in the program could participate as volunteers (no remuneration), for a tax credit (to be applied against local taxes) or be paid a small salary. Many participated. Most volunteered, some took a tax credit, and a few opted for the salary.

The program became popular and spread to other Westville town services (library, recreation center, after school sports, and other activities). The cost savings reduced educational and other town expenses. Tax increase requests reduced in frequency and magnitude. Seniors became more integrated in the community. They felt less of a need for separate senior programs and so those demands decreased. What remained of them was adequately met by a portion of the town's cost savings.

Westville succeeded in transforming a win, lose, compromise situation to one in which the vast majority of citizens were happy. It did so by reframing an "or" situation into an "and" one using the principle of complementarity.

We will return to complementarity later in this chapter where it will be used with the controlled–influenced–uncontrolled framework to illuminate the complementary relationship between differentiation and integration, and how complexity can be harnessed through thoughtful intervention in processes that are beyond our direct control.

2.4.2 Interdependence

Classical systems thinking has focused on analysis and analysis has focused on problems in which the variables are independent or may reasonably be assumed to be so. When the variables are, indeed, independent this is a powerful framework since it allows the systems engineer to logically decompose a system into its variables and examine the response of the system to changes in each variable, one at a time [6].

A common pitfall is not recognizing that a system may contain slack between its variables and it is the slack which allows the analyst to treat the variables as though they are independent *but only in a number of circumstances* or *up to a point*. The performance of the variables in a system can be improved until the slack among them is used up at which point the variables become interdependent and improvement in one comes only at the expense of others.

Enterprises or other large, complex organizations may run into interdependence as an issue when engaged in quality or other system-wide improvement campaigns. In these situations it is customary to gauge improvement as performance increases in multiple variables (shorter development time, better cost containment, increased customer satisfaction, reduction in product defects, improved safety, etc.). Often, these campaigns reach a plateau, stall in their progress, or even regress as measured by the variables of interest. When this happens it is usual to think that the organization needs to be re-energized in the improvement campaign and, of course, this may be the case. However, plateauing may be a symptom that the enterprise or system—in its current form—has reached its potential. It has used up all its slack. Further progress can only come from a redesign of the organization or operation of the enterprise or system.

Example

The trolley example is an illustration of a situation in which further progress required a redesign of the system. As long as the design of the overall trolley system kept the driver and conductor on the trolley, it created an interdependence between driver performance and conductor performance that traded improvements in one at the expense of the other. Moving the conductor from the trolley was—in essence—a redesign of the system that provided additional slack between the two performance variables.

2.5 A Framework for Harnessing Complexity

Earlier in this chapter an operational definition of enterprise was introduced that is based on the notion that the elements or variables in our world could be partitioned into those we control, those we influence, and those beyond our control. A consequence of the U.S. government evolving toward an enterprise model is that each of us, as participating actors in our enterprise, are losing the ability to control many elements while at the same time we are gaining in our potential to influence many others. Systems engineering, therefore, is becoming more and more about the ability to influence in our increasingly complex environment.

2.5.1 A Systems View of Development

Development* is a purposeful transformation toward higher levels of differentiation and integration at the same time. Development is an evolutionary process by which a social system (e.g., individual, team, community, enterprise) increases its ability or value (as gauged by performance, effectiveness, impact, influence, profitability, etc.). *Differentiation* refers to deviations among entities that are apparently similar and *integration* refers to the similarities among things that are apparently different. The former emphasizes tendencies toward increased complexity, variety, and autonomy while the latter tends toward increased order, uniformity, conformity, and collectivity [6].

The basic idea is the seemingly opposite tendencies of differentiation and integration complement each other to create *innovation*, which is both a destination and a journey or process.

The discussion that follows in the next several paragraphs will step through Figure 2.5 using an enterprise as an example, but it should be kept in mind that this developmental model is applicable to individuals and teams as well as enterprises.

---

* This development model is quite general and can be applied to individuals, teams (large or small), systems, and enterprises as defined previously. Our focus will generally be on teams, systems, and enterprises which will be generically referred to as "organizations." The other terms will be used where appropriate.

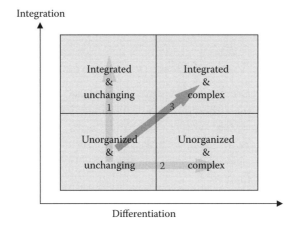

Figure 2.5 A systems view of development.

An enterprise in the lower left quadrant of Figure 2.5 is static and unorganized. Change is infrequent or nonexistent and the piece-parts in there are unorganized into any semblance of a coherent whole. Movement or evolution is generally possible, of course, so the question becomes one of direction.

One evolutionary path is to organize all the elements and processes that currently exist in the enterprise, that is, integrate straight up to the upper left quadrant, as indicated by arrow 1 in Figure 2.5. The result is organized simplicity: a highly integrated organization which knows how to do things very well in essentially one way. Change in the way things are done is difficult to effect. The organization is rigid and inflexible in how it responds to problems, opportunities, and other changes in its environment.

Another possibility for evolution is toward unorganized complexity, as indicated by arrow 2. This is an organization that proliferates many ideas for accomplishing activities or developing products (differentiation) but never integrates them into a coherent process or product. Frequently, there is an element of innovation in this differentiation, but what keeps it from being true innovation is that the good ideas never get realized. Examples of unorganized complexity are some of the early Internet inventors who produced numerous elegant and novel approaches to solving Internet technology problems but never put them together into a coherent, financially profitable business package.

The last possibility for evolution is toward integrated complexity, as indicated by arrow 3. This is a movement toward complexity and order at the same time. This is an organization that proliferates many new ideas about its processes, products, and solution approaches and then selects and integrates the best of them into subsequent versions or entirely new incarnations. Some organizations develop elaborate and detailed 10-year plans and then execute to that plan. An integrated, complex organization is not like that: instead, it reinvents itself through continual differentiation and integration to higher levels of value and performance. This continuous reinvention provides adaptability to environmental changes.

2.5.2 A Systems View of Development—Examples

Complex entities such as individuals, teams, or enterprises do not develop in straight lines, as suggested by the arrows in Figure 2.5; nor are specific developmental outcomes necessarily guaranteed when an individual, team, or enterprise embarks on a developmental program. So, how do complex entities develop? The following examples motivate a general developmental framework which will be elaborated in the remainder of this chapter.

Example

John is a capable performer who wanted to improve his value and impact at work. In collaboration with his supervisor and mentor, John decided to achieve his goal by improving his communication skills, technical currency, and visibility within his organization.

John looked into several opportunities for improving his communication skills including a variety of courses on effective writing and effective briefing preparation. He enrolled in one of each. John also joined a debating club. He soon realized that he communicated adequately in prepared situations and so he dropped the courses and increased his involvement in the debating club. John integrated his debating club experiences into his on-the-job extemporaneous discussions.

John explored different ways of increasing his technical currency: evening classes at a local university; e-Learning; and in-house courses at his place of employment. His frequent business travel affected the evening class schedules. The e-Learning environment was too unstructured for John. The in-house courses provided the right balance of flexibility for his travel schedule and structure, so John focused on that approach to improve his technical currency.

John was known and well regarded in his business unit. But it was a small and established part of the company's overall operations. John diversified his individual work portfolio so that he and his work became more widely known in the company. He also sought out and contributed to initiatives perceived as vitally important to the company's future. As a result, John got visibility at higher organization levels.

Example

Jane runs an engineering consulting group that is geographically divided between two locations. One part of the group is collocated with the client and the other is located at a facility 400 miles away.

The client was generally pleased with the consulting services provided but began to question the contributions of the part of the group that was physically separated from them.

Jane took several actions to remedy this perception. She began having weekly teleconference meetings that included the client and group members at both locations. She made a point of having group members away from the client location lead and brief activities to the client. Jane also rearranged some of the work so the away location had sole staffing and product responsibility for it.

These actions succeeded in changing client understanding of the away location staff and perception of their value to the existing contract. It also led to an unexpected, positive consequence: the client contracted with Jane's company for a different consulting activity based on the abilities and experience of the remote location staff.

Example

Linda is the GM of a large and diverse business enterprise. Each division, by itself, has been successful in its niche and each has a reputation for innovating within its product area and market. The divisions' tendency, however, is to innovate incrementally in fairly predictable steps along their separate, various product lines.

Linda understands the importance and value of this "directional" innovation but she sees enormous potential for transformational innovation that develops fresh, groundbreaking ideas for new products and markets by associating concepts and technologies of one division with those of another division [11].

Linda's vice presidents pay all the right lip service to her vision but they never seem to get around to putting together any creative cross-division proposals. They are not opposed to Linda's idea: they are each just too busy managing their own successful operations. Linda has held focus group sessions and off-sites in an attempt to generate some ideas and enthusiasm. She even considered a major reorganization to break down interdivisional barriers and better align operations with transformational innovation. But Linda understands that such a reorganization risks undoing the current organization's incremental innovation ability, which is also very important to future business.

Linda decided to leave the current organizations in place to add a small but important business process. Approval for full-scale development of a division's new product now requires the usual documentation and briefings and, in addition, the division is required to arrange and present an agreed, funded, cross-division collaborative proposal aimed at transformational innovation.

With this small change, Linda linked success within a division to collaboration across divisions. In effect, her new policy changed the rules of success for the divisions and reshaped the interaction patterns of her vice presidents and the divisions they lead to place a premium on collaboration. Linda cannot predict the exact details of the cross-divisional collaborations (nor would she necessarily want to), but she can assure that they will happen if the vice presidents and their divisions wish to remain successful. The strategy became so successful that it was copied by other GM in the company.

2.5.3 A General Development Framework for Complex Systems

The above three very different examples presented the development of a complex entity or system (individual, team, enterprise). In each example, we can see movement toward the integrated complexity depicted in Figure 2.5. How did it happen?

Clearly, each of the central characters in the examples *intervened* in their complex situations in ways that helped bring them to a future they were seeking.

What was the nature of these interventions? The first example concerned an individual who explored a *variety* of opportunities for developing his communication skills and technical currency before selecting specific approaches that served his needs. The second example illustrated how a line manager solved a perception problem concerning part of her team by changing the *interaction* patterns between them and her client. The last example illustrated a GM *selecting* a strategy* to improve innovation, which worked so well it was copied across the corporation.

While simple, these examples illustrate the key elements of a guiding developmental framework for any complex adaptive system[†]: variation, interaction, and selection. These are interlocking concepts that can generate productive actions in complex situations that cannot be controlled but can be influenced (see Figure 2.4). The framework will help us ask and answer the question, "what interventions in our complex system are likely to bring us to a future we would prefer?"[‡] This moves us from being passive observers in complex situations to active participants who *manage* variation, *shape* interactions, and *make* selections to guide, even accelerate[§] improvement over time.

What do these three processes do? *Variation* produces raw material for adaptation.** *Interaction* makes or changes the rules and strategies agents play by and their interaction patterns. *Selection* promotes adaptation.

---

* A *strategy* is the way an agent responds to its surroundings and pursues its goals. The usage includes deliberate choice in the sense of business strategy and patterns of response that pursue goals with little or no deliberation.
† A *complex adaptive system* is a complex system (refer footnote *) that contains agents or populations that *seek* to adapt. An *agent* is an entity that has the ability to interact with its environment, including other agents. It can respond to what happens around it and can do things more or less purposefully. A *population* is a grouping of entities (e.g., agents, strategies) that have a common attribute, affinity, or bond (e.g., population of business managers, population of coworkers).
‡ The framework provides a systematic way to analyze complex systems that suggests useful questions and illuminates promising possibilities for action and it can clarify relationships among seemingly separate issues. Analyzing complex adaptive systems within this or any other framework does not assure the ability to produce or predict *specific* outcomes. But it can foster an increase in value, impact, or performance of the system over time. It provides a basis for inquiring where leverage points and significant trade-offs of a complex system may occur and it suggests what kinds of situations may be resistant to policy interventions and when small interventions are likely to have large effects.
§ The idea of manipulating a complex system through regimens to *accelerate* improvement comes from discussions with and writings of M. L. Kuras (The MITRE Corporation) [12].
** *Adaptation* is said to occur when a selection process leads to an improvement according to some measure of success.

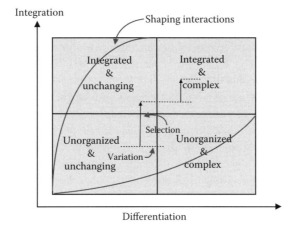

Figure 2.6 A general development framework for complex systems.

What questions do these processes help answer? A variation–interaction–selection framework clusters the discussion of change mechanisms in a complex system on three central and connected questions[*]:

- What is the right balance between variety and uniformity in a system?
- What (or who) should interact with what (or who) and when?
- What should be maintained or proliferated and what should be eliminated?

Much of the intellectual content of this variation–interaction–selection framework comes from seminal work done by Robert Axelrod and Michael Cohen (who, together with Arthur Burks and John Holland, comprise the original members of the BACH group) [1].

The perspective that is presented in this report is one which overlays Axelrod and Cohen's variation–interaction–selection construct onto Gharajedaghi's systems view of development, as depicted in Figure 2.6. Gharajedaghi provides the destination. Axelrod and Cohen provide mechanisms for moving toward it.

2.5.4 Variation

Variation produces the raw material for adaptation [1].[†] Variety is not only desirable to development of a complex adaptive system, but also essential. And it is essential not only in the popular survival-of-the-fittest sense in which a "best-of-breed"

---

[*] Although not mentioned explicitly, the change mechanisms may be in response to changes in the environment.
[†] *Variation* produces *artifacts* which are objects that are used by agents (e.g., a systems engineering process). Artifacts can have important properties (e.g., location or capabilities). Artifacts usually do not have their own purposes or powers of reproduction.

artifact or agent is identified and selected from among a set of "also rans" which are then eliminated. That which is not currently best may be a critical resource for the future of a complex adaptive system. For example, without variety, the introduction of a new threat, such as a parasite or virus, can wreak havoc whether in crops, computers, or other monocultures.

But to take advantage of what has already been learned, some limits must be placed on the amount of variety or diversity in the system. Otherwise, there is the risk of expending resources generating options for achieving a goal and never actually choosing and using one. These factors and trade-offs were at work in the example of the individual who explored a variety of opportunities for developing his communication skills and technical currency before selecting specific approaches that served his needs.

The actions available to shape the behavior of a complex system often work not just by accommodating variety, they also work by actually increasing or decreasing the variety of agents in a population (e.g., Internet standards, product designs under consideration for development). Variety turns up repeatedly in complex systems as a crucial factor in their development. But it is not as simple as saying that homogeneity is bad and variety is good. Homogeneity can be very useful indeed, as the following example shows.

Example

Consider a C2 Enterprise built on a layered architecture in which Internet protocol (IP) is a chosen point of convergence for all network implementations.* The standardization of IP is a form of homogeneity imposed on the enterprise. Differentiation is lost at that particular level of the architecture and innovation is—if not lost—seriously affected because the inertia of a widely accepted and adopted standard can only be overcome by the most compelling reasons. But that same standardization enables innovation to flourish above and below the standard and is therefore a net gain for the enterprise.

Therefore, the key question in variation surrounds choosing the right balance between variety and uniformity.

2.5.4.1 The Exploration–Exploitation Trade

There is a fundamental trade-off between exploration and exploitation in a complex adaptive system. The former creates types that are likely to be substantially novel. The latter creates agents or strategies similar to types that have already been successful. What are ways of thinking about and managing that balance in an enterprise?

For the purpose of this discussion, an enterprise capability is defined to have the following characteristics. It involves contributions from multiple elements,

---

* Electronic Systems Center Strategic Technical Plan (STP) v 2.1.

agents, or systems of the enterprise. It is generally not knowable in advance of its appearance. Technologies and their associated standards may still be emerging and it may not be clear yet which will achieve market dominance. There may be no identifiable antecedent capability embedded in the cultural fabric of the enterprise and thus there is a need to develop and integrate the capability into the social, institutional and operational concepts, systems, and processes of the enterprise.

The personal computer emerged as a replacement for the combination of a typewriter and a hand-held calculator, which were firmly embedded in our social, institutional and operational concepts, and work processes. The personal computer is a system or system capability, not an enterprise capability, by this definition. But the Internet is an enterprise capability. Its current form could not possibly have been known in the 1980s. Its technology has been emerging and continues to do so. More fundamentally, there was no identifiable antecedent capability embedded in the cultural fabric of our society before the Internet's emergence, nor were there associated problems and solutions to issues such as identity theft, computer viruses, hacking, and other information security concerns.

Enterprise capabilities evolve through emergence, convergence, and efficiency phases as suggested by a stylized s-curve similar in its essentials to Rogers' diffusion of innovation curve [13] which later influenced Moore's technology adoption curve [14]. Emergence is characterized by a proliferation of potential solution approaches (technical, institutional, and social). Many of these potential solutions will represent evolutionary dead-ends and be eliminated (convergence) through market-like forces. This is followed by a final period (efficiency) in which the technology is integrated and operationalized to such a degree that it becomes invisible to the humans, institutions, and social systems that use them.

Enterprise capabilities evolve through emergence, convergence, and efficiency phases whether or not an enterprise (or society) has intervention processes in place to actively manage them. Interventions, whether purposeful or accidental, can alter the shape of the evolutionary curve, the speed at which evolution progresses, and the level of evolutionary development achieved. Table 2.1 summarizes characteristics of each phase of enterprise capability evolution and ideas for shaping the exploration–exploitation balance.

2.5.5 Interaction

Interaction among agents shapes the creation and destruction of variety and produces the events that drive attribution of credit [1]. While thinking about the variety of agents and their strategies one is led to the question of the right balance between variety and uniformity. This balance is achieved partially through interaction which seeks to answer, "what (or who) should interact with what (or who) and when?" Interaction patterns and their influences are all around us, as illustrated by the familiar example below.

Table 2.1 Guiding Enterprise Evolution

Enterprise Capability Phase Characteristics	Examples, Rules of Thumb, and Anecdotes
Emergence	
• When there are no clear solutions or multiple, emerging solutions.	• Reward proliferation of potential solutions. Example: DARPA Grand challenge to accelerate evolution of critical autonomous robotic vehicle technology.
• When extensive or long-term use can be made of a solution.	• Operating systems and some applications (e.g., target tracking and identification) are among the longest lived elements of IT.
• When there is low risk of catastrophe from exploration.	• Modular enterprise architectures that provide market-like choices for different service layers.
Convergence	
• When there are multiple, adequate solutions with little performance differentiation.	• Narrow the solution space by providing rewards or incentives to programs and contractors that achieve common solutions through collaboration.
• When maintaining multiple solutions impairs enterprise performance or cost.	• Solutions are not prescribed from above.
Efficiency	
• When there are a small number of mature solutions that dominate the market. • When the probability of technology change is low.	• Reward exploitation of existing solutions • Reward use of common solutions • Solution is not specified from above.

Source: Adapted from Rebovich, G. Jr. *Proceedings of the 2007 1st Annual IEEE Systems Conference*, April 2007.

Our individual networks of current acquaintances have a strong local bias because it is convenient to work, shop, attend school, and become affiliated with civic and religious institutions near home.

More fundamentally, and beyond manipulating variety and uniformity, interactions help shape the outcome space within which a complex system develops as depicted in Figure 2.6. Events of interest within a complex social system arise (or do not) from the interactions of its agents with each other and with artifacts, as illustrated by the example below.

Example

The 1961 French military putsch against Charles de Gaulle, precipitated by his policies on Algeria, failed before it could get off the ground. None of the officers had taken more than a passing notice when thousands of transistor radios were issued to French troops several weeks before de Gaulle's Algeria policy announcement. The radios were regarded as a harmless comfort for the troops aimed at relieving their boredom. But de Gaulle used them to broadcast directly to the troops his reasons for the Algeria policy. Though the policy was wildly unpopular with most career soldiers in leadership roles, de Gaulle succeeded in convincing thousands of conscripts of the wisdom of his choice. This one-on-many interaction between de Gaulle and his troops, over the heads of their intermediate leaders, broke the back of the revolt before it started [16].

Interaction patterns shape the events in which members of a complex social system become directly involved and they provide the opportunity for spreading and recombining of types that lead to creation and destruction. The events drive processes of selection and amplification that ultimately change the frequency and variety of agent types. Interaction patterns help determine what will be successful for the agents and the system and this, in turn, will help shape the dynamics of the interaction patterns themselves.

There are numerous methods, rules, and strategies for shaping interaction patterns. In turn, these interaction patterns among agents shape the creation and destruction of variety and produce the events that drive attribution of credit. Next, we examine how selection itself works and how it feeds back onto variety and interaction.

2.5.6 Selection

To this point, the discussion has been on mechanisms that create and destroy types and processes and structures that govern interaction among agents. The discussion now turns to how selection can be employed to promote adaptation. This involves making decisions on which agents or strategies should be proliferated and eliminated [1].

If one wants to design a system that is able to explore new possibilities while being able to exploit what has already been achieved, biological evolution provides an important benchmark. Of course, selection in a complex social system need not operate in the same way as evolutionary biology. But, natural selection provides a well-studied and familiar example of how selection can work. It requires three things.

- A means to retain the essence of the agent
- A source of variation
- Amplification or changes in the frequencies of types

Interestingly, biological evolution demonstrates that adaptation can be achieved without agents or anyone else having any understanding of how the system works. This has implications for human social systems. They, too, evolve with or without purposeful intervention.

Compared to more directed methods of achieving adaptation, natural selection has disadvantages since it occurs at the level of the agent. Whenever it is feasible to attribute success to something more specific than an entire agent, there is the possibility of selecting strategies, rather than whole agents.*

Example

Knowing that quinine-related compounds reduce malaria allows a strategy of spreading them through the world instead of waiting for many generations for natural selection to breed malaria-resistant humans, particularly since the main antimalarial solution nature makes the carrier susceptible to sickle-cell disease. [1]

When attribution is sufficiently precise it can be advantageous to make numerous copies of a good strategy in a short time instead of reproducing complete agents. The latter would be difficult, costly, or perhaps even impossible to accomplish. Agent selection and strategy selection share the need to make copies that retain effective adaptations, to incorporate variation for further adaptation, and to amplify the success and cull the failures that do occur. But the two methods differ at the level at which they operate—and selection at the two levels can work very differently.

Example

Selection of one contractor from a population of competing firms (agent selection) can have quite different dynamics from selecting among a population of technical approaches proposed by a single contractor (strategy selection).

The view advocated here is one in which trade-offs are made in an evolutionary context and considers factors such as innovation versus exploitation and variety versus uniformity. Note that the trade-offs may go beyond concerns surrounding the immediate decision (technical approach for a particular project, in the above example). For example, depending on the immediate decision, there may be implications concerning the survival or extinction of contractors and, hence, the variety of them available to do similar projects in the future.

A design for an adaptive system of selection must address four issues:

- Defining success criteria
- Determining level of selection (agents or strategies)

---

* Recall that agents have strategies. For example, an employee helps a coworker in the hopes that the latter will reciprocate; someone needing a loan asks a friend to help; a nation seeking to promote favorable norms leads by example.

- Attribution of credit (for success or failure)
- Creating new agents or strategies

In application, these design issues do not necessarily separate neatly into four distinct factors. The central question in selection is, "which agents or strategies should be copied or destroyed?" The answer to the question is strongly intertwined with answers to two other major questions: "what is the right balance between variety and uniformity?" (variation) and "what should interact with what and when?" (interaction).

2.6 Application to an Enterprise

This chapter has presented a framework for harnessing complexity in a world with many diverse, mutually adapting players, where the emerging future is difficult to predict. Increasingly, these are the situations systems engineers find themselves in. What follows is a brief summary of the ideas presented and some thoughts about how they might be applied in the design and evolution of improved enterprise organizations, strategies, and processes.

2.6.1 Purposeful Questions

Systems engineering has always been about asking good questions, answering them, and following their implications. What follows is a series of questions that can help the systems engineer harness the complexity of a particular system or enterprise.

2.6.1.1 Systems Thinking Questions

- What is my enterprise? What elements of it do I control? What elements do I influence? What are the elements of my environment that I do not control or influence but which influence me?
- How can a balance be achieved between optimizing at the system level with enabling the broader enterprise, particularly if it comes at the expense of the smaller system?
- How can analytic and synthetic perspectives be combined in one view to enable alignment of purposes across the enterprise? Would a change in performance at the subsystem level result in a change at the enterprise level? If so, how, and is it important? How would a new enterprise level requirement be met and how would it influence systems below it?
- How can the solution space of a seemingly intractable problem be expanded by viewing in its containing whole?
- How can complementary relations in opposing tendencies be viewed to create feasible wholes with seemingly unfeasible parts? How can they be viewed as

being separate, mutually-interdependent dimensions that can interact and be integrated into an "and" relationship?
- Is interdependence of variables in a system or enterprise hidden by slack? Is the inability to make progress in one variable except at the expense of others an indication that slack among them is used up? Can a redesign of the system or enterprise remove interdependence or provide additional slack?

2.6.1.2 Harnessing Complexity Questions

- What interventions in a complex system are likely to bring it to a future we would prefer? How can variation be managed, interactions be shaped, and selections be made to guide and accelerate improvement over time? What is the right balance between variety and uniformity? What (or who) should interact with what (or who) and when? What should be copied (or proliferated) and what should be eliminated?

2.6.1.2.1 Variation

- What is an appropriate balance between variety and uniformity?
- Can variety be created by combining portions of strategies or agents already in use to introduce new types that have some degree of correlation with the enterprise's other conditions?
- Can variety be created by relaxing or reversing assumptions about a problem or a solution?
- How do errors occur in current processes? Does the variety that results ever offer potential value?
- Where and how can homogeneity be used? Can homogeneity (e.g., convergence on a standard) be used to enable innovation to flourish elsewhere in the enterprise?
- How can premature convergence be prevented? Are the types available in a population likely to be the best possible? Are environmental conditions changing so that what is best now may not be best in the near future?
- What is the right balance between exploration and exploitation? Is exploration especially valuable because improvements can be widely applied or used for a long time? Is there a risk of disaster from trying a bad strategy?

2.6.1.2.2 Interaction

- What are events of interest whose likelihood should increase or decrease?
- What are the current patterns of interactions among types?
- How can interaction of agents with each other and with artifacts be used to alter the likelihood of events of interest in a complex system? What (or who) should interact with what (or who) and when?

2.6.1.2.3 Selection

- How can selection be used to promote adaptation?
- Are selection trade-offs being made in a context that goes beyond an immediate decision? Does the trade-off consider factors important to future decisions, like variety versus uniformity and exploration versus exploitation?

2.6.2 Examples

It is the nature of a complex social system or enterprise that some of its important attributes do not scale from the system or SoS level. What follows are examples of enterprise-level thinking and processes that may not have well-developed (or needed) system or SoS-level antecedents.

2.6.2.1 Portfolio Management

A technologically sophisticated enterprise whose capital is knowledge and the ability to apply it looks at the future from the perspective of: "what hard questions or problem areas do we want to be working on in the next five to fifteen years?" and "how do we position ourselves to increase our likelihood of being selected for and succeeding in them?" This is a form of the basic question, "what will bring us to a future we would prefer?"

The answer is similar to the familiar personal question "what will bring me to a financial future I would prefer?" It is sound portfolio management. And it is portfolio management with a view that looks beyond the short-term performance or current value of the individual instruments that comprise the portfolio. It requires the right balance between different investment types. As in personal portfolio management, that balance can and does change over time due to factors such as stage of life, other obligations and expenses (college tuition, car payments, unexpected catastrophic asset loss), and change in life goals. Viewed in this way, managing the balance or the whole is clearly more than the summation of managing the parts of the portfolio.

There will always be a need to look at and manage the immediate and short-term performance of the separate current programs within an enterprise portfolio. But it is future-oriented hard problem areas that focus our attention on managing a portfolio in a way that moves the enterprise toward various preferred futures.

Consider the following future problem areas from the perspective of enterprise portfolio management.

Example: The Next Generation of U.S. AEW&C System

What is the right amount of variety in the AEW&C enterprise portfolio when considering the next generation of U.S. AEW&C system? Does the current portfolio provide a mix of technical problems and experiences that will be applicable to the next generation U.S. AEW&C, including those surrounding choice of platform, sensors, information systems, communications, and networks? If not, how can that variety be created? Should an attempt be made to redefine engineering roles and technical focus of current work? Should new work be brought in that requires the assumption of less coveted task engineering roles instead of a general systems engineering and integration role if it provides the opportunity to acquire intellectual capital in a critical future technology?* Should some programs in the portfolio be divested? If the portfolio itself cannot be changed in a way that brings in opportunities to acquire critical intellectual capital, are there ways of changing AEW&C enterprise interaction patterns to increase the likelihood of acquiring it? For example, can exchange programs be initiated with other program offices [e.g., Navy Hawkeye Global Information Grid (GIG)] in which intellectual capital is acquired via periodic meetings or engineer exchange programs? Is the population of AEW&C enterprise staff being viewed primarily through a lens of current problems? Or is the population of staff being assessed and modified through training and hiring to assure the right mix of technology skills and domain experience for addressing the next generation of AEW&C? What is the right balance between servicing current programs and positioning for the future?

Example: The Next Generation of Air SA Systems

The next generation of air SA systems has other solution possibilities beyond AEW&C options consisting of different combinations of airborne platforms, onboard sensors, mission system capabilities, and communications. Alternatives include space-based and unmanned aerial vehicle (UAV)-based sensors and different options for mission crew location. As a result, the variety in technology and mission experience required expands to include space-based radars, UAVs, UAV-based sensors, control mechanisms for UAVs, and communications between a remotely located mission crew and space-based or in-theater air SA assets. These are long-term possibilities than the next generation of AEW&C. How much intellectual capital in these areas is needed now? How much is needed later? Does it need to exist in the AEW&C enterprise or can it reside in a broader enterprise portfolio? If the latter, what should be the interaction patterns between them and the AEW&C enterprise? How should the interactions evolve over time?

Example: The Next Generation of SA Systems

The next generation of SA systems goes beyond air SA to include ground SA and perhaps other domains. The variety in technology and mission experience needed encompasses both airborne moving target indicator (AMTI) and ground moving target indicator (GMTI) areas. Portfolio management questions arise concerning program, technology focus, and staff mix, as they did in considering the next generation of AEW&C and air SA systems.

---

* Note that the portfolio perspective permits a view of a low-impact/high-value activity as one that needs no growth to high impact as a single program as long as it serves an enterprise-level high-impact future need.

But there are new and different questions, as well. Interesting interactions arise when thinking of solutions to problems that have been addressed separately in the past. For example, current AMTI (e.g., AWACS) and GMTI (e.g., Joint STARS) capabilities have completely separate sensors, mission systems, communication systems, and platforms. Current day tactics, techniques, and procedures for the two capabilities are incommensurate in a number of important respects (e.g., on-station orbits, sensor information exchange with other network participants, and its timing). What does a solution look like in which both AMTI and GMTI missions are both well-served at the same time and not traded off against each other in a zero-sum game? How can the principle of complementarity be used to interpret these opposing tendencies as mutually-interdependent dimensions that can interact and be integrated in a new way of thinking about SA? Are there opportunities for transformational innovation at the intersection of the AMTI and GMTI domains—domains that have largely evolved along separate paths? Can AMTI and GMTI information be used to develop new and useful representations of interactions or relationships among objects in different domains? Does this then change the nature of SA? Does it suggest unimagined cross-domain system operator roles, mission applications, and decision-support aids? Do these new areas need to be explored? By whom?

2.6.2.2 Evolving Enterprise Processes and Practices

Enterprises that develop products have an interest in measuring the progress of product development. From a single program or product perspective, the focus of incremental system development is primarily concerned with the ultimate goal of successfully fielding the full capability of a specific system or capability. As a result, it is common in system acquisitions to use intermediate measurements to give insight into questions such as "is the program on course?" and if not "what are we doing wrong and how do we correct it?"

In developing products or capabilities, particularly highly complex ones, intermediate measurements are also the basis of *build-a-little, test-a-little* incremental acquisition approaches. Their focus on developing working, useful intermediate capabilities more readily leads to the successful deployment of a full capability.

In an enterprise that develops many products, systems, or capabilities, particularly over an extended time, there is an opportunity to examine intermediate development criteria to determine whether they accurately predicted development success and how they could be improved for future developments. The focus of this perspective is on improving the intermediate measurement criteria used to track the progress of product, system, or capability development not on the status of a particular product in development.

Surprises (results different—better or worse—than expected) can help revise success criteria so that, in future developments, attribution of credit is able to produce more accurate results, provided one can correlate factors that were observable or predictable in the short run with the surprise. The question becomes, "what observable criteria were often high or low when the result was better or worse than expected?" This is an important shift from a focus on "what predicted the outcome?" to "what predicted the surprise?" that is, the deviation of what occurred from what was expected.

This kind of process allows the enterprise to accumulate a store of increasingly accurate criteria for attributing credit to a variety of intermediate positions in developing products, systems, or capabilities. The enterprise thus evolves its acquisition processes, learning from its own experiences. The enterprise becomes collectively more capable in a way that enables managers of future programs to make increasingly more intelligent decisions throughout the development cycle.

2.7 Postscript

What is the relationship between classical system engineering as exemplified by the International Council on Systems Engineering (INCOSE) model and the evolutionary development framework for complex systems in this chapter? Is there any? Are there more than one? There is no simple answer to it, but the discussion in this chapter suggests some interrelationship and at least a partial nature of it.

The INCOSE model produces systems, subsystems, and SoSs. For this discussion, call them all "products." There are many developers producing products which perform similar roles or functions. And there is variation or differentiation across these similar products. So, one view is that classical system engineering produces artifacts that provide variety for the processes of interaction and selection that result in improved future products. To use the AEW&C example again, the current market of AEW&C systems provides the variety for the next generation.

But there are other artifacts that come out of executing a classical system engineering process like the INCOSE model. Improved system engineering processes do. And there is variety in these processes which fuels the evolutionary model that produces improved future processes. So, clearly the classical model feeds the evolutionary model.

It works the other way, as well. Changes occur in well-defined, deterministic processes, whether introduced by mistake (e.g., misunderstanding of a system engineering process by an inexperienced employee) or by design (e.g., deliberate attempt to innovate). And so evolution is found in and part of the very fabric of the classical system engineering process.

The relationship between the classical system engineering and the evolutionary developmental framework is akin to the Yin Yang* symbol which suggests that they are complementary elements. Both are needed to form the whole and each contains an aspect of the other (Figure 2.7).

---

* The Yin Yang is also called the Tai-Chi symbol. The Tai-Chi is from I-Ching, a foundation of Chinese philosophy.

Figure 2.7 Yin Yang symbol.

Acronyms

AEW&C	Airborne Early Warning and Control System
AMTI	Air Moving Target Indicator
C2	Command and Control
GIG	Global Information Grid
GM	General Manager
GMTI	Ground Moving Target Indicator
INCOSE	International Council on Systems Engineering
IP	Internet Protocol
IT	Information Technology
SA	Situation Awareness
SoS	System of Systems
UAV	Unmanned Aerial Vehicle
WWW	World Wide Web

References

1. Axelrod, R. and Cohen, M. D., *Harnessing Complexity: Organizational Implications of a Scientific Frontier*. New York, NY: Basic Books, 2000.
2. Rebovich, G. Jr. Systems thinking for the enterprise: New and emerging perspectives. *Proceedings of 2006 IEEE International Conference on Systems of Systems*, Los Angeles, CA, April 2006.
3. Joseph K. DeRosa et al. *Evolving Systems Engineering at MITRE*. Bedford, MA: The MITRE Corporation, August 2007.
4. Malone, T. W., *The Future of Work*. Boston, MA: Harvard Business School Press, 2004. (Malone argues a similar view. A key element of his thesis is that the cost of communication has dropped dramatically, thus precipitating major social changes.)
5. Senge, P. M. and Sternman, J. D., Systems thinking and organizational learning: Acting locally and thinking globally in the organization of the future. In *Proceedings of the 1990 System Dynamics Conference*, 1990.

6. Gharajedaghi, J., *Systems Thinking: Managing Chaos and Complexity*. Boston, MA: Butterworth Heinemann, 1999.
7. Ackoff, R., *Systems Thinking and its Radical Implications for Management*. IMS Lecture/Boston, February 15, 2005.
8. Net Centric Implementation Framework, v1.0.0, 17 Dec 04, *Net Centric Enterprise Solutions for Interoperability*, Defense Information Systems Agency, Washington, DC, 2004.
9. Ackoff, R., From mechanistic to social systemic thinking. *System Thinking in Action Conference*, November 1993.
10. Parry, S. B., The quest for competencies, *Training Journal*, 33 (7), 48–56, 1996.
11. Johansson, F., *The Medici Effect*. Boston, MA: Harvard Business School Press, 2004. (Johansson differentiates between two types of innovation—directional and intersectional.) The former represents incremental development and the latter represents transformational development which requires the removal of *high associative barriers* between different cultures in an organization.
12. Kuras, M. L. and White, B. E., Engineering enterprises using complex systems engineering (presentation), in *1st Annual System of Systems Engineering Conference*, Johnstown, PA, 14 June, 2005.
13. Rogers, E. M., *Diffusion of Innovation*, 5th edition. New York, NY: Free Press, 2003.
14. Moore, G. A., *Crossing the Chasm*. New York, NY: Harper Collins, 2002.
15. Rebovich, G. Jr., Engineering the enterprise. *Proceedings of the 2007 1st Annual IEEE Systems Conference*, April 2007.
16. Forsyth, F., *The Day of the Jackal*. New York, NY: Bantam Books, pp. 20–21, 1985.

Chapter 3

Pilots and Case Studies

Kimberly A. Crider

Contents

3.1 Background 65
3.2 Summary of Key Findings 66
    3.2.1 The ESE Processes Are Useful in the Practice
          of Complex Systems Engineering..67
    3.2.2 Stakeholder Analysis Is a Logical Sixth ESE Process.
          Several Pilots and Case Studies Showed That an Important
          Enabler of ESE Is Active Stakeholder Participation........................67
    3.2.3 ESE Does Not Follow a Linear Path 67
    3.2.4 Effective ESE Requires a Deliberate Investment
          of Resources above the Scale of Individual Systems 67
    3.2.5 Context Is Key and Architecture Can Help 68
    3.2.6 ESE Must Consider the Numerous and Complex
          Interdependencies of People, Process, and
          Technology That Result When Innovation
          Is Introduced within an Enterprise 68
    3.2.7 ESE Requires New Sociocultural Skills 68
3.3 Enterprise Engineering Systems Processes 68
    3.3.1 Capabilities-Based Engineering Analysis..69
    3.3.2 Enterprise Architecture 69
    3.3.3 Enterprise Analysis and Assessment 69
    3.3.4 Strategic Technical Planning 69
    3.3.5 Technology Planning 70
3.4 Overview of Pilots and Case Studies 70
3.5 ESE Pilots—Summary of Findings 70

3.5.1 Pilot 1 71
    3.5.1.1 Objective 71
    3.5.1.2 Enterprise Outcome 71
    3.5.1.3 Enterprise Processes Applied 71
    3.5.1.4 Approach 71
    3.5.1.5 Findings 71
3.5.2 Pilot 2 72
    3.5.2.1 Objective 72
    3.5.2.2 Enterprise Outcome 72
    3.5.2.3 Enterprise Processes Applied 72
    3.5.2.4 Approach 72
    3.5.2.5 Findings 73
3.5.3 Pilot 3 74
    3.5.3.1 Pilot Objective 74
    3.5.3.2 Enterprise Outcome 74
    3.5.3.3 ESE Processes Applied 74
    3.5.3.4 Approach 74
    3.5.3.5 Findings 74
3.5.4 Pilot 4 77
    3.5.4.1 Objective 77
    3.5.4.2 Enterprise Outcome 77
    3.5.4.3 Enterprise Processes 77
    3.5.4.4 Approach 77
    3.5.4.5 Findings 78
3.5.5 Pilot 5 78
    3.5.5.1 Objective 78
    3.5.5.2 Enterprise Outcome 78
    3.5.5.3 Enterprise Engineering Processes Targeted 79
    3.5.5.4 Approach 79
    3.5.5.5 Findings 79
3.5.6 Pilot 6 80
    3.5.6.1 Objective 80
    3.5.6.2 Enterprise Outcome 80
    3.5.6.3 ESE Processes Applied 80
    3.5.6.4 Approach 80
    3.5.6.5 Findings 80
3.6 ESE Case Studies—Summary of Findings 81
    3.6.1 Case Study 1 81
        3.6.1.1 Objective 81
        3.6.1.2 Approach 81
        3.6.1.3 Findings/Lessons Learned 82
    3.6.2 Case Study 2 83
        3.6.2.1 Objective 83

		3.6.2.2	Approach	84
		3.6.2.3	Findings/Lessons Learned	84
	3.6.3	Case Study 3		85
		3.6.3.1	Objective	85
		3.6.3.2	Approach	85
		3.6.3.3	Findings/Lessons Learned	85
	3.6.4	Case Study 4		86
		3.6.4.1	Objective	86
		3.6.4.2	Approach	86
		3.6.4.3	Findings/Lessons Learned	86
	3.6.5	Case Study 5		87
		3.6.5.1	Objective	87
		3.6.5.2	Approach	87
		3.6.5.3	Findings/Lessons Learned	87
	3.6.6	Case Study 6		88
		3.6.6.1	Objective	88
		3.6.6.2	Approach	88
		3.6.6.3	Findings/Lessons Learned	88
	3.6.7	Case Study 7		89
		3.6.7.1	Objective	89
		3.6.7.2	Approach	89
		3.6.7.3	Findings/Lessons Learned	89
	3.6.8	Case Study 8		90
		3.6.8.1	Objective	90
		3.6.8.2	Approach	90
		3.6.8.3	Findings/Lessons Learned	90
	3.6.9	Case Study 9		91
		3.6.9.1	Objective	91
		3.6.9.2	Approach	91
		3.6.9.3	Findings/Lessons Learned	91
3.7	Final Recommendations			91
References				92

3.1 Background

In October 2004, the MITRE Corporation former Center for Air Force Command and Control (C2) Systems, established an enterprise systems engineering (ESE) focus group, based on the recognition that systems engineering was advancing beyond traditional approaches in order to more effectively address the complex nature of today's systems, their acquisition and development, and their target operating environments. At the same time, complex systems theory was evolving and seemed to align with new systems engineering techniques that were yielding improved results in delivering operational capability [1]. The ESE focus group was formed to document these new approaches and techniques, and advance the practice and theory of systems engineering to address complex, enterprise-level issues.

The first year's activity produced an internal nine volume series in the theory and practices of ESE as well as a number of related documents and journal articles. A set of five ESE processes was defined based on first principles in the theory of complexity and enterprise evolution. The first principles include the use of traditional systems engineering (TSE) techniques, as well as sociocultural aspects of engineering systems that require new methods to address the evolution of complex systems within a complex enterprise [2,3]. The ESE processes that emerged from the first principles involve the use of capability-based outcomes to drive engineering analysis, the development of enterprise architectures (EAs) to inform engineering analyses and assessments, and the implementation of strategic technical plans and technology planning efforts to produce systems that will be integrated and interoperable across the enterprise.

In October 2005, the focus group turned to conducting a set of pilots and case studies to explore the techniques used in applying the derived ESE processes. Each pilot investigated the applicability of one or more of the ESE processes to tackle an existing enterprise challenge or fulfill a critical enterprise capability need. Each case study looked retrospectively at individual programs to critically evaluate how ESE concepts may have played a role in shaping program outcomes. In this way, the theory and current state of the practice of ESE were linked.

This chapter provides a synthesis of the findings and lessons learned in supporting enterprise objectives. The chapter begins with a summary of the key findings from the overall effort and a brief description of the ESE processes, which provided the context for the analysis. The chapter then synopsizes the individual results from each pilot and case study. It concludes with a combined set of recommendations to evolve the understanding of ESE and the application of ESE processes and concepts to meet the needs of system engineers and their most critical and complex challenges.

3.2 Summary of Key Findings

The pilots and case studies illuminate the current state of ESE across Air Force C2 System programs, and offer insights into other projects. It was concluded that *the five ESE processes provide the basis for a practice to support enterprise-level systems engineering* and that *there is a need for a sixth process for engineering an enterprise— Stakeholder Analysis*—which is especially important in contexts that involve a broad, diverse set of stakeholder interests. The pilots and case studies further revealed that *successful ESE goes beyond establishing a core set of processes, and requires new skills* that emphasize collaboration and the ability to establish the enterprise context for all stakeholders, balancing competing and cooperative interests, and other behaviors that can impact enterprise outcomes; *new perspectives* and the resources necessary to continuously scan the external environment and address the interplay of political, operational, economical, and technical factors in the development and delivery of enterprise capability; and *a new understanding of likely use cases* through which ESE process steps may be applied, recognizing that ESE processes are not necessarily applied sequentially, but in varying combinations depending on the enterprise context and objective. A more detailed summary of findings is provided below.

3.2.1 The ESE Processes Are Useful in the Practice of Complex Systems Engineering

The pilots and cases studies validated that the five original ESE processes and their associated toolkits provide useful guidance to inform ESE. They offer a methodology to facilitate evolutionary development and delivery of enterprise capability in a complex environment of people, processes, and technology.

3.2.2 Stakeholder Analysis Is a Logical Sixth ESE Process. Several Pilots and Case Studies Showed That an Important Enabler of ESE Is Active Stakeholder Participation

Understanding stakeholder equities in the broad context of enterprise outcomes is essential to achieving consensus and resolving conflict. See Ref. [4] for details of stakeholder analysis.

3.2.3 ESE Does Not Follow a Linear Path

The pilots and case studies showed that ESE process steps do not necessarily follow a sequential implementation, but often are performed in combination or parallel depending on the nature of the enterprise objective or challenge to be solved. A set of ESE "use cases" derived from a common set of enterprise challenges, with guidance on how to put together ESE process steps to achieve enterprise objectives in each sample case, would further assist ESE practitioners.

3.2.4 Effective ESE Requires a Deliberate Investment of Resources above the Scale of Individual Systems

Foremost among these are support for up-front capabilities-based engineering analysis (CBEA), strategic technical planning (STP), active collaboration among key stakeholders (stakeholder analysis) and agreement on enabling infrastructure (technical, business, and operational) to achieve enterprise outcomes.

3.2.5 Context Is Key and Architecture Can Help

The application of ESE processes should be done within an established context that is based on enterprise objectives. This context is used to inform programs of their role in supporting enterprise outcomes. This includes guidance on the set of ESE activities each program must implement, giving due consideration to program resources, and a collective understanding of the value proposition for applying enterprise process(es) at the program and enterprise level. Many studies found that architecture is an effective tool for communicating intent and synchronizing stakeholders. However, the level of detail and partitioning of the so-called EAs (more appropriately called federated architectures), varied widely. Architecture was a powerful tool, but the right partitioning and scale to support net-centric operations is not yet clear.

3.2.6 ESE Must Consider the Numerous and Complex Interdependencies of People, Process, and Technology That Result When Innovation Is Introduced within an Enterprise

Technology alone did not produce new enterprise behaviors (operational capability). It had to be shaped by the operational vision embodied in capabilities-based analysis and assessment. Choices of candidate technologies were informed by cues from the enterprise environment (both commercial and military). Winning technologies were selected through a competitive process. The evolutionary forces of variety, shaping, and selection are driven by the interplay of people, processes, and technology.

3.2.7 ESE Requires New Sociocultural Skills

Finally, we see throughout many of the pilot and case study results that to effectively carry out the engineering of an enterprise and the complex systems involved requires not only new engineering processes, but also new skills (e.g., leadership/strategic vision, conflict management, ability to balance cooperation and competition, and coalition building) to address the operational, business, and technical processes that influence enterprise outcomes, and foster convergence on the outcomes themselves, as well as the enabling infrastructures, design tenets, and technologies needed to achieve those outcomes. Enterprise solutions depend on technical, social, and cultural variables. In general, they cannot be separated.

3.3 Enterprise Engineering Systems Processes

TSE processes are used to organize the work routines of an engineering organization to provide order and increase the efficiency by which systems are produced and delivered. However, today's systems are being conceptualized and designed for more functionality and much higher degrees of integration to support increasingly complex interactions among people and technology at an enterprise scale. Today's systems are also increasingly being called on and designed for a future in which the operational environment will be characterized by ephemeral threads of functionality that are put together to serve a need now and then go away just as quickly. As a result, the processes used to plan, develop, and deliver those systems have become complex. Engineering systems at this scale, therefore, must consider how best to combine the interdependencies of people and technology, balance the exploitation of today's best approaches and practices, and the exploration of tomorrow's untested technologies that may be superior, and ultimately improve the productivity and effectiveness of the enterprise as a whole [2]. These principles are at the heart of the five ESE processes developed to facilitate engineering in today's environment. A brief description of each is given below.

3.3.1 Capabilities-Based Engineering Analysis

CBEA operationalizes the goals and vision of the enterprise. It focuses on the development of enterprise capabilities, grouping capabilities into portfolios, and defining the strategies to achieve those capabilities by addressing the interrelationships among the portfolio systems. CBEA also involves the development of capability-based roadmaps to support planned evolutionary development of systems, technologies, and new acquisitions to fulfill capability needs.

3.3.2 Enterprise Architecture

EA captures the enterprise vision, strategy, and implementation approach, describing the interdependencies at play. EA comprises the steps for developing a set of products to characterize an enterprise; its goals, effects, and properties; its component systems and elements; and the interactions of people, process, and technologies among its components.

3.3.3 Enterprise Analysis and Assessment

Enterprise analysis and assessment (EA&A) helps shape the environment that causes competing options to flourish or perish by continuously characterizing progress toward enterprise goals.

3.3.4 Strategic Technical Planning

STP sets the technical strategy for the enterprise; establishes the balance between standards and competing technologies.

3.3.5 Technology Planning

Technology planning is an approach to explore enterprise concepts by assessing technical innovation and integration opportunities in the marketplace.

We will see in the following pilot and case study summaries that these ESE processes interact with each other to support various ESE activities. Figure 3.1 presents one way to envision the interrelationship of these processes.

3.4 Overview of Pilots and Case Studies

Each of the ESE pilots and case studies was chartered to address the relevance and applicability of one or more of the five ESE processes to a complex enterprise challenge or objective. What follows is a short summary of objective, approach, key findings, and recommendations from each.

3.5 ESE Pilots—Summary of Findings

The ESE pilots represent an exploratory activity in that they were conducted specifically to evaluate the utility of the ESE processes as well as deliver operational

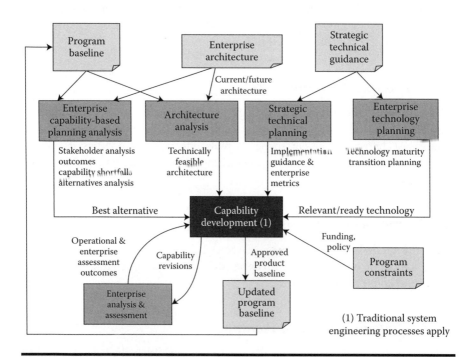

Figure 3.1   Interrelationship of ESE processes.

capability to the users. The processes informed the pilots while the feedback from those conducting the pilots informed the processes.

3.5.1 Pilot 1

3.5.1.1 Objective

The purpose of this pilot was to determine the applicability of any of the ESE processes to facilitate integration of a military enterprise that crossed traditionally separate application domains.

3.5.1.2 Enterprise Outcome

Improve the ability to monitor, assess, plan, and execute operations across a distributed joint cross-domain environment.

3.5.1.3 Enterprise Processes Applied

Technology planning, EA, CBEA, EA&A, and STP.

3.5.1.4 Approach

The pilot mapped the steps commonly followed for integrating systems to achieve enterprise capability to one or more ESE process steps. The system integration steps involved in this pilot were: (1) identify user needs for information sharing; (2) establish underlying data and services interoperability requirements; (3) assess current technologies available against desired capability attributes; (4) develop experimentation to explore candidate approaches; and (5) plan for system modification and implementation into a program baseline. Mapping of these steps to the ESE processes was used to identify how ESE could be applied in a systematic way to inform enterprise integration.

3.5.1.5 Findings

ESE process toolkits provide useful information to guide and inform the development of complex enterprise-wide systems. The pilot showed that ESE process steps can be applied directly to the systems integration needs of the cross-domain enterprise, and provide meaningful guidance to inform and shape the complex systems integration to achieve enterprise objectives. In this case, the most useful process steps were those that facilitated rapid, iterative implementation, and maximum stakeholder involvement—two key tenets associated with the complex systems engineering of the cross-domain enterprise. For example, given the complex nature of an integrated cross-domain enterprise environment, early consensus among stakeholders centered on the need for a flexible, open-enterprise system that would allow for technology reuse and adaptation to take advantage and meet the needs of the dynamic enterprise environment in which the systems would be developed and operated. The importance of reuse also drove a flexible design framework, which could satisfy multiple stakeholder objectives to recoup prior technology investments wherever possible, and provided an opportunity for rapid implementation of iterative capability overtime. ESE toolkits which defined ESE approaches for technology planning, EA&A, CBEA, and STP provided guidance that helped identify the importance of technology reuse and ways to effectively apply technology to cross-domain integration needs. Stakeholder analysis and continuous characterization of stakeholder needs were also critical ESE principles in this pilot effort that helped clarify important stakeholder needs up front, on which reuse and other technology attributes could be assessed.

ESE does not follow a linear path. While there was definite alignment between the ESE processes and enterprise integration in this study, the pilot team found that the ESE processes did not necessarily follow a linear, sequential flow. In fact, there was a good deal of moving across ESE processes to effectively inform the enterprise integration effort. This suggested an opportunity to develop a set of ESE use cases for a common set of enterprise challenges, with guidance on how to put together the ESE process steps in various ways to achieve enterprise objectives in each case.

3.5.2 Pilot 2

3.5.2.1 Objective

The purpose of this pilot was to inform the ESE processes and continue to apply ESE techniques (especially EA&A) to establish data format, content, sizing, performance, and user needs for an enterprise-wide capability to produce and consume a specific type of sensor data. By applying ESE processes and techniques to this effort, the pilot team hoped to advance techniques for applying ESE to other enterprise challenges.

3.5.2.2 Enterprise Outcome

Engineer an enterprise-level capability that extends the availability of this type of sensor data to a wider set of users.

3.5.2.3 Enterprise Processes Applied

EA&A.

3.5.2.4 Approach

The pilot team focused on the applicability of EA&A process in this study, given the numerous working group activities conducted in the spirit of EA&A (e.g., operational assessments of new technology/applications, technical experimentation, modeling and simulation, incremental development and delivery, community of interest (COI) collaboration, standards evolution, and use of a common infrastructure for analysis). The review resulted in a comparison of EA&A principles to enterprise goals for producing and consuming the sensor data and identified matches, gaps, changes required, points of confusion, and suggested additions needed for EA&A. In addition, the pilot team considered the remaining ESE processes, highlighting both applicable and problematic aspects of each. The pilot validated the need for TSE processes such as requirements analysis, augmented by complementary ESE principles, such as defining capability outcomes, leveraging stakeholder communities, and using experimentation (to provide continuous characterization and adaptation) to evolve the required capability.

3.5.2.5 Findings

ESE works best when stakeholders can converge on broad enterprise outcomes. In this pilot, the role of the COI was important not only to help identify specific data sharing needs and mechanisms, but also to provide a forum for stakeholders to come together on a shared set of common objectives—what the complex systems engineering regimen [5] refers to as outcome spaces. The pilot showed that early and continuous identification and engagement of stakeholders is a critical aspect of ESE, regardless of the enterprise issue. Additionally, getting stakeholders to converge on a common enterprise objective can provide focus and synergy to achieve a common enterprise effect within an established outcome space, not disregarding other enterprise needs or consequences that may have to be addressed in parallel or mitigated in the process. In this case, the COI framework and use of EA&A provided the opportunity to formulate broad outcome spaces that appealed to many enterprise participants, and also created a forum for interaction management [6]. These constructs proved useful in encouraging stakeholder participation and gaining leadership endorsement in sensor production and consumption capability definition, architecture development and analysis, and STP to guide its development.

Is ESE art or science? The pilot showed that the application of ESE is in many ways more art than science. As noted earlier, ESE emphasizes the ability to combine the interactions of people, process, and technology at a given time to achieve enterprise outcomes. In doing so, the paths toward enterprise outcomes often emerge as a result of those interactions and cannot be predetermined, or even repeated. ESE practitioners require added skills and techniques, therefore, to assess stakeholders and their needs (via stakeholder analysis), sense potential outcomes (via CBEA and technology planning), and influence positive interactions via ongoing experimentation and characterization (via EA&A and STP). It was suggested by the pilot team that ESE might better be characterized by a set of attributes and also that it is not necessarily distinct from TSE. More practical examples of how to apply ESE to various cases are needed to clarify the techniques for applying the art and science of ESE, at whatever degree is required to achieve enterprise objectives.

3.5.3 Pilot 3

3.5.3.1 Pilot Objective

This pilot assessed the application of technology planning process to ongoing work program enterprise integration efforts.

3.5.3.2 Enterprise Outcome

Enterprise engineering process improvement.

3.5.3.3 ESE Processes Applied

Technology planning.

3.5.3.4 Approach

The pilot leveraged the work of an enterprise integration team (EIT) engaged in improving its technology planning process and related efforts. In particular, emphasis was being placed on identifying specific, key windows of opportunities to mature the process and its application to work program enterprise integration needs. In doing so, the team also reviewed the multiple, varied sources for technology planning ideas, including collaborative relationships (industry, academia, nonprofits) and experimentation venues, and compiled a list of cross-organization needs as well as identified enterprise technology topics.

3.5.3.5 Findings

The pilot resulted in a number of lessons learned and recommendations to guide technology planning in ways that would broaden its scope from TSE approaches, and increase its impact at an enterprise level. As described below, the sociocultural aspects of technology planning include dealing with the complex interdependencies of people and processes interacting with technology, which yields uncertainty, competition, and evolution in the enterprise outcomes—all of which are important components of ESE thinking and practice.

3.5.3.5.1 ESE Must Address the Complex Nature of a Systems' Operational Environment as Part of Technology Analysis and Assessment

The pilot team observed that traditional and enterprise-level systems engineering approaches to technology planning differ in the nature of the environment within which technologies will be implemented. As an example, for technology planning to be an effective part of an ESE strategy, technology solutions must be addressed within the context of the enterprise in which they will be used, considering the numerous and complex interdependencies of people, process, and technology that will result in new, yet-to-be-known enterprise behaviors. These emergent behaviors will likely impact the technology solutions and drive selection in unexpected directions.

3.5.3.5.2 Technology Planning Must Assume Uncertainty, Not Only in the Target System Environment, But Also in the Technologies Themselves

Another related lesson learned from this pilot is that technology planning will always involve some level of uncertainty. A primary reason for this uncertainty is that all current investments in technology simply cannot be known at any point in time. Therefore, it is important to establish strong networks of communication between technology developers, operators, and financers to open as many avenues as possible to technological innovations that may meet capability needs. Even if a technology's contribution to a given problem were known, the complexity of the environments in which the technology is to be implemented may result in emergent behaviors that negate or reinforce the technology's benefit. The impact of technologies at an enterprise level cannot be fully known. The use of ongoing experimentation, based on realistic scenarios, with active participation of all stakeholders, is an important part of increasing awareness of enterprise behaviors that emerge in real-world operations and the role a particular technology plays in those environments.

3.5.3.5.3 Technology Planning Must Be Lightweight and Flexible

The pilot team observed that while a generalized approach to technology planning could be defined, there is no step-wise procedure that works in every case. The approach to technology planning must be flexible to allow the opportunity to take advantage of new technologies as they become available, assess emergent behaviors of an enterprise and its technology components, and conduct ongoing CBEA and adjustments, and enable rapid transition to satisfy immediate capability needs while longer-term solutions continue to be evaluated. Similar to CBEA, technology planning needs to be an iterative, collaborative process that emphasizes exploratory analysis based on capability needs, but also serves to inform capability needs and enterprise planning overtime.

Good technology will emerge through competition. Competition is a cornerstone of technology development because it fosters innovation based on rewards. An important question raised by the pilot team, in this case, is how to define and provide proper incentives and competition that enables technology innovation to address enterprise capability. Given the complexities associated with how enterprises operate, desired outcomes often require coordination and collaboration among entities to shape and influence behaviors toward enterprise objectives. The use of rewards and incentives therefore often need to stimulate cooperation and competition to address enterprise needs. Finding opportunities to encourage innovation based on competition among collaborative groups therefore becomes an important part of enterprise technology planning.

From these observations, the EIT pilot team developed a set of recommendations for how to improve the technology planning process.

3.5.3.5.4 Focus Technology Planning on Capability Gaps

The pilot affirmed the need to use the CBEA process to derive both system level and enterprise-level capability needs. It should also account for user concerns and bottoms-up program inputs. This analysis must be done at the beginning, middle, and end of the technology planning process to inform its decision makers about which technologies benefit which capabilities at different levels of an EA. The ongoing capability gap analysis and the role that technologies play in satisfying enterprise needs is part of an important complex systems engineering principle that calls for continuous characterization of an enterprise and the technologies that contribute to an enterprise's behaviors and outcomes.

3.5.3.5.5 Look to Stakeholders to Find the Cues in the Technology Environment

The pilot validated the need to conduct stakeholder analyses early and often throughout the technology planning process to identify the major funding opportunities in defense department research domains, technology research COIs, investment trends in commercial industry, opportunities to converge with/influence commercial industry roadmaps, and the effects of different technology solutions on meeting different stakeholder needs.

3.5.3.5.6 Facilitate Innovation Networks

The pilot suggested the need for technology planning to encourage a strategy and new roles for technology planners to act as strategic technical leaders within the technology planning process in order to facilitate sharing of capability needs and innovative solutions between users and providers of new technologies. These strategic technical leaders might work with industry, research laboratories, and other technology developers to encourage innovation through competition and other incentives, in order to advance the availability of solutions to meet enterprise capability needs. The innovation network is a way to reach more broadly throughout the enterprise.

There is a need to expand the technology planning process to include the development of a strategy for technology transition. The strategy and the associated process steps should include development of success criteria driven by enterprise parameters (i.e., what enterprise needs must be met in the implementation and operation of a technology or set of technologies before transition can be deemed a success?).

3.5.4 Pilot 4

3.5.4.1 Objective

This pilot used the CBEA process as a basis for building a framework to evaluate the capability-based contribution of systems participating in an enterprise. The capability framework developed in this pilot would ultimately be used to define an integrated master schedule (IMS) of planned, funded, and fielded capabilities necessary to accomplish desired capabilities in the next 5 years.

3.5.4.2 Enterprise Outcome

Enterprise engineering process improvement is established with an ultimate goal of providing the basis for an enterprise-level mechanism to exploit opportunities and minimize risk.

3.5.4.3 Enterprise Processes

CBEA.

3.5.4.4 Approach

The pilot defined five key steps to develop an IMS using CBEA:

- Align program plans with strategic direction, using a master capabilities list to describe program capabilities.
- Conduct end-to-end capability analysis by mapping program capabilities to a theater-level mission scenario, and identify gaps, risks, and critical dependencies, or explore a controlled reaction to environmental changes.
- Identify investment and divestiture opportunities across programs or for programs supporting multiple missions.
- Address interoperability and net centricity by considering relevant COI concerns/objectives and alignment of technology needs across related platforms.
- Support program planning and advocacy—use program capabilities and associated criticalities/dependencies to facilitate consistent funding analysis and planning.

3.5.4.5 Findings

Effective enterprise planning begins with capabilities analysis. The pilot team affirmed the criticality of capability-based analysis in driving enterprise planning. Recognizing that a defense department desires a capability-driven requirements process and yet, far too often, programs remain focused on system-level functional descriptions, the pilot team developed a framework for evaluating program contributions to missions of current or potential interest. Those contributions were then formulated into a set of capability statements linked to specified outcomes (i.e., missions/mission effects). The use of the capability framework enabled the pilot team to take a bottoms-up approach to establishing capabilities based on program contributions to one or more outcome spaces, and establish a baseline to frame capability portfolios for the enterprise. From here, the pilot team was then able to begin an exploratory analysis of the capabilities established, using the CBEA process and capability framework to assess end-to-end capability solutions, investment and divestiture opportunities, interoperability and net-centricity factors, relevant communities of interest and key participants, alignment of technology needs across related platforms, program planning/advocacy, and consistent financial analysis. The framework and underlying CBEA process thus proved critical to providing a basis for enterprise analysis based on capability outcomes versus individual system-level analysis.

Convergence on enterprise outcomes facilitates opportunity management and program planning. Use of the capability framework in this pilot not only reinforced the importance of capabilities to drive enterprise planning, but also emphasized a key principle from the complex systems engineering regimen to identify outcome spaces and encourage cooperation in order to achieve collective action in enterprise decision making. Ultimately, the capability framework implemented in this pilot will drive program investment and divestiture decision making. Therefore, getting stakeholders to agree up front on the outcome spaces (i.e., mission needs) most critical to the enterprise provides a basis for collaborative assessment, opportunity management, and program planning.

3.5.5 Pilot 5

3.5.5.1 Objective

The purpose of this pilot was to explore the applicability of the set of ESE processes to define, document, and execute an engineering process for evolving a C2 enterprise.

3.5.5.2 Enterprise Outcome

Identify and correct integration shortfalls in a C2 Enterprise that were hampering user community effectiveness.

3.5.5.3 Enterprise Engineering Processes Targeted

All currently defined ESE processes.

3.5.5.4 Approach

This is a program responsible for delivering cross-enterprise services and other infrastructure solutions to enable integration of multiple missions across a military enterprise. The program has defined an enterprise integration rhythm of activities, which involves a set of processes to continuously identify capability shortfalls, analyze alternate solutions, and field and sustain successful solutions for the multimission enterprise. The pilot team compared the set of ESE processes against the multimission enterprise integration processes to identify alignment and opportunities for improvement.

3.5.5.5 Findings

ESE processes must address not only technical systems engineering factors, but also the confluence of operational, business, and technical processes and activities that influence enterprise outcomes. The pilot found the current set of ESE processes tend to be product or technology focused, and do not sufficiently define steps and techniques for addressing business and operational factors. For example, EA should be applied to the business process architecture as well as the enterprise systems architecture. Trade-off analyses with cost–benefit criteria and use of business incentives to drive enterprise-level outcomes should be included as part of the ESE toolkits. The pilot team recommended ESE process toolkits be expanded to address nontechnical aspects of ESE impacting the achievement of enterprise objectives.

An important first step in the application of ESE is developing a strategic context for implementing ESE to achieve enterprise outcomes. ESE process toolkits should emphasize the importance of strategy. The pilot team identified the need for the ESE process toolkits to establish a strategic context at the outset of an ESE endeavor. Strategy defines how an engineering process is to be implemented in an enterprise, recognizing stakeholder needs and objectives, funding and requirements flow, and other sociocultural-political factors that guide enterprise operations.

Enterprise collaboration is critical. It builds networks of knowledge sharing and understanding among enterprise participants so that enterprise outcomes are more broadly accepted and synergistic efforts can be applied to exploring capabilities, technologies, and opportunities. In considering the strategic context for ESE, enterprise engineers must recognize and promote enterprise collaboration as critical investment.

ESE processes do not follow a sequential pattern, but intersect and interact depending on the enterprise issue being addressed. In enterprise integration, the flow of activities from capability planning to program planning cuts across the ESE processes at various points. More work is needed to identify process touch points and hand-offs, especially when applied to enterprise problems like how to achieve systems integration on an enterprise scale, how to develop a capability roadmap that aligns program plans and technology initiatives to deliver operationally defined capability outcomes, or how to assess cost–benefit of specific investments at an enterprise level.

3.5.6 Pilot 6

3.5.6.1 Objective

This pilot applied the CBEA process to aid in developing a roadmap for integration and evolution of a global enterprise operation.

3.5.6.2 Enterprise Outcome

Update a global enterprise portfolio with an enterprise-level capabilities definition and roadmap for the next-generation evolution.

3.5.6.3 ESE Processes Applied

CBEA, STP.

3.5.6.4 Approach

Using the CBEA process, the pilot team met with enterprise stakeholders to clarify capability needs and specify outcome spaces. The team framed the capability areas and explored technology concepts and concept validation tools to discover technological opportunities. An integrated process team was formed to conduct technology assessments and review technologies against the STP. Various evolutionary strategies were developed. Approved concepts were included in the enterprise portfolio roadmap, and related to the EA.

3.5.6.5 Findings

Stakeholder analysis is an essential ESE step. While the pilot team found the CBEA process useful overall in providing a framework to support capability planning and roadmap development, the team highlighted stakeholder analysis as an essential step to characterize the enterprise and its boundaries, its constituents, their roles, objectives, and levels of influence on enterprise outcomes. This step was deemed crucial to ensure proper validation of needs and funding support.

Extend CBEA to address cross-enterprise or cross-operational concepts opportunities. In many cases, acquisition programs are developing systems with functionality to serve multiple enterprises. That was the case for this particular pilot team, which found common capability needs across military and civil agency enterprises that had similar missions. The ability to conduct CBEA across enterprises adds complexity to all aspects of enterprise integration. The pilot team noted that additional examples of cross-enterprise analysis and solutions development are needed to assist in dealing with integration needs at this level.

3.6 ESE Case Studies—Summary of Findings

The ESE case studies represent a retrospective look at programs that utilized common sense heuristics in dealing with enterprise challenges. The activity identified lessons learned, assessed the degree to which the fundamental principles of ESE were used even before they were articulated, and conjectured how ESE processes may have helped if they had been available.

3.6.1 Case Study 1

3.6.1.1 Objective

This case study explored how ESE methods and principles were applied in an initiative to synchronize operating concepts for a shared mission across two global enterprises from inception through prototype development, experimentation, implementation, and transition to field operations.

3.6.1.2 Approach

The process used in this initiative grew out of an effort to improve a military mission thread by enabling better information flow and collaboration across two enterprises—one military and the other civil. An ad hoc group comprised of mid-level engineering and operational managers from across the stakeholders was formed to manage this cross-domain enterprise integration activity. Capability shortfalls in the mission thread were identified and solutions to address them were developed and rank ordered. Experts in each of the domains and systems—a COI—developed strategies for how existing field capabilities and technologies emerging from laboratories could be used to realize operational capability improvements. Much of what was needed to achieve significant improvements in interoperability between the two enterprises already existed, was planned to be delivered, or at least was already identified as a requirement for future development.

The group then aligned the priorities of the various systems and, in a few instances, identified new solutions to provide the glue that enabled the systems to better work together in the mission thread. The approach leveraged existing functionality to produce a new or enhanced operational capability in a relatively short time and at a fraction of the cost of a stand-alone, new solution.

The group's methodology began to take shape with a first-block implementation to address a specific scenario. This included an experiment featuring key elements of the first block, a series of rapid acquisition program proposals to obtain funding to assist with fielding of the first block, seed money to develop experiment proposals, and later, several experimental capability transition proposals. All the activities worked together to deliver the envisioned first-block capability—a federation of military and civil solutions focused on a specific mission thread.

3.6.1.3 Findings/Lessons Learned

Effective ESE is as much about product as it is about process.

In this case study, the methodology proved to be a fast, inexpensive way to implement net-centric communications across the enterprise, as well as information exchanges that improved mission management effectiveness in the chosen scenario. Thus, the methodology was both a product and a process.

The product was the first block of cross-enterprise capability. The process was a repeatable methodology for identifying, advocating, managing, and fielding cross-domain/enterprise integration initiatives. The initiative demonstrated the viability of a process that accelerates the fielding of select solutions across two enterprises with similar but not identical missions. The process used capabilities-based planning, experimentation and analysis, architectures, and technology planning to establish an enterprise roadmap and garner the funding to deliver the needed capability. A next step for this initiative is to institutionalize the approach with a team that accelerates the fielding of solutions in a repeatable fashion.

To effectively engineer an enterprise solution requires the use of coalitions of key stakeholders who influence enterprise outcomes. The methodology involved teaming arrangements to foster collaboration in developing and fielding a cross-enterprise capability. The team created a framework to converge on enterprise outcomes and ensure delivery of operational capability. It built a COI comprised of operational and technical experts from the participating domains, commands, programs, and associated systems which defined seams that needed to be closed and capabilities that needed to be implemented among military and civil stakeholders. The COI influenced and monitored the execution of system development programs to minimize delivery gaps. In addition, the COI provided expertise that identified quick-reaction program alternatives and near-term solutions to operational problems. It served as an environment for continuous interface between acquisition programs and operational communities to ensure that the user needs were met.

Successful ESE is based on convergence of key enterprise capability needs and a flexible approach to address those needs that leverages available technology, rapid funding, and acquisition approaches. Key to the methodology was a focus on specific, clearly defined mission threads, and analysis of which capability improvements could be achieved by using assets in new or better ways, or making simple, less expensive modifications to the existing systems. The team was kept small and agile, and focused on high-return operational capabilities. Not everything can be addressed by this process but for mission threads in this initiative it provided a way of getting improved capability in the hands of the operational user quickly and cheaply.

3.6.1.3.1 ESE Processes

The six ESE processes provided a useful framework to guide the development and implementation of an operations synchronization process across two enterprises and the delivery of a first-block product set.

3.6.1.3.2 Enterprise Architecture

The vision, strategy, and implementation were documented in an architecture.

3.6.1.3.3 Strategic Technical Planning

The initiative leveraged systems that adhered to evolving STP tenets.

3.6.1.3.4 Enterprise Evaluation and Assessment

The initiative inserted capability into a joint live experiment to determine if the operational user liked it and, if so, to transition it to an operational capability.

3.6.1.3.5 Capabilities-Based Engineering Analysis

Initially translated a capability need into a cross-domain solution set for near-term execution and later, linked the overarching process to emerging capability planning processes.

3.6.1.3.6 Technology Planning

The initiative leveraged new technologies being developed by one enterprise and provided transition opportunities for each. A follow-on advanced technology demonstration was to facilitate technology insertion to improve the capability introduced.

3.6.1.3.7 Stakeholder Analysis

This emerged as an important sixth process with the formation of a COI comprised of developers, users, acquisition experts, and technology experts from the military and civil domains that could identify capability shortfalls and guide potential solutions.

3.6.2 Case Study 2

3.6.2.1 Objective

This case study is about a program to improve an enterprise's ability to develop and distribute information on objects of interest in a military operational environment.

The study highlights the use of innovative development and integration strategies to provide that capability.

3.6.2.2 Approach

This program was intended to provide an enterprise capability that improves interoperability by improving the sharing of information on objects of interest across a military enterprise. Each participating node in the information-sharing network implements a set of common functions and processes to produce a consistent understanding of objects of military significance across a specified geographical region. The program provided the common functions and processes through the use of model-driven architecture tools and methods, and produced a representation of the needed information production and distribution functions. The engineering effort was comprised of military, government civilian, nonprofits, and commercial contractors provided by each military service.

3.6.2.3 Findings/Lessons Learned

The following findings and lessons learned from the case study highlight the applicability of ESE to this situation awareness improvement development program.

3.6.2.3.1 Leadership Is a Critical Component in the Achievement of Enterprise Capability

The continuous interaction and influence of leaders in the complex environment within which the capability is developed was an essential ingredient of the process to create, deliver, and make the capability operational. All stakeholders actively engaged in developing the enterprise capability.

3.6.2.3.2 Clarity of Vision in the Face of a Complex Objective

The ability to create a clear vision of the program objective made a complex endeavor tractable.

3.6.2.3.3 Flexibility and Consensus in Achieving the Vision

While the vision articulated was straightforward, its implementation was a new and relatively complex way of acquiring military enterprise functionality that involves both technical and human interactions on multiple scales. The path to achieving it was discovered through the confluence of stakeholders, emerging technologies and nascent business, and engineering processes. Achieving a consensus about the enabling infrastructure, design tenets, and a high-level functional architecture and guiding precepts provided a common baseline with flexibility to evolve overtime. The planned use of early product releases, known as capability time boxes, and ongoing demonstrations and experiments provided infrastructure and events to explore operational uses and benefits and coordinate test and validation efforts. These improved both the ongoing development and stakeholder confidence in the emerging capability.

3.6.3 Case Study 3

3.6.3.1 Objective

This case study describes an approach to data sharing based on a COI model and touch points of the model to ESE processes of EA, EA&A, CBEA, and technology planning. The study provides recommendations for applying ESE to other COI efforts.

3.6.3.2 Approach

The mission of the initiative directly mandated enterprise engineering, for example, define the business rules for seamless operations and expose data assets to be reused as information services. COI products include data schemas, services definitions, a mapping of COI services to systems, which produce or consume data described by the COI, as well as scripts for orchestrating COI services within the affected programs that developed the systems. A COI roadmap was produced which describes the timeline, technical approach, and investment strategy for migrating systems and programs to the common schema and services described by the COI, so information sharing could more effectively enable cross-enterprise missions and objectives. A program for developing a net-centric enterprise capability, which integrates databases, servers, client workstations, local area networks, and computer software into an open, scalable, network-centric single architecture was identified as the primary enterprise program to support the COI services. The case study evaluated how well the implementation of the COI services into this net-centric enterprise capability was accomplished, and what improvements could be made to facilitate enterprise engineering approaches for the formalization of enterprise information, and the implementation of architectures and governance approaches to migrate enterprise products into development programs.

3.6.3.3 Findings/Lessons Learned

The engineering of enterprise information sharing works best when stakeholders converge on critical capabilities and supporting services to meet the common enterprise objectives. A key ESE principle is the need to employ systems thinking when defining and building services to meet enterprise-wide needs. This requires an upfront analysis of key stakeholders, their role in the enterprise and the capabilities about which they are most concerned. Where consensus for specific capabilities cannot be achieved, ESE calls for convergence on agreed outcome spaces in which more detailed capabilities can be further refined over time. It requires methods to analyze and assess alternatives directed toward improving the effectiveness and efficiency of cross-enterprise operations with outcomes that provide value to stakeholders. It stresses the need to balance the political, operational, economical, and technical dimensions of delivering enterprise capability.

3.6.4 Case Study 4

3.6.4.1 Objective

This was a laboratory and field assessment of a service-oriented architecture (SOA) as a technical paradigm to support net-centric operations throughout an enterprise. The case study addressed the relevance of a demonstration of net centricity to ESE and lessons that could be applied to other programs.

3.6.4.2 Approach

This case study examined whether the use of SOAs would improve enterprise performance sufficiently to justify widespread adoption of them across the elements of the enterprise. The objectives emphasized the operational and technical benefits of SOAs to support a time-critical mission. The specific SOA capabilities were:

- Orchestrate and configure IT services on-the-fly
- Interoperability of independently developed services
- Ensure performance of services is maintained through military networks
- Support common infrastructure services such as failover, discovery, network time, and domain name service
- Demonstrate reuse of legacy systems

The architecture employed services from a number of military systems, development programs, test beds, and commercial companies. A scenario was used that required a dynamic alignment of systems and services that mimicked actual mission operations.

3.6.4.3 Findings/Lessons Learned

The demonstration provided a basic confirmation of the availability of several SOA technical capabilities to support the enterprise attributes needed for net-centric operations. It also provided a better understanding of the challenges of ESE.

3.6.5 Case Study 5

3.6.5.1 Objective

This was a case study in using cross-Service collaboration to build architectures, identify shortfalls, and provide an enterprise capability. ESE processes analyzed were CBEA, EA&A, EA, and technology planning.

3.6.5.2 Approach

This case study resulted from a cross-Service analysis of a specific time-critical mission thread. Seven interoperability points were identified where relatively simple changes to the flow of information between the Services could substantially improve the prosecution of the mission thread.

Using a capabilities-based approach, the concept in this case study was derived from an analysis of the operational activities and information flows each Service follows to manage information specific to the time-critical mission in a cross-Service operation. A high-level architectural analysis revealed several similarities across the Services. Each Service has similar tools for prosecuting the mission, and each Service also experiences difficulties finding, verifying, and checking status of objects of interest. The result was a need to develop a capability to share information more efficiently and effectively, eliminate operator-intensive merging of information and resolution of conflicting information, and manual assessments of mission success. This capability called for a common information format, and the use of a Web-based service to enable efficient means to describe, post, and access target information.

3.6.5.3 Findings/Lessons Learned

CBEA, EA, and technology planning had direct applicability in responding to this enterprise capability need. CBEA was applicable since the enterprise-level issue and its manifestation in a mission was the driving factor. The use of EA products helped articulate the operational challenge and expressed the nature of an infrastructure to enable cross-Service awareness that expedited cross-Service coordination and response. Technology planning and STP precepts guided the team to adapt Web-services technology to provide common interface definitions across systems and Services that enabled better interoperability, reduced manual coordination, and provided a more consistent approach to mission execution at all levels of the operational enterprise.

The case showed the need to conduct an early stakeholder analysis. Use of an ESE environmental profiler tool [7] could have helped to clarify the role of key stakeholders and the need for their early involvement.

3.6.6 Case Study 6

3.6.6.1 Objective

This case study concerned a directory service for an enterprise that provides information about and access to planned, developing, and fielded information services for enterprise systems developers and engineers. Its intent was to improve system and enterprise acquisition by facilitating information service reuse and SOA development. The case study identified relevant ESE techniques and processes used in the design, development, and execution of the directory service as a tool for enterprise capability development, as well as the lessons learned in the implementation of it.

3.6.6.2 Approach

The directory service was meant to assist in enterprise portfolio management by identifying and providing information on available services to support enterprise needs. From an engineering and development standpoint, the directory service helps enterprise architects assess the readiness of specific services, understand the feasibility and risks associated with implementing the service(s), compare alternative service choices, and use the service(s) in an SOA. It also might be used in capabilities-based planning to map operational threads to services, and thereby help identify SOA risks. By encouraging information service reuse, the directory service facilitates SOA development and interoperability, and ultimately reduces design, development, and implementation costs as systems that work together to support a capability benefit from consistent use of key enterprise services.

3.6.6.3 Findings/Lessons Learned

To be effective, ESE must address not only the technical dimensions of an enterprise, but also the operational, economical, cultural, organizational, and political dimensions of an enterprise. This case study revealed that while the directory service was designed to enable more agile acquisition and systems engineering, its implementation suffered without an enterprise business model and service-support strategy to encourage its use. There were no service-level agreement guarantees for directory service users. The directory service allowed for tracking of services, but there was no single agency that had authority for the phasing of a service from concept to implementation to retirement, or monitoring of services to assure standards conformance. As a result, would-be users of the directory service did not see an advantage in using it. In fact, without a comprehensive and consistent support strategy in place, program managers and engineers were reluctant to use the services provided by the directory service repository. The directory service was designed to be a repository for enterprise information services. However, incentives to encourage its use were not provided. It was as easy for program managers and engineers to develop information services as needed for their own systems to avoid the technical and operational risks associated with the lack of support agreements. Moreover, there were no cultural, organizational, or political factors driving the use of the directory service. No enterprise governance structure or recognized authority was in place to guide information service development and use, or influence a new development and engineering paradigm that emphasized use of the directory service. Therefore there was no reason for a change in behavior by systems engineers or managers that would advance directory service use. The case study demonstrates the need for well thought-out strategies that address all dimensions of an enterprise—technical, operational, political, cultural, and organizational—when delivering capability designed to encourage broad use and to facilitate integration among enterprise players.

3.6.7 Case Study 7

3.6.7.1 Objective

The objective of this case study was to identify the relationship and applicability of the ESE processes to an integrated capabilities readiness and risk-assessment process.

The capabilities readiness and risk-assessment process identifies and prioritizes capability gaps based on an analysis of key mission threads that underpin one or more major enterprise concepts of operations. Integrated capabilities readiness and risk-assessment process assessments are conducted annually and result in a set of capability objectives, which establishes specific tasks and actions to achieve enterprise capability overtime. The integrated capabilities readiness and risk-assessment process and the resulting capability objectives are a principle component of an enterprise's capability planning process to guide future development and research in a way that most effectively meets enterprise capability needs.

3.6.7.2 Approach

This case study reviewed the integrated capabilities readiness and risk-assessment process to identify the relationship and relevance to ESE.

3.6.7.3 Findings/Lessons Learned

The ESE processes reviewed were EA, EA&A, and CBEA. While applicable to the integrated capabilities readiness and risk-assessment process, these ESE processes focus primarily on technical analysis and need to be expanded to include methodologies for addressing and balancing financial, operational, and political dimensions of enterprise alternatives. Each was found to be strongly aligned to the integrated capabilities readiness and risk-assessment process. Enterprise architecting provides a framework to conduct assessments to support capability objective solutions identification and analysis. Key architecture products are those that help describe the operational activities associated with a mission thread that may be analyzed as part of an integrated capabilities readiness and risk assessment.

EA&A and capabilities-based engineering are also very closely aligned to the integrated capabilities readiness and risk-assessment process. The integrated capabilities readiness and risk-assessment process is based on capabilities planning and analyses using EA&A techniques to evaluate alternative approaches to achieving capability across enterprise systems. The study revealed that the current ESE processes lack a methodology to define the broader "trade space" (to include technical, economical, operational, political factors for engineering enterprise capability). Current methods tend to focus primarily on technical dimensions of ESE. More tools and techniques are needed that provide insights to the programmatic and cost trade-offs associated with evaluation of enterprise alternatives.

3.6.8 Case Study 8

3.6.8.1 Objective

This case study evaluated the value of an enterprise system framework into the enterprise, its SOA infrastructure including a service management process, and its patterns for integrating toward an adaptive system of systems.

3.6.8.2 Approach

This case study evaluated an enterprise system for providing support services to enterprise personnel in deployed and garrison locations.

3.6.8.3 Findings/Lessons Learned

The enterprise system framework in this case study is an example of how to apply ESE effectively to provide enterprise capability. The following are a few examples of what went right in developing and implementing the enterprise system framework.

STP for the enterprise system framework provided a systematic approach to integrating enterprise goals into individual systems. The enterprise goal for this enterprise system framework was to rapidly field and deploy enterprise capability and make it available to users worldwide. STP was used to establish integration standards to enable interoperability and scalability for an application to join the framework, and be accessible to users wherever they are.

The enterprise system framework used an EA to describe its services, standards, and integration approach. Applications to be integrated to the enterprise system framework can be planned for by reviewing its well-documented architecture and integration procedures.

EA&A, CBEA, and technology planning allowed the enterprise system framework program to continually characterize the effectiveness of infrastructure and applications to satisfy enterprise needs. The enterprise system framework used EA&A effectively to identify opportunities to provide new capabilities or technology in an evolutionary manner, thus reducing costs and technical risks, and targeting improvements that would provide the most overall value to the enterprise.

Another success of the enterprise system framework was its utility in building consensus around the enabling infrastructure. The enterprise system framework is an enterprise infrastructure that provides services to deliver and manage capabilities for end users. By gaining consensus around the infrastructure, the enterprise system framework can be used to exert more control over the phasing of services and evolution of capability overtime.

3.6.9 Case Study 9

3.6.9.1 Objective

This case study was about a program to deploy multiple capabilities and technologies to its enterprise via an incremental development process. It focuses on the relevance of ESE processes and techniques to the identification, maturation, and assessment of critical technologies, core technology modernization planning, incremental development planning, and impacts on technology planning due to changing threats.

3.6.9.2 Approach

Compare case study technology planning experiences with ESE technology planning process to identify best practices and suggest process refinements.

3.6.9.3 Findings/Lessons Learned

ESE process description must set the enterprise context for programs. Technology planning should provide a context for individual programs to interpret the intent of the process activities and needs from an enterprise perspective. This includes guidance on the minimum set of specific activities that programs must implement, giving due consideration to program resources. Programs require guidance on questions such as: which technology planning activities must be implemented and which are optional; how should the planning activities be implemented; and who are the stakeholders and what are their roles and responsibilities.

The value proposition of enterprise process(es) at both the enterprise and program levels must be clear. This would facilitate better buy-in at the program level, allow programs to make the connection with the enterprise level, and could result in needing less detailed guidance at the program level to implement the process.

3.7 Final Recommendations

As a result of the pilot and case studies, the following steps are recommended to further explore the practice of ESE.

Stakeholder analysis: Stakeholder analysis arose as an important first step in many ESE activities. Therefore, a formal process should be defined for stakeholder analysis, along with accompanying use case examples.

ESE use cases: Many cases showed that ESE processes do not follow a linear path, but rather interact with each other in ways that depend on the enterprise objective or problem to be addressed. A set of use cases could be defined that would suggest how to apply ESE processes to achieve common enterprise objectives (e.g., STP, capabilities-based planning/road-mapping, or enterprise systems integration). The archetypical use cases should be documented as part of the ESE process toolkits and continually reviewed and updated.

ESE tools: Update ESE toolkits with specific tools developed across the enterprise. Many of the cases reinforced the utility of existing ESE tools or suggested the development of others to support ESE processes. These tools and their application should be formalized as part of existing ESE toolkits and made available for use across the enterprise.

Education and training in ESE should focus on ESE tools, process, and skills development. The cases showed that ESE processes and tools must be accompanied by the development of new skills that emphasize collaboration building, ability to balance cooperation and competition, and the ability to recognize and shape the confluence of political, operational, technical, and economic factors influencing enterprise-level systems engineering. Education and training programs should be developed and deployed that foster these skills for both technical engineers and program managers.

The ESE focus group effort has reinforced the importance of applying enterprise thinking and the value of having a defined set of principles, processes, and tools available to guide engineering activities toward enterprise outcomes. While this effort has been successful in highlighting the need and applicability of ESE, the discipline is still in the early stage of a cultural change toward a more formal approach to enterprise thinking and behavior as part of the systems engineering discipline.

The pilots and case studies reviewed in this report show that the ESE principles and processes developed to date are relevant to the complex demands for enterprise capability. The results of this analysis shows the need to expand and deepen the set of processes and tools to include stakeholder analysis and an ESE environmental profiler tool as added components of an ESE discipline.

References

1. DeRosa, J. K., Rebovich, G., and Swarz, R. "An enterprise systems engineering model." *INCOSE 2006 International Symposium*, Orlando, FL, July 2006.
2. DeRosa, J. K. and McCaughin, K. "Process in ESE." *INCOSE 2006 International Symposium*, Orlando, FL, July 2006.

3. Swarz, R. and DeRosa, J. K. "A framework for enterprise systems engineering processes." *International Conference on Software and Systems Engineering*, Paris, France, December 2006.
4. McCaughin, K. and DeRosa, J. K. "Stakeholder analysis to shape the enterprise." *NECSI International Conference on Complex Systems*, NECSI, Boston, MA, 2006.
5. Kuras, M. L. and White, B. E. "Engineering enterprises using complex-system engineering." *Proceedings of 15th International Symposium, INCOSE*, Rochester, NY, July 2005.
6. Gharajedaghi, J. *Systems Thinking: Managing Complexity and Chaos*. Boston, MA: Butterworth Heinemann, 1999.
7. Stevens, R. *Engineering Megasystems: The Challenge of Systems Engineering in the Information Age*, CRC Press, June 2010 (expected).

Chapter 4

Capabilities-Based Engineering Analysis

Steven J. Anderson and Michael J. Webb

Contents

4.1	Introduction	97
	4.1.1 Capabilities Perspective of the Enterprise	98
	4.1.1.1 Complex Contextual Issues	98
	4.1.1.2 Need for Responsive Modernization	99
	4.1.2 Capabilities Focus in the Department of Defense	100
	4.1.2.1 Capabilities-Based Planning	100
	4.1.2.2 Capabilities-Based Planning—The Defining Characteristics	101
	4.1.3 The Need for Capabilities-Based Engineering Analysis	103
4.2	An Overview of Capabilities-Based Engineering Analysis	105
	4.2.1 Embodying Complex System Strategies	105
	4.2.1.1 Modularity	105
	4.2.1.2 Exploratory Modeling and Analysis	106
	4.2.1.3 Adaptive Evolutionary Planning	106
	4.2.2 Capabilities-Based Engineering Analysis—The Analytical Model	107
	4.2.2.1 Terminology	108
	4.2.2.2 Principles	108
	4.2.2.3 Processes	109
	4.2.3 Varied Implementations	112

96 ■ Enterprise Systems Engineering

- 4.3 Purposeful Formulation 113
  - 4.3.1 Specify Outcome Spaces 113
    - 4.3.1.1 Stakeholder Analysis 115
    - 4.3.1.2 Describe Desired Outcomes 116
    - 4.3.1.3 Specify Conditions 116
    - 4.3.1.4 Derive Task Structure 117
    - 4.3.1.5 Develop Metrics 117
  - 4.3.2 Frame Capability Portfolios 118
    - 4.3.2.1 Enterprise Architecture 119
    - 4.3.2.2 Bound Relevant Structures 120
    - 4.3.2.3 Identify Interdependencies 121
    - 4.3.2.4 Describe Enterprise Context 121
    - 4.3.2.5 Establish Reference Portfolio 121
- 4.4 Exploratory Analysis 122
  - 4.4.1 Assess Performance and Cost 122
    - 4.4.1.1 Prepare for Analysis 122
    - 4.4.1.2 Assess Baseline Performance 125
    - 4.4.1.3 Evaluate Costs 125
    - 4.4.1.4 Bound Solution Space Performance 126
    - 4.4.1.5 Identify Gaps and Opportunities 127
    - 4.4.1.6 Special Challenges for Command and Control Analysis 128
  - 4.4.2 Motivations for Change 128
  - 4.4.3 Explore Concepts 129
    - 4.4.3.1 Generate Alternatives 129
    - 4.4.3.2 Develop Concept 130
    - 4.4.3.3 Forecast Impact 131
    - 4.4.3.4 Assess Feasibility 131
- 4.5 Evolutionary Planning 131
  - 4.5.1 Examine Evolution Strategies 132
    - 4.5.1.1 Establish Partnering 132
    - 4.5.1.2 Create Vision 134
    - 4.5.1.3 Develop Strategy 134
    - 4.5.1.4 Plan for Contingencies 135
    - 4.5.1.5 Refine Portfolio Plans 135
  - 4.5.2 Evaluate Enterprise Impacts 136
    - 4.5.2.1 Identify Technical Impacts 136
    - 4.5.2.2 Identify Capability Impacts 137
    - 4.5.2.3 Identify Resource Impacts 137
  - 4.5.3 Concept Decision 138
    - 4.5.3.1 Purposes 138
    - 4.5.3.2 Analytical Information 139
    - 4.5.3.3 Possible Outcomes 139
    - 4.5.3.4 The Need for Tools 139

    4.5.4 Develop Portfolio Roadmap 140
        4.5.4.1 Integrate Plans 140
        4.5.4.2 Map Evolution Pathways 141
        4.5.4.3 Describe Evolution 142
4.6 Joint Capabilities Integration and Development System 144
    4.6.1 Top-Down, Concepts-Centric Approach 144
    4.6.2 The Analysis Process 144
        4.6.2.1 Functional Area Analysis 145
        4.6.2.2 Functional Needs Analysis 146
        4.6.2.3 Functional Solution Analysis 146
    4.6.3 Joint Capabilities Integration and Development System
          Analytical Output 147
    4.6.4 A Critique of Joint Capabilities Integration and
          Development System 147
4.7 Air Force Capabilities-Based Planning 148
    4.7.1 CONOPS-Based Approach 148
    4.7.2 Capability Categories 149
        4.7.2.1 Functional 149
        4.7.2.2 Operational 150
    4.7.3 Air Force Analytical Process 150
        4.7.3.1 Functional Area Analysis 151
        4.7.3.2 Functional Needs Analysis 151
        4.7.3.3 Functional Solution Analysis 153
    4.7.4 Air Force Analytical Output 153
    4.7.5 Timing 155
4.8 Summary and Conclusions 155
Bibliography 155

4.1 Introduction

The increasingly complex, dynamic, and uncertain nature of today's world has led many enterprises to design and manage their organizations as systems of capabilities and outcomes, as opposed to collections of processes and reporting relationships (Haeckel, 1999). By organizing assets as systems of capability modules, an enterprise can engage challenges and opportunities more dynamically, and at the most fundamental level, such an organization can be managed as a purposeful complex adaptive system (Cohen and Axelrod, 1999).

    Capabilities-based engineering analysis (CBEA) is an engineering approach that supports such an enterprise strategy by focusing on the development and evolution of capabilities, rather than specific programs or functions. As a component of enterprise systems engineering (ESE), CBEA analyzes the technical options associated with alternative plans for achieving capabilities; it is the analytical framework that supports capabilities-based planning (CBP), programming, and acquisition in a systemic approach to the purposeful evolution of the enterprise. The basic motivation for CBEA is to enable an enterprise to develop and evolve its critical capabilities in a highly complex and dynamic environment to bring superior value to its customers.

This chapter outlines the guiding principles and key analytical activities of CBEA for researchers and practitioners engaged in ESE. In order to understand the context and purpose of CBEA, it is first necessary to examine the reasons for focusing on capabilities in enterprise management.

4.1.1 Capabilities Perspective of the Enterprise

Economic theory has increasingly focused on the concept of *capabilities* as the defining characteristic of an enterprise, and this view has propagated through the fields of business and management. Indeed, in its firm-oriented incarnation, evolutionary economics (Nelson and Winter, 1982) is often referred to as *the capabilities view* (Foss, 1993; Langlois, 1992). In this view, an enterprise is conceptualized and distinguished in terms of its unique capabilities, and these capabilities are considered to be the most potent source of competitive advantage, growth, and success of an enterprise (Collis, 1994). Business and management literature has highlighted the success of a capabilities approach (Coyne, 1986; Ghemawat, 1986; Grant, 1991; Hall, 1993; Stalk et al., 1992; Williams, 1992).

Capability-based business strategies are founded on the notion that distinctive capabilities provide the strategy platform that underlies long-term survival and success, and such strategies emphasize the need to constantly evaluate an enterprise's strengths and weaknesses from a capabilities perspective. Such enterprise strategies typically include (Haeckel, 1999) the following:

1. Organizing assets and capabilities as a system of modules that can be dynamically employed for in pursuit of specific outcomes
2. Propagating purpose, bounds, and essential structure throughout the enterprise
3. Distributing power and responsibility
4. Managing interactions rather than the actions of modular capabilities

4.1.1.1 Complex Contextual Issues

Central to the capability view is a primal focus on deploying enterprise resources to achieve end-user effects or outcomes (Haeckel, 1999). However, the focus on purposeful effects is complicated by the complexity of the enterprise and its environment. Figure 4.1 provides a high-level illustration of an enterprise and its complex contextual issues.

The complex contextual issues shape the efficacy of the enterprise. For example, in evolutionary economic theory (Nelson and Winter, 1982), organizational capabilities and decision rules evolve over time through deliberate problem-solving and

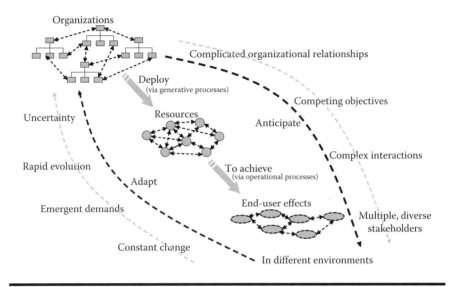

Figure 4.1 Complex context of the enterprise.

random events. An economic analogue of natural selection operates as the market determines which firms are viable and not viable in the real-world difficulties of complexity, uncertainty, and bounded rationality. Interestingly, the authors conclude that attempts at long-range optimization and control of technological advances will lead not to efficiency but to inefficiency.

4.1.1.2 Need for Responsive Modernization

Researchers emphasize the need for an enterprise to invest in capabilities rather than functions or business units (Stalk et al., 1992). Chandler (1990) emphasizes the need to focus on the integration, coordination, and maintenance of organizational capabilities. Chandler views the maintenance of capabilities to be as difficult as their creation, as changing technologies and markets constantly pressure capabilities toward obsolescence. One of the more critical tasks of top management is to modernize organizational capabilities and to integrate them into a coherent, unified organization—so that the whole becomes more than the sum of its parts. A continuing emphasis on modernizing organizational capabilities, in turn, provides the source dynamic for the continuing survival and growth of the enterprise.

As the pace of change continues to increase and the nature of that change becomes increasingly discontinuous, the *business-as-efficient-mechanism* paradigm is rapidly yielding to the *business-as-adaptive-system* organizing principle of preference. In this paradigm, the concept of *dynamic capabilities* becomes a key concern of the enterprise. *Dynamic capabilities* are typically viewed as the ability to integrate, build, and reconfigure competencies to address rapidly changing environments (Adner and

Helfat, 2003; Eisenhardt and Martin, 2000; Teece et al., 1997; Winter, 2003). CBEA is intended to serve as an adaptable approach to achieving dynamic capabilities.

According to Wade and Hulland (2004) information system resources can take on many of the attributes of dynamic capabilities, and thus may be particularly useful to firms operating in rapidly changing environments. Even if information system resources do not directly lead the firm to a position of superior sustained competitive advantage, they may nonetheless be critical to longer-term competitiveness in unstable environments they help it to develop, add, integrate, and release other key resources over time.

4.1.2 Capabilities Focus in the Department of Defense

The capabilities perspective of the enterprise is not limited to commercial firms; the United States Department of Defense (U.S. DoD) has also adopted such a view. With the end of the Cold War and its focus on conflict with the Soviet Union, the United States sought a more adaptive approach to military planning. The 1992 *National Military Strategy* (NMS) and *CJCS Posture Statement* argued for a capabilities-based force posture to deal with a wide range of poorly understood threats. The new direction initiated a move away from a few canonical threat-based scenarios and established a central role for uncertainty in planning.

4.1.2.1 Capabilities-Based Planning

The 2001 Quadrennial Defense Review (QDR) promulgated CBP "that focuses on developing the general wherewithal to fight successfully in a wide range of scenarios …." Arguing that the United States could not know the exact origin of threats, the QDR focused instead on the idea of anticipating the kinds of capabilities that an adversary might employ and the capabilities that the U.S. military would need to deter and respond against a diverse and uncertain set of security threats.

On the basis of this perspective, the Air Force (AF) established its Capabilities-Based Review and Risk Assessment (CRRA) in 2002. The CSAF articulated a vision for CBP and programming: "Our goal is to make warfighting effects, and the capabilities we need to achieve them, the drivers of everything we do" (Jumper, 2002).

While only recently emphasized, the concept of CBP in the DoD is not completely new. RAND developed an approach (objectives-based planning) to the subject in the 1980s; extracts from an early work (Kent, 1989) appear very relevant today:

> "In sum, the United States critically needs a force-planning process that focuses on the building blocks of operational capability rather than on hardware." (p. 1)
>
> "The Services should focus on demonstrating the relationship of strategies to operational tasks … [and] demonstrate how accomplishing these tasks … will achieve stated operational objectives." (p. 15)

> "A statement of the specific operational capabilities expected from a particular force element requires an end-to-end operational concept for accomplishing specific operational tasks." (p. 21)
>
> "Resources should be allocated on the basis of the best overall effect in achieving operational objectives and in winning campaigns and wars." (p. vii)

4.1.2.2 Capabilities-Based Planning—The Defining Characteristics

CBP has been defined as "... planning, under uncertainty, to provide capabilities suitable for a wide range of modern-day challenges and circumstances, while working within an economic framework." (Davis, 2002) Recent implementations of this concept in DoD, for example, Joint Capabilities Integration and Development Systems (JCIDS) and the AF CRRA, are based on some common key elements.

1. Analysis is focused on outcomes, usually in terms of achieving effects, vice inputs such as particular programs, platforms, or activities (functional areas).
2. The ability to achieve outcomes must be evaluated in a wide range of diverse and stressing operational circumstances.
3. The planning process must address the multiple dimensions (e.g., doctrine, organization, training, materiel, leadership, personnel, and facilities [DOTMLPF]) required to effect a capability and consider the full range of alternative solutions available to meet warfighting needs.
4. The concept of modularity is critical in managing multiple capabilities, focusing on the ability to dynamically integrate diverse resources to achieve different outcomes (categories of capability).

Figures 4.2 and 4.3, taken from the *2004 Joint Defense Capabilities Study*, highlight the DoD corporate view of CBP. While CBP has often been contrasted with *threat-based planning*, several authors have appropriately identified the contrast as a fallacy, because CBP is also very much concerned about threats and scenarios, it simply does not depend completely on one or two canonical scenarios (Davis, 2002; Troxell, 1997):

> "... serious capabilities planning and operations planning require the concreteness that comes with considering specific scenarios—either real or credibly constructed. Further the enthusiasm and focus needed to generate good ideas or to worry about problems creatively are enhanced when the scenario being considered is either real or obviously a relevant surrogate." (Davis, 2002, p. 8)

102 ■ *Enterprise Systems Engineering*

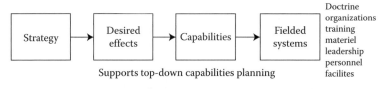

Figure 4.2 DoD view of the need for CBP.

The innovation in CBP, therefore, is not the irrelevance of threat, but the importance of assessing and managing risk across a much more diverse and uncertain set of threat conditions. This view was reaffirmed in the recent QDRs.

In summary, the DoD capabilities-based approach to planning emphasizes the importance of assessing and managing risk across very diverse and uncertain problem sets, with consideration and competition among a broad range of alternative solution approaches (i.e., DOTMLPF) constrained by cost and resource constraints. Military planners have instantiated processes to address these concerns; Figure 4.4, for example, outlines recent Joint and AF CBP processes. Though the processes appear intricate and complex, policy guidance (cited in Sections 4.6 and 4.7) is available which assigns roles and responsibilities for a hierarchical, strategy-to-task "family" of joint operations concepts and analyses. The need for engineering analysis pervades

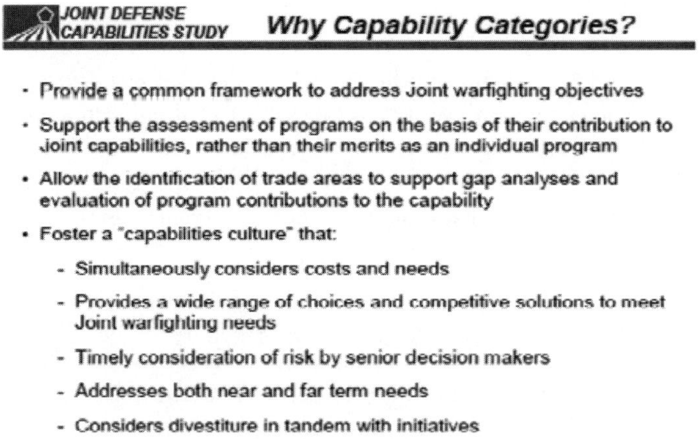

Figure 4.3 The need for capability categories.

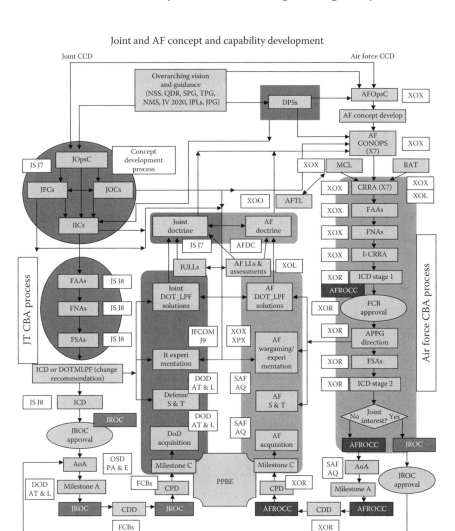

Figure 4.4 CBP processes.

these processes, and this chapter outlines key considerations and approaches to serve as an analytical framework for that work.

4.1.3 The Need for Capabilities-Based Engineering Analysis

From an enterprise perspective, the scope and complexities of CBP, programming, and acquisition require a disciplined and robust analytical construct. The capabilities perspective in ESE fosters a change in focus from a program or platform-centric approach to a capability-centric design that facilitates horizontal integration of a distributed, mission-oriented enterprise. CBEA supports this perspective by analyzing enterprise capabilities, linking capabilities to portfolios of resources and processes, and identifying possible reconfigurations to achieve effective adaptive responses. Partitioning the enterprise into capability categories, with attendant capability portfolios, helps to decompose the problem to a manageable degree,* but intricate complications inevitably persist; for example,

1. Multiple systems/programs typically contribute to a particular capability.
2. A particular system/program can contribute to multiple capabilities.
3. There are multiple ways and means to provide a particular capability.
4. Interdependencies among systems/programs entail complex, cascading effects.
5. The enterprise must coevolve multiple capabilities (multiple complex portfolios).
6. The environment is uncertain, dynamic, and complex.

To help enterprise decision makers adjudicate risks through their policy and resource allocation decisions, analysts must shape and resolve these enormous complications to the greatest extent possible. In this regard, CBEA supports the capabilities-based perspective across a broad range of activities and concerns.

1. Technical inputs to CONOPS refinements
2. Assessments of enterprise functional performance, identifying capability gaps, overlaps, and possible improvements
3. Cost estimates and risk assessments
4. Identification of critical issues for in-depth analysis
5. Reengineering of mission processes
6. Technology proposals for cross-cutting capability improvements
7. Investment strategies for cross-cutting capability portfolios
8. Capability roadmaps that capture comprehensive time-phased views of modernization
9. Assessments of programmatic status and progress against the roadmaps and recommendations for programming changes
10. Proposals for selected risk-reduction development and demonstration activities that are identified as critical tasks for the development of new capabilities
11. Recommendations on potential experimentation and/or test approaches
12. Assistance in the development of capability requirements documents, acquisition strategies, test strategies, and supportability strategies
13. Process and analytical recommendations for planning, executing, and reporting analyses and cost-effectiveness studies of capabilities

---

* This is not the same as reductionism, where one decomposes a system and works on the pieces in isolation from one another, which can be dangerous, as the interactions among the pieces and the whole and with their environment continually changes in a complex system.

Given the complexity of ESE challenges, the goal of CBEA is not necessarily an optimal design, but an adaptive, robust evolution of the enterprise. The following sections describe the principles and practices that facilitate such an effort.

4.2 An Overview of Capabilities-Based Engineering Analysis

The CBEA analytical approach is founded on the premise that capabilities must be viewed as the result of complex adaptive systems with multidimensional interdependencies. CBEA is aimed at measuring the world in terms of end-user outcomes, and it explicitly represents the profound uncertainties involved in assessing such issues in an increasingly complex and dynamic environment. The conceptual elements of CBEA are derived from modern approaches to managing complexity.

4.2.1 Embodying Complex System Strategies

The complex systems paradigm has strongly influenced modern enterprise management theory, and numerous researchers have studied the implications of complexity for enterprise strategies (e.g., Beinhocker, 1999; Brown and Eisenhardt, 1998; Cohen and Axelrod, 1999; Connor, 1998; Haeckel, 1999; Kelly and Allison, 1999; Lissack and Roos, 1999; MacIntosh and MacLean, 1999; Sanders, 1998; Waldrop, 1992). In the context of capabilities analysis, complexity results from the interconnectedness of capabilities, the social relationships within the enterprise (Barney, 1991) and from cospecialized assets, that is, assets which must be used in connection with one another (Teece, 1986). To address the complexity of enterprise management, CBEA is founded on a few key strategies from complex systems theory.

4.2.1.1 Modularity

Inspired by "The Architecture of Complexity" (Simon, 1962), many researchers advocate the adoption of modularity as a design principle to manage complexity (Baldwin and Clark, 2000; Sanchez and Mahoney, 1996; Schilling, 2000). The adoption of modular design principles for all aspects of the enterprise—organization, resources, processes, effects—entails the creation of semiautonomous modules or units with stable and visible rules for communication and interaction. As long as the integrity of inter-module interaction is preserved, module designers are free to engage in local (e.g., within module) adaptation and innovation. Modularity enables parallel innovation efforts while maintaining conformance to interface specifications.

Partitioning the enterprise into capability categories, with attendant capability portfolios focused on critical end-user effects, is a strategy to manage complexity. However, as noted above, important interactions across the enterprise are likely to remain to some degree. For example, while two systems may be weakly coupled structurally, they may be highly coupled functionally. Schaefer (1999, p. 325) concludes that, "it would seem unlikely that a firm could ever hope to uncover an optimal modular design partition …."

4.2.1.2 Exploratory Modeling and Analysis

Enterprise planning problems are typically characterized by enormous uncertainties that should be central considerations in the design and evaluation of alternative courses of action (COAs). A key approach to addressing pervasive uncertainty is *exploratory modeling and analysis* (Bankes, 1993; Davis and Hillestad, 2001; Lempert et al., 1996). The objective of exploratory modeling and analysis is to understand the implications of highly uncertain problem and solution spaces to inform the choice of strategy and subsequent modifications. In particular, exploratory modeling and analysis can help identify strategies that are flexible, adaptive, and robust.

CBEA uses exploratory modeling and analysis to examine enterprise capability issues in the broadest possible context of scenarios, conditions, and assumptions, readily complementing the technique of planning with multiple scenarios (Beinhocker, 1999; Courtney, 1997; Epstein, 1998; Schoemaker, 1997). Relevant models and analysis methodologies should be able to reflect hierarchical decomposition through multiple levels of resolution and from alternative perspectives representing different aspects of an enterprise. Ideally, the analytical agenda should be able to examine the fitness of enterprise responses to a great variety of possible futures.

4.2.1.3 Adaptive Evolutionary Planning

The complex dynamics of enterprise relationships suggest that enterprise evolution depends critically on factors other than global intention and design. Complex systems have the internal capacity to change in unpredictable ways that cannot be described by optimization planning approaches. Researchers point to the limits of predictability and conclude that reliance on a single evolutionary strategy is inappropriate (Beinhocker, 1999; Pascale, 1999; Waldrop, 1992). Instead, the goal should be to develop a collection of strategies to facilitate ready adaptation.

As an extension of its exploratory modeling and analysis, CBEA considers a broad set of risk factors (e.g., cost, schedule, performance, technology, etc.) to envision alternative evolutionary paths. This implies planning for multiple options and adapting strategies as scenarios unfold. Preparing for multiple contingencies is a key element of an adaptive planning process (Haeckel, 1999) emphasizes that creating an enterprise culture of adapting and responding is paramount to survival and success.

4.2.2 Capabilities-Based Engineering Analysis—The Analytical Model

On the basis of the cited complex system strategies, CBEA is best viewed as a modular approach to analysis. Figure 4.5 identifies the basic modules and flow of CBEA. All of the modules interrelate, but there are three major phases: (1) Purposeful Formulation; (2) Exploratory Analysis; and (3) Evolutionary Planning. The modularized schema is not a standardized process, but a collection of interrelated analytical activities that can be assembled as needed to support CBP, programming, and acquisition. The succeeding sections follow this top-level conceptual categorization and describe each of the three module groupings in more detail.

The modular representation of CBEA conveys four important points:

1. There are important dependencies among the activities
2. CBEA is typically not a linear process, but conducted iteratively
3. CBEA may be conducted for different reasons and by different stakeholders, and thus, entry points into the process vary
4. A particular analysis may not include all modules. Because of the breadth and complexity of the issues, CBEA is typically an extensively collaborative effort

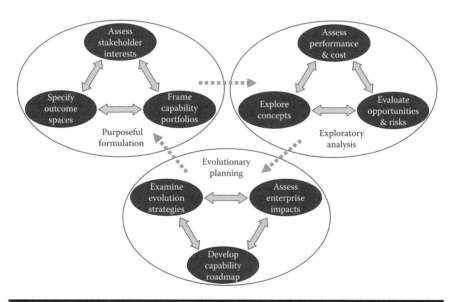

Figure 4.5   CBEA. *Note*: **For readers familiar with the JCIDS, the CBEA approach described in this chapter represents a broader and more generic analytical construct. However, the primary analytical processes of JCIDS can be readily mapped to the CBEA framework, especially the purposeful formulation and exploratory analysis activities; the evolutionary planning activities go beyond the scope of JCIDS analyses.**

4.2.2.1 Terminology

Semantics for capabilities-based approaches vary among practitioners, but the following definitions seem most useful and consistent among applications in the DoD; they are from the *Joint Concept Development and Revision Plan [JCDRP]* (2004)

- *Capability*: The ability to achieve an *effect* to a *standard* under specified *conditions* through multiple combinations of means and ways to perform a set of *tasks*.
- *Effect*: An outcome (condition, behavior, or degree of freedom) resulting from tasked actions.
- *Standard*: The minimum proficiency required in the performance of a task; for mission-essential tasks of joint forces, each task standard is defined by the joint force commander and consists of a measure and criterion.
- *Condition*: A variable of the environment that affects performance of a task.
- *Task*: An action or activity based upon doctrine, standard procedures, mission analysis or concepts that may be assigned to an individual or organization.

Implicit in the definition of *capability* is a focus on end-user operational effects. Authors typically use the word *outcome* to name such an end-user effect, and thus, a *capability* can be regarded as the ability to achieve an outcome (Haeckel, 1999). Additionally, authors use the word *function* in different ways, referring variously to behavior, task accomplishment, organizational components, or the ability to achieve effects (Chandrasekaran and Josephson, 1996; Gharajedaghi, 1999).

This chapter remains consistent with the basic JCDRP definitions outlined above and avoids use of the term *function*, because of its ambiguity. However, this chapter does use the word *structure* in a distinctive manner, referring to all the constituent inputs of a system or enterprise and their interrelationships. Thus, in the DoD context, all relevant DOTMLPF issues are referred to collectively as *structure*. The dynamics (actions) of these structural elements, actual or modeled, is referred to as *behavior*, and effects are the resulting operational outcomes. These meanings are consistent with the systems analysis framework of Gharajedaghi (1999).

4.2.2.2 Principles

On the basis of the foregoing context and motivations, CBEA is founded on a key set of analytical principles.

1. Focus on purposeful outcomes (desired operational effects) of the enterprise end users, vice inputs such as particular programs, platforms, or activities. It is in this sense that CBEA is referred to as a *top-down* approach. In the DoD, warfighters and their mission areas are the forcing functions for the

development of information technology and the collaborative processes that serve to promote innovation.
2. Frame a *portfolio perspective* as a means of partitioning the problem and solution spaces in terms of capabilities (*capability partitions or modules*).
3. Approach issues holistically, considering all aspects of structure, behavior, and effects; consider a full range of alternative solutions available to provide a capability.
4. Examine the complex networks of interdependencies. Such interdependencies exist across all the fundamental dimensions of analysis (structure, behavior, and effects), in multiple aspects, and at different levels of hierarchical description. (CBEA seeks to trace and bound these complex interactions.)
5. Explicitly bound the profound uncertainties attendant to complex adaptive system problems; all assessments must be accompanied by rigorous risk analysis.
6. Pursue an adaptive evolutionary approach to planning to position the enterprise to effectively respond to changes as they occur.
7. Seek to assess and balance the evolution of capabilities within resource constraints by focusing on critical risks and opportunities across the enterprise in a wide range of diverse and stressing operational circumstances.

4.2.2.3 Processes

Figure 4.6 depicts the CBEA modules in a more traditional process view as activities conducted in the context of other ESE processes. This process decomposition illustrates the content and logic of CBEA activities from the customer-focused formulation of the capability problem space to a time-phased plan for solutions. The following paragraphs provide an overview of the CBEA process activities, with the subparagraph numbers corresponding to the numbered process steps of Figure 4.6.

The purposeful formulation activities of CBEA establish the framework for analysis in a participative process that engages stakeholders as purposeful systems (Ackoff and Emery, 1972).

1. *Specify Outcomes Spaces.* On a basis of a thorough review of stakeholder needs and objectives, the analytical process must specify the relevant outcome spaces, that is, the operational goals, contexts, and conditions for which solutions must be designed. A significant part of the CBEA approach is to stimulate analysts to consider a wide set of possible scenarios and conditions against which solution options must be evaluated. In each context, stakeholder needs and objectives should be formulated in terms of desired effects, identifying outcome-based metrics as analytical criteria.
2. *Frame Capability Portfolios.* Given the outcome space descriptions, the capability portfolios must be bounded and described. A capability portfolio includes all the structural elements that must cooperate to provide the desired

110 ■ *Enterprise Systems Engineering*

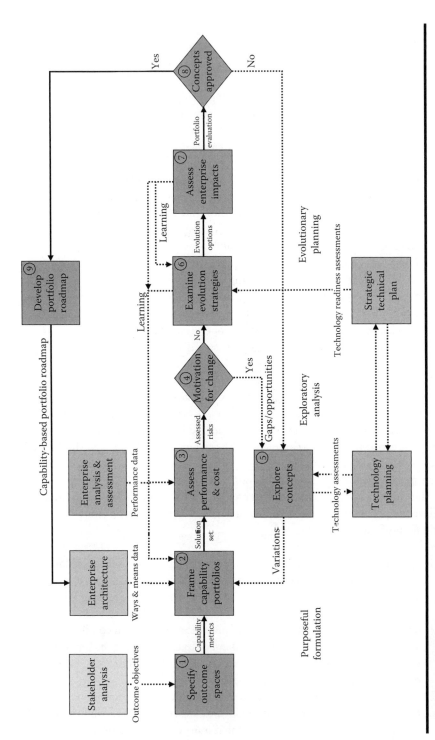

Figure 4.6 CBEA in the context of ESE.

operational outcomes in a capability area. As multiple capability areas may be involved in a particular analysis, multiple portfolios need to be defined. Additionally, analysis typically requires a time-phased view of these portfolios as they change over time.

The exploratory analysis activities of CBEA evaluate the performance and cost of portfolio options over the full range of formulated contexts and generate new concepts for possible improvements.

3. *Assess Performance and Cost.* The performance and cost of portfolio options must be assessed over the broad range of formulated contexts and conditions. The focus of performance evaluation is an assessment of the risk in achieving required capabilities—the ability of concepts and systems to achieve the desired outcomes. The analysis must trace effects through the enterprise toward end-user intended effects (outcomes) and identify key drivers in performance and cost. The results of the performance evaluation should focus on an assessment of capability risks and opportunities, emphasizing the identification of critical capability drivers, capability gaps, and possible areas of significant improvement.
4. *Motivation for Change.* The risk and opportunity assessments resulting from the foregoing capability analyses must be evaluated for their significance. The basic issue to decide is whether the performance and cost assessments motivate the search for alternative solutions. Such motivations typically include mission shortcomings (gaps), excessive costs, or hypothesized opportunities for significant performance improvements. A decision to explore alternative solution concepts may proceed in parallel with continued portfolio planning, as an iteration of the CBEA process may be linked to a time-dependent management process; in this case concept exploration activities would be included in the portfolio evolution strategies.
5. *Explore Concepts.* This activity represents a comprehensive effort to identify possible changes in the enterprise architecture (EA; in all solution dimensions, i.e., DOTMLPF) that will improve capabilities (or reduce cost). The key motivation of concept exploration is to identify feasible ideas that offer the promise of significant performance improvement, cost savings, or risk reduction. The focus on capabilities, vice existing solutions, should facilitate proposals for new ways and means to accomplish a mission and thus, potentially foster new transformational capabilities. CBEA emphasizes competition among alternative concepts, with cost considerations serving as critical constraints in defining feasible solutions.

CBEA evolutionary planning activities focus on developing flexible, robust, and adaptive approaches to the development and fielding of solutions within the context of capability portfolios and the broader enterprise.

6. *Examine Evolution Strategies.* Central to this effort is the examination and integration of alternative evolution strategies aimed at synchronizing component structures and behaviors under different possible contingencies; different time-phased cost and performance profiles are developed for different evolution paths. Such strategies must typically integrate (or at least de-conflict) component program planning, budgeting, and decision making.
7. *Assess Enterprise Impacts.* Beyond a single capability portfolio, plans must be assessed for their impacts (technical, capability, and resource) on other capability portfolios and the broader enterprise. Evolutionary planning activities are conducted at various compositional levels (e.g., program, portfolio, and enterprise), typically distributed among various organizations. Different stakeholders may manage programs that contribute to a particular capability, and a program may well support multiple capabilities. Such complicated interrelations present challenges for evolutionary planning, and as in all CBEA modules, establishing partnering relationships across the enterprise is critical.
8. *Concepts Approved.* This concept decision point represents the key managerial evaluation and selection of capability solution alternatives, constrained by fiscal realities and acceptable degrees of risk. The information and supporting analyses are tailored and focused on the issues that will most influence the decisions. The requirements for analysis must be balanced with the need for imposing as minimal an impact as practicable on delaying decisions. The analysis must, however, be adequate to characterize the decision trade-space (to include risk) for use by senior decision makers when deciding among alternative COAs.
9. *Develop Portfolio Roadmap.* A capability portfolio roadmap documents the analysis, planning, and decisions for the future of a capability area. A roadmap is used as a fundamental capability planning and management tool to enable the development of materiel solutions which meet warfighter needs. A roadmap provides an integrated, time-phased, and fiscally informed plan that assists in conducting capability assessments, guiding systems development, and defining investment strategies. A roadmap is also used as a basis for aligning resources and as an input to strategic guidance documents, program development, and program reviews.

4.2.3 Varied Implementations

CBEA provides an approach to analysis that can be used by all stakeholders throughout the enterprise to focus on end-user capabilities (with an enterprise perspective). Given the different stakeholder purposes and management responsibilities, various adaptations of the CBEA will be executed. Even within one stakeholder group there can be various motivations for such analyses. For example, operational users may have various reasons for initiating an analysis

1. Actual operational shortcomings
2. Perceived future needs
3. Broad evaluation of a mission area or functional area
4. Examination of a proposed operational concept
5. Fiscally driven changes to programmed resources
6. Potential opportunities provided by technical innovations

Such different motivations have implications for what a particular application of CBEA should emphasize. For example, a CBEA based on an actual (urgent) operational failure may spend relatively little time on foundational issues, whereas an application based on a perceived future operational need will likely require considerable work in purposeful formulation. The CBEA process modules should be adapted, as necessary, to the nature of the problem to be addressed.

The following three sections of this chapter detail activities in the three phases of CBEA: (1) Purposeful Formulation; (2) Exploratory Analysis; and (3) Evolutionary Planning. The process decompositions and descriptions are intended to convey the basic concepts and relationships of the analytical framework, not to provide an algorithmic checklist for the conduct of analysis. As noted above, the process modules must be adapted to the purpose and context of a particular analytical project.

4.3 Purposeful Formulation

As in any problem-solving or planning methodology, formulating the problem to be solved is a fundamental concern in CBEA. The view presented in this chapter relates two process modules in this purposeful formulation phase: *specify outcome spaces* and *frame capability portfolios*. However, there are also important input processes that, while not formally parts of the CBEA process, provide needed information: *stakeholder analysis* and *EA*. This section describes the core analytical issues and process decompositions for each of these activities, as depicted in Figure 4.7.

4.3.1 Specify Outcome Spaces

Formally defining the problem domain and concepts to be examined is a foundational step in framing a CBEA. As with any analysis, CBEA must begin with a problem statement to be examined. For example, in JCIDS, a Joint Integrating Concept (JIC) will contain a description of the military problem, as well as the central idea of the concept of operations. Given this general problem description,

114 ■ *Enterprise Systems Engineering*

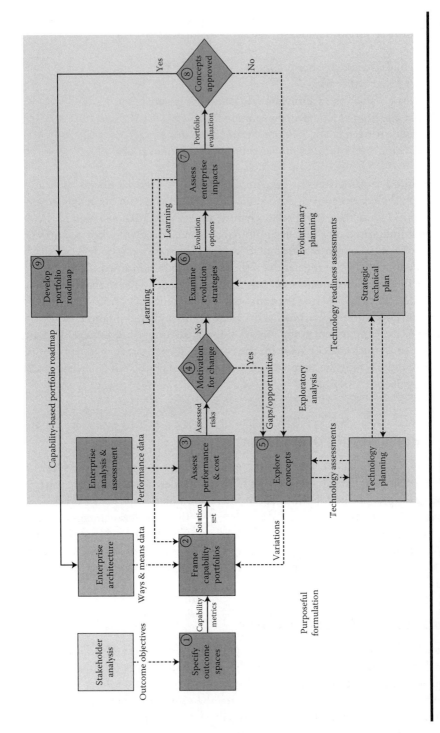

Figure 4.7 Purposeful formulation steps of CBEA.

outcome spaces detail the contexts, conditions, and measures to be used to analyze the problem. Figure 4.8 provides a sample approach to this work.

4.3.1.1 Stakeholder Analysis

As a holistic analytical approach, CBEA should be founded on the identification of key stakeholders, an assessment of their interests, and the ways in which those interests influence the analysis process. Stakeholders are simply the organizations or groups with interests in a policy, program, or project. In the DoD context, a wide range of stakeholders may be relevant to a particular CBEA, to include: users, acquirers, developers, testers, trainers, sustainers, researchers, financiers, and governors/regulators. In CBEA, preeminent focus is directed at end-user operational interests and their outcome space objectives (needs and goals).

Stakeholder analysis shapes the analytical agenda by identifying the roles and goals of different groups and helping to formulate appropriate forms of engagement with these groups. Different stakeholder roles and responsibilities lead to different analytical needs. Ideally, there should be a large degree of commonality among analytical efforts in a particular capability area. For example, measures of effectiveness (MOEs) for a particular outcome-effects construct could be used by all stakeholders in a capability area, from research to end users. Such commonality facilitates the CBEA process and collaborative decision making among all enterprise stakeholders.

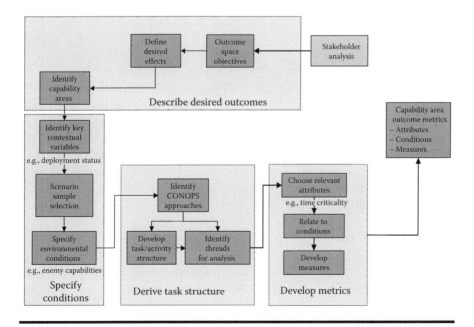

Figure 4.8 Specify outcome spaces.

In particular, a stakeholder analysis can be used to

1. Identify the goals and interests of stakeholders.
2. Identify conflicts of interests between stakeholders, to help manage such relationships.
3. Assess the capacity of different stakeholders and stakeholder groups to participate in the analysis.
4. Assess the appropriate type of participation by different stakeholders.

This stakeholder information is needed to manage, interpret, balance, and process stakeholder input. This effort is foundational in a CBEA effort, as stakeholders generally provide the purpose, information, resources, and authority to conduct the analysis. It is important to identify these areas as early as possible to ensure that the stakeholders are engaged and that their interests are considered from the outset.

4.3.1.2 Describe Desired Outcomes

Typically derived from an end-user statement of outcome objectives (in terms of capability gaps or mission goals), the analytical framework must be guided by an understandable and actionable description of the problem, in terms of the ability to produce certain operational effects and the capability areas relevant to these effects. Outcome objectives are achieved by creating effects, *and the abilities to achieve those effects are the capabilities that are the basis of operational effectiveness analysis.*

4.3.1.3 Specify Conditions

Formal descriptions of outcomes are not completely meaningful until they are embedded in operational contexts and conditions. Selecting scenarios and conditions are often the most important scoping steps in a CBEA, as they provide

1. The context to assess the capabilities associated with the mission area
2. The means to connect the assessment topic to the existing strategic guidance
3. The means to test the concept against the breadth of the defense strategy
4. The spectrum of conditions for analysis

The DoD Analytical Agenda process provides a comprehensive set of scenarios that have already been approved, coordinated, and populated (to varying degrees) with data on both friendly and enemy intentions and capabilities. Furthermore, all the scenarios in the DoD Analytical Agenda have been mapped to established threat categories for policy makers.

The idea of sampling is very important in scenario selection; the goal is to select a comprehensive, but manageable set. There are quantitative methods available to help choose a sample. One useful approach is to take a quick look and examine all of the Analytical Agenda scenarios (there are approximately 30 of them, most with multiple variations) and suggest combinations that are comprehensive and analyzable within the time and resources available.

Identifying particular environmental conditions (in the broad, systems theory sense of *environment*) can be used to supplement, or in lieu of specific scenarios. For some analyses, strong *ceteris paribus* assumptions can be made to simplify the formulation, and a set of well-selected conditional assumptions can be used to provide the context for analysis. It is, of course, critical that these foundational assumptions be made explicit as part of the analytical framework.

4.3.1.4 Derive Task Structure

On the basis of the scenarios and desired capabilities, an operational task structure must be defined, typically based on operational plans/concepts and doctrinal guidance. The task structure must also be allocated to a force structure (current and/or future) to represent the *ways* and *means* to achieve capabilities

1. Employed force elements
2. Time phasing
3. Sequencing of tasks and dependencies among tasks
4. Logistics support required, for example, basing, transport, allied cooperation, and so on

The timing, concurrency, and sequence of operational activities should be identified, and dependencies among the operational activities identified. The relationship between operational nodes should be identified as information exchanges and as inputs or outputs of operational activities. When available, scenarios or operational threads should be developed utilizing subject matter experts (SMEs) to understand alternative approaches for executing the mission.

4.3.1.5 Develop Metrics

Defining the attributes (dimensions) and criteria for measuring performance is another critical step in formulating the problem. Attributes are "a quantitative or qualitative characteristic of an element or its actions." (CJCSM 3170.01E, 2005, p. GL-5) For example,

1. Operating capacity
2. Deployment and employment rates
3. Degree of interoperability
4. Survivability
5. Accessibility in varying environments

Attributes capture a concept's intent for judging the utility of alternative concepts, and provide a starting point for evaluation criteria. Comparing the scenarios, objectives, and task structure to attributes available in the operational concepts facilitates measure selection. Few attributes have obvious pass–fail criteria associated with them, and instead are probably better represented by a continuum of values. A representation of payoffs versus metrics is a useful construct, as it allows evaluation of a continuum of outcomes. Also, expressing evaluation criteria in a common scale will facilitate tradeoffs later in the CBEA.

This activity should define operationalized outcome metrics as analytical criteria for the capability area. Such measures must be operationalized in two senses: (1) Operational definitions should exist for the metrics (i.e., the process for measuring is known and feasible); and (2) the metrics should relate to enterprise end-user operations (in achieving effects)—MOEs. (A particular artifact can be characterized by measures of performance, but those measures should be assessed for their contribution to MOEs, translated through process performance.)

Ideally, there should be a large degree of commonality among analytical efforts in a particular capability area. For example, MOEs for a particular outcome-effects construct could be used by all stakeholders in a capability area, from research to end users. Such commonality facilitates the CBEA process and collaborative decision making among all enterprise stakeholders.

4.3.2 Frame Capability Portfolios

This step identifies, characterizes, and relates systems necessary to achieve objective operational effects and describes the structures and behaviors of solution alternatives. A capability portfolio includes all the structural elements that must cooperate to provide the desired operational outcomes in a capability area. As multiple capability areas may be involved in a particular analysis, multiple portfolios may need to be defined. Additionally, analysis typically requires a time-phased view of these portfolios as they change over time. An EA construct can provide the framework to facilitate this process, providing the information repository needed to describe the *as-is*, *to-be*, and related enterprise transition plans. As CBEA focuses on tracing structural changes through behavior to operational outcomes, even small-scale subenterprise changes may require broad-scope capability portfolio information.

Framing the capability portfolios in architecture views provides the means to understand and manage complexity, including

1. A standardized approach that is reproducible, unconstrained by past mission experience, and independent of the personalities executing it
2. A structured approach for supporting analysis, primarily through comparing and contrasting existing or planned relationships

3. An integrated approach for linking operational concepts and needs to the providing systems and their associated technical standards

Figure 4.9 highlights the basic analytical activities of this step.

4.3.2.1 Enterprise Architecture

EA is another core ESE process, and its information is critical in providing a framework for CBEA. An EA data repository

1. Serves as a strategic information asset that describes the enterprise design, the information necessary to achieve the intended effects, the necessary technologies, and the transitional processes for implementing new technologies.
2. Represents an integrated view of the enterprise across various dimensions (needed to analyze and assess complex interrelations of enterprise structure, processes, and effects).
3. Provides the information and tools on which to base the analytical support to decision making (for all stakeholders).
4. Provides shared vocabulary and views facilitating communication and understanding.
5. Typically includes a baseline architecture (*as-is*), target architecture (*to-be*), and transition plan (*roadmap*).
6. Links current material systems and programs to warfighting capabilities.

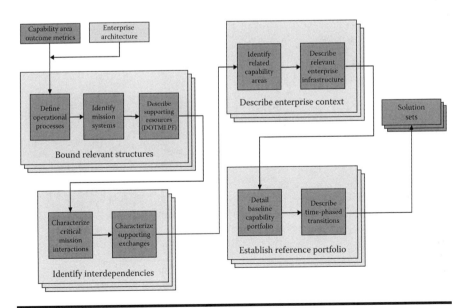

Figure 4.9 **Frame capability portfolios.**

120 ■ *Enterprise Systems Engineering*

The reader can refer to Chapter 6 (Architectures for Enterprise Systems Engineering) of this book for more information about the roles of EA in analysis.

4.3.2.2 Bound Relevant Structures

This step translates operational tasks into the functions that must be performed by systems, applications, personnel, and combinations thereof. The step also identifies the preferred sequence in which the functions are to be performed to deliver the mission capability. The data flow (inputs and outputs) among functions should be identified to depict how the operational information is processed, generated, and exchanged. The functional decomposition should generally be conducted down to the high-level system interfaces. It is important to emphasize that, as depicted in Figure 4.10, the portfolio architectural information can be structured hierarchically to trace relationships from ways and means to effects.

Different portfolio alternatives can be reflected in the planned architecture to generate instances of the objective architecture. The planned architecture provides the basis for integrating each of the alternatives recommended for analysis into a

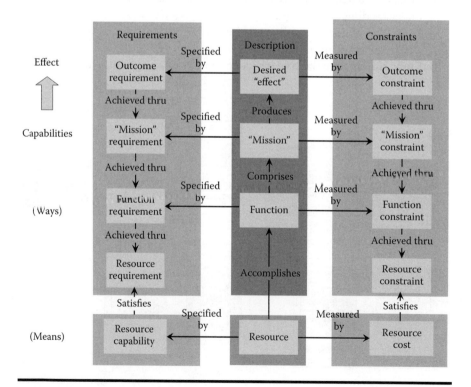

Figure 4.10 Relating structure to effect.

new "objective" architecture. Using the planned architecture as a tool, the impact of enterprise changes can be assessed. Each option likely has different cost, schedule, performance, and risk implications that must be assessed.

4.3.2.3 Identify Interdependencies

A critical aspect of portfolio analysis (and hence formulation) is to understand the interdependencies among the elements of the portfolio. This step of the formulation phase must identify the critical mission interactions and physical interfaces among the systems, as well as, internal and external platform networking and communication mechanisms. The relevant data include the relationships among operational activities, information exchanges among organization elements, functions allocated to system, application or humans, and the platforms, facilities, systems, and interfaces that provide the capability. In addition, data accessed from data stores, such as databases, should be depicted as necessary data in order to accomplish the transformation of operational inputs into operational outputs. This analysis will likely result in second- and third-tier allocations between portfolio programs, and the level of detail must be adapted to the purpose at hand.

4.3.2.4 Describe Enterprise Context

The relation of the capability portfolio to other capability areas and their interactions must also be understood. Networks within and among systems must be identified to facilitate integration and interoperability analyses. Such information is necessary to assess whether the portfolio has been bounded properly such that the systems necessary to execute the capability are included (as an integrated and interoperable system). If integration and interoperability issues are identified, then alternatives should be pursued (unless cost prohibitive, or technologically infeasible) to resolve the deficiencies.

4.3.2.5 Establish Reference Portfolio

This step documents the purposeful formulation phase of analysis, identifying the systems (either in-service or in acquisition) that contribute to the capability, the interrelationships between those systems, and the planned changes to the portfolio over time. When complete, the planned architecture represents the set of mission-related systems that are intended to provide the specified capability. If multiple portfolios are involved in providing the desired capability, then the architecture of each portfolio should be captured. Solution alternatives (generated in *exploring concepts*) are depicted as modifications to this reference portfolio. This reference portfolio and the proposed modifications are the objects of evaluation and comparison throughout the remainder of the CBEA process.

4.4 Exploratory Analysis

The objective of exploratory analysis is to understand the implications of highly uncertain problem and solution spaces to inform the choice of strategy and subsequent modifications. CBEA uses exploratory modeling and analysis to examine enterprise capability issues in the broadest possible context of scenarios, conditions, assumptions, and parameter values. The analysis must typically be conducted to examine hierarchical decomposition through multiple levels of resolution and from alternative perspectives representing different aspects of a capability area.

On the basis of the formulation described in the previous section, exploratory analysis entails the analysis of performance, the identification of possible improvements, the determination of costs, and the assessment of mission risks. The view presented in this chapter describes three basic steps in this phase of analysis: (1) assessing performance and cost; (2) deciding if there is motivation for change in the solution set; and (3) exploring new solution concepts. This section describes the core analytical issues and process decompositions for each of these activity modules, as depicted in Figure 4.11.

4.4.1 Assess Performance and Cost

The performance and cost analysis assesses the ability of the current and programmed capabilities to accomplish the tasks under the full range of operating conditions and to the designated standards that the purposeful formulation phase identified. It also serves to further define and refine the integrated architectures, with cost and performance as guiding criteria. Using the tasks and metrics identified for the outcome spaces as primary input, the performance analysis generates an assessment of risks in accomplishing the tasks and possibly identifies capability gaps or excesses that require resolution. This analytical activity may well be informed by other relevant assessments, as conducted in the Enterprise Analysis and Assessment (another major ESE process).

The results of the performance evaluation should focus on an assessment of capability risks, emphasizing the identification of critical capability drivers, capability gaps, and excesses. The analysis process will typically incorporate both quantitative and qualitative input, addressing the issues that concern decision makers with transparent analytical methodologies that are appropriate to the subject matter. Figure 4.12 illustrates a sample process for conducting an assessment of performance and cost.

4.4.1.1 Prepare for Analysis

Preparation for this analysis includes the standard steps of deciding on an analytical approach and gathering necessary data. A central aim of the DoD Analytical Agenda is to facilitate this activity by making such data readily available for major

Capabilities-Based Engineering Analysis ■ 123

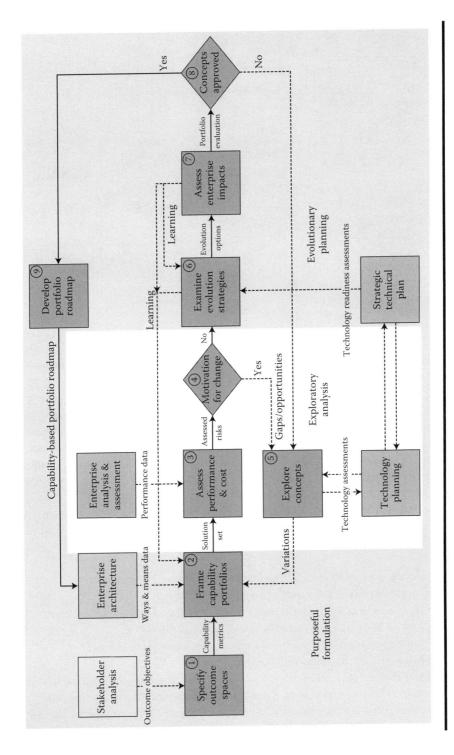

Figure 4.11 Exploratory analysis steps of CBEA.

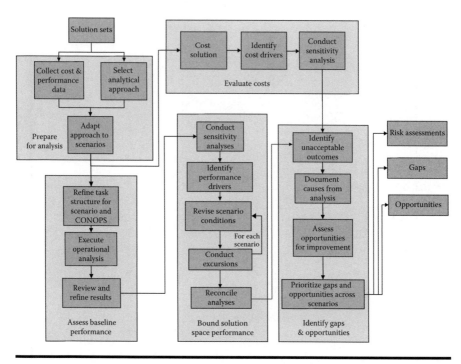

Figure 4.12 Assess performance and cost.

studies. Since DODI 8620.2, *Implementation of Data Collection, Development, and Management for Strategic Analyses*, was published in 2003, it has become much easier to get information on both U.S. and enemy capabilities for modeling purposes. The general approach to data collection should be to obtain as much as possible from current studies; this facilitates leveraging efforts that have already been scrutinized and should reduce data collection efforts.

The issue of choosing an analytical approach involves basic issues such as analytical objectives, study scope, and resources available. The following is a set of questions to consider in evaluating analytical approaches:

1. Can the approach evaluate the doctrinal approaches?
2. Can the approach estimate the specified MOEs?
3. Can the approach represent the specified scenarios, tasks, and functions?
4. Does the approach represent the correct warfighting scope?
5. How large a team does the analytical approach require to execute?
6. How long will the approach take to execute?
7. Does the approach require construction of a set of special-purpose models? If so, how difficult will it be to win community acceptance of these models?
8. Is the approach agile enough? Can it quickly assess a large number of alternatives (U.S. and enemy CONOPS, scenarios, and capabilities)?

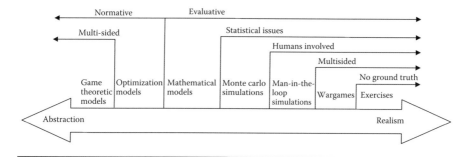

Figure 4.13 Modeling taxonomy.

A variety of modeling approaches can be used in exploratory analysis. Figure 4.13 conveys a common taxonomy of such modeling approaches.

The availability of a particular tool or methodology (or the statement that it is validated) should not drive the analytic approach. The approach must fit the problem, not vice versa.

4.4.1.2 Assess Baseline Performance

The baseline performance analysis assesses the mission effectiveness of the identified portfolio option, given the formulated contexts and conditions. The focus of performance evaluation is an assessment of the risk in achieving required capabilities—the ability of concepts and systems to achieve the desired outcomes. The analysis must trace effects through the enterprise toward end-user intended effects (outcomes) and identify key drivers in performance and cost.

Any technical alternative has structure, behavior, and effects that must be examined for their contributions to operational effects. For example, an effectiveness analysis of command and control (C2) needs to trace active processes, information and decision flows, and the systems that enable operational activity effectiveness. Regardless of how narrowly bounded a particular enterprise may be defined, its effects can propagate broadly and often persistently throughout larger enterprises and beyond (in time and space).

4.4.1.3 Evaluate Costs

A consideration of cost is required throughout the CBEA processes, and this step is aimed at providing a portfolio life-cycle cost estimate to the fidelity appropriate for the level of analysis. Sometimes ignored in operational capability studies, cost always represents a real-world constraint; consequently, the resource implications of proposed solutions should always be considered. The cost of ownership of proposed alternatives should be analyzed so that total life-cycle cost of ownership can be utilized in making informed decisions. Life-cycle cost of ownership includes the cost to develop, acquire, operate, support, and dispose of an alternative, including the related manning, training, and logistics infrastructure. Typically, the following categories of cost (or savings) must be characterized

1. Developmental costs
2. Acquisition costs
3. Recurring operating/sustainment costs
4. Facility or infrastructure costs

For existing programs, gathering cost estimates is typically a matter of contacting the program office or OSD/PA&E's Cost Analysis Improvement Group (CAIG). Demonstration programs or new concepts are more difficult; in these cases, it is probably necessary to conduct a separate cost analysis to estimate the required resources. The required fidelity of such cost estimates will depend on the type of analysis being conducted. For example, rough order of magnitude (ROM) estimates may well suffice for early concept development assessments.

In costing capability portfolios, various cases should be considered

1. Best obtainable solution if costs are unconstrained
2. Best solution that neither increases nor decreases total costs
3. Best solution that achieves some specified decrease in costs

However, there are also questions of feasibility

1. Does the portfolio provide a special-purpose or general-purpose solutions?
2. Is the portfolio generally consistent with current investment trends?
3. Are there realizable nonmateriel solutions?

Note: The reader can reference extensive source material on DoD cost analysis procedures.

4.4.1.4 Bound Solution Space Performance

This step provides an iterative assessment of different variations of the baseline portfolio under a wide range of scenarios and conditions. The process should seek to identify key attributes of a solution set that most significantly influence capability performance in terms of purpose, tasks, and conditions. It should also identify the metrics that a proposed solution set improves or degrades. The mission risks associated with the capability portfolio should be documented and evaluated to determine the probability of the risk occurring and the resulting consequences of that risk occurrence. Unacceptable risks should be evaluated and the cause of the risks identified. Possible resolutions should be identified by iterating the analysis.*

In evaluating mission capability, the process must "... indicate critical components of the overall capability, that is, components whose failure will cause the system to fail ... This is not a standard decomposition into subordinate missions and tasks (although there may be considerable overlap) ... [it is] organized by purpose of the components ... and by how critical the components are, not by a desire for logical completeness or a desire to cover all of the physical systems involved in the operation ..." (Davis, 2002, p. 35). These critical components then must be afforded detailed focus.

4.4.1.5 Identify Gaps and Opportunities

The results of the cost and performance analysis must be examined for the adequacy of proposed solution sets and opportunities for improvement. The results to be examined typically include:

1. Scenarios considered
2. Alternative CONOPS considered
3. Estimated results and costs of executing the CONOPS, in terms of the developed metrics
4. Results which appear to be unacceptable according to current strategic guidance
5. Reasons for the unacceptable results
6. Operational needs that result from those reasons

These results should be examined to evaluate the acceptability of the solution set concerning the portfolio configuration, cost, and the risk in achieving the level of capability established by strategic guidance. Deficiencies (gaps) in these aspects and opportunities (hypothesized improvements from DOTMLPF changes) to significantly improve upon them should be identified and prioritized. Prioritization becomes difficult when assessing multiple scenarios across the breadth of the defense strategy. The list of needs will usually contain a few things that are common to all scenarios, so their pervasiveness probably makes them a high priority. On the other hand, some needs that are critical in one situation may be irrelevant in others; the priority of such needs depend on the likelihood and possible consequences of their relevant scenarios.

Often, the strategic guidance contains priorities; the best prioritization scheme is one that can be directly traced to the strategic guidance. At this stage of the analysis prioritization may not be critical, as solutions and costs must be further

---

* Enterprise uncertainty management including the pursuit of opportunities should also be considered (White, 2006).

investigated. It is critical, however, to provide the linkage from needs to estimated operational outcomes for each scenario, in terms of the MOEs. This allows senior decision makers to consider both the likelihood of the scenario occurring and the consequences of failure, which are the major components of risk. It also allows them to perform their own calculus in terms of tradeoffs among MOEs.

4.4.1.6 Special Challenges for Command and Control Analysis

C2 is a domain in which the effects are traditionally difficult to quantify, and with the development of new concepts such as *network-centric warfare* and *effects-based operations*, the importance of this challenge will increase. The difficulties in measuring C2 contributions to warfighting are well documented:

> "None of the relationships between cognitive dominance and other operational objectives are well understood ... some of these relationships "are subject to gross uncertainties, are poorly understood, or are so dependent upon highly variable situations that evaluation is extremely difficult" (Pirnie and Gardiner, 1996, p. 21).
>
> "... [in regards to C2] a limited capability exists to allow an assessment of the value that a new capability might bring to an operation. Consequently, the Air Force finds itself with an almost infinite list of 'could do's' with limited means for determining what it 'should do' (USAF Scientific Advisory Board, 1996, p. 13)."

In addition, the emerging importance of network-centric concepts complicates the problems of attributing capability and deciding resource allocations. There is, of course, real synergy in *the network*—improvements in target location accuracy, timeliness of actionable time-critical targeting and threat information, tracking performance against stressing targets, and so on—however, the means of rigorously quantifying the impact on mission accomplishment remain a challenge.

4.4.2 Motivations for Change

On the basis of the foregoing assessment of gaps and opportunities, a key decision must be made as to whether new solutions should be explored. The risk and opportunity assessments resulting from the foregoing capability analyses must be evaluated for their significance. The basic issue to decide is whether the performance and cost assessments motivate the search for alternative solutions. Such motivations typically include

1. Mission shortcomings (gaps)
2. Unacceptable risks in achieving mission outcomes

3. Willingness to accept increased risks in a capability area (to allocate resources to other areas)
4. Excessive costs
6. Hypothesized changes (in DOTMLPF) that could provide significant performance improvements

A decision to explore alternative solution concepts may proceed in parallel with continued portfolio planning, as an iteration of the CBEA process may be linked to a time-dependent management process; in this case concept exploration activities would be included in the portfolio evolution strategies.

4.4.3 Explore Concepts

This module represents efforts to identify possible changes in the EA (in the broadest sense, DOTMLPF in military terms) that will improve capabilities (or reduce cost). It identifies potential mitigating DOTMLPF COAs, assesses potential for COAs to address capability gaps and excesses, and identifies the COA(s), which best enhances the needed capabilities. The focus on capabilities, vice existing solutions, should facilitate proposals for new ways and means to accomplish a mission and thus, potentially foster new transformational capabilities. CBEA emphasizes competition among alternative concepts, with cost considerations serving as critical constraints in defining feasible solutions (e.g., cost as an independent variable). The approaches identified should include the broadest possible range of possibilities for providing decisive improvements.

Concept exploration analysis will likely require activities in other modules of CBEA, and the process should be supported by a robust, analytical process which incorporates innovative practices—including best commercial practices, collaborative environments, modeling and simulation (M&S), and electronic business solutions. Capability requirements and option sets should be architecture based to ensure the necessary interoperability and appropriate linkages to other capabilities. The integrated option sets developed may cut across multiple platforms, programs, and/or portfolios. The goal is a balanced, cost-effective, integrated solution, within an acceptable level of risk. Figure 4.14 outlines the general approach to concept exploration.

4.4.3.1 Generate Alternatives

During this phase, operational user representatives and engineers use an iterative process to identify a number of potential solution sets, that is, alternative methods of allocating functionality and performance among the systems, which achieve a war fighting capability, and apply heuristics based on experience and lessons learned from similar problem domains to limit the solution set to the most promising alternatives.

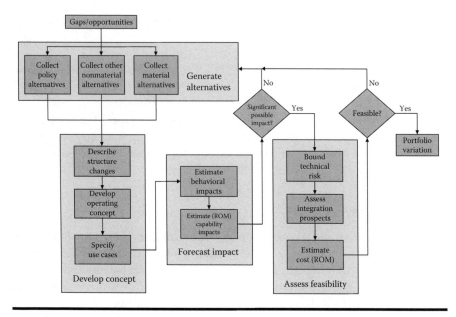

Figure 4.14 Explore concepts.

Some strategies to assist the search for solution concepts include:

1. Working with diverse stakeholders to search for solutions
2. Emphasizing cost-effective common technologies, interfaces, components
3. Integrating capability requirements through an architecture-based enterprise approach
4. Collaborating with industry

The ESE process of technology planning helps identify emerging technologies that may have significant impact on the level of capability delivered by the portfolio. Approaches should be identified to mature the technologies and to develop appropriate technology transition plans for the realization and fielding of the desired capability improvements.

4.4.3.2 Develop Concept

Candidate alternatives for satisfying the capability must be described to a level of detail that facilitates an initial assessment of the possible capability impact, feasibility, and cost (at rough orders of magnitudes). Key elements of such a description include: the proposed structural (DOTMLPF) changes, the supporting concept of operations, and specific use cases to be used as illustrative sample scenarios. The focus in developing the concept is to initially conceive a concept design that is resilient to satisfying competing demands such as: adaptable and scaleable to support evolutionary or spiral development; provides effective capability in multiple contexts; and addresses other constraints (e.g., manpower, training, cost, and schedule). The goal is to establish a preliminary, but sound architectural framework for considering the possible change.

4.4.3.3 Forecast Impact

For each materiel or nonmateriel approach (or combination of approaches) and the associated DOTMLPF and policy implications (to the extent that they can be identified), an initial assessment should be made of the possible impact that the new approach may have on the capability area. The motivation here is to pursue only alternatives with possible significant mission impacts. Follow-on analysis will assess the performance, but some options may be filtered at this preliminary level.

Among possible new concepts may be a *game-changing capability*, or a completely different way of conducting operations. Finding a game-changing capability introduces the need to provide new employment concepts, as there will likely be no doctrinal guidance on how to make the best use of something totally new. A potential game-changing capability is a good candidate for experimentation, because it will take actual field work to confirm or deny the utility of the idea, as well as discover good ways to employ it.

4.4.3.4 Assess Feasibility

Despite the preliminary nature of concept development, an assessment of solution feasibility should be conducted before the alternative is further analyzed. The motivation is to consider key programmatic risk factors from the outset (e.g., technological maturity, technological risk, integration and interoperability issues, supportability, and affordability). These considerations must be addressed using the best data available, representing largely a heuristic *sanity check* at this point in the process. Promising solution concepts, assessed in detail for performance and cost, are further developed as prospective solution sets in the evolutionary planning processes.

4.5 Evolutionary Planning

CBEA evolutionary planning activities focus on developing flexible, robust, and adaptive approaches to the development and fielding of programs within the context of capabilities portfolios and the broader enterprise. Central to this effort is the examination and integration of alternative evolution strategies aimed at synchronizing component structures and behaviors under different possible contingencies, as different time-phased cost and performance profiles are developed for different evolution paths. Such strategies must typically integrate (or at least deconflict) component program planning, budgeting, and decision making.

A capability portfolio is comprised of a set of programs that contribute to a capability. While project management usually focuses on the cost, schedule, and performance of a specific program, portfolio evolutionary planning directs attention at a more aggregated level. Its primary use is to identify, select, finance, monitor, and maintain the appropriate mix of projects and initiatives necessary to achieve capability goals and objectives. It involves the consideration of the aggregate costs, risks, and returns of all projects within the portfolio, the various tradeoffs among them, and their impact on the larger enterprise.

Figure 4.15 highlights the CBEA evolutionary planning activities discussed in this section.

4.5.1 Examine Evolution Strategies

This module focuses on the evolution of a program within the context of its capabilities portfolios. An evolution strategy represents an integrated plan aimed at synchronizing the evolution of component structures (i.e., DOTMLPF) to support the purposeful evolution of the program, the capability portfolio, and the enterprise. An evolution strategy must typically

Allocate capabilities and associated budget to component programs.
Synchronize component program planning, budgeting, contracting, configuration management, testing, delivery, and deployment.
Support planning, requirements management, budgeting, configuration management, risk management, and testing processes for component programs.

Evolution strategies address the system performance allocations and interface relationships among the portfolio systems and establish capability increments and fielding plans based on the planned evolutionary development of systems, new acquisition programs entering the portfolio, and emerging technologies from science and technology investments. Figure 4.16 outlines activities relevant to this examination of evolution strategies.

4.5.1.1 Establish Partnering

The enlargement of focus from programs to portfolios typically requires a shift in culture, process, and responsibility (for identifying and delivering operational value) from individual program offices to stakeholders throughout the enterprise. Evolutionary planning activities are conducted at various compositional levels (e.g., program, portfolio, and enterprise), typically distributed among various organizations. Different stakeholders may manage programs that contribute to a particular capability, and a program may well support multiple capabilities. Such complicated interrelations present challenges for evolutionary planning, and as in all CBEA modules, establishing partnering relationships across the enterprise is critical.

Capabilities-Based Engineering Analysis ■ 133

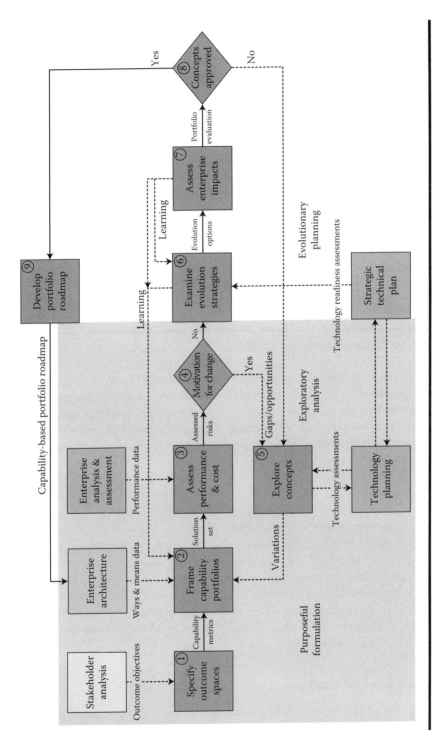

Figure 4.15 Evolutionary planning steps of CBEA.

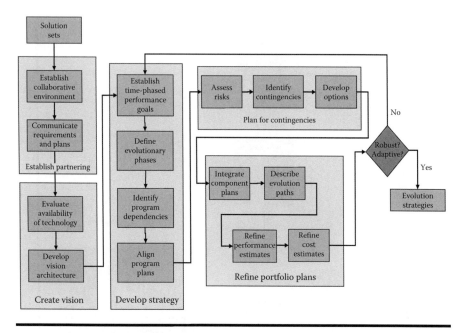

Figure 4.16 Examine evolution strategies.

Many organizations manage and maintain their portfolios by leveraging their strategic planning, budgeting, procurement, and investment control processes. This helps provide the necessary governance and incentive structure to ensure that portfolio management is integrated throughout the organization. This ensures the stages and steps used to formulate, manage, and maintain portfolios are consistent and repeatable.

4.5.1.2 Create Vision

The vision for an evolution strategy provides a long-term view of how the capability architecture will be evolved over time, the level of capability achieved at specified time increments, and the fielding of new systems, enhancements, or upgrades that provide the capability increments. This analysis requires the ability to assess the rate of technology maturation supporting capability increments and to identify areas for science and technology investment that further support the capability evolution. The vision provides a description of the *to-be* architecture and a general description of the transition plan.

4.5.1.3 Develop Strategy

The objective of this activity is to develop and maintain an integrated portfolio schedule that identifies the important time-phased interdependencies between portfolio programs that are essential to fielding fully integrated and interoperable systems that deliver the required capability. Important interdependencies include such things as establishing and demonstrating intersystem behavioral criteria, interface requirements and designs, and shared database requirements and designs. This activity provides for updating the integrated portfolio schedule based on changes to individual program schedules due to progress variations experienced during program execution, or as a result of PPBE decisions that affect program funding. This activity identifies the important relationships between portfolio programs and evaluates the impact of changes to one program on its related portfolio programs.

The nature of evolutionary development dictates that a portfolio will be made up of a diverse set of legacy and newly developed systems. Additionally, each capability increment may bring unique manning and tactics, techniques, and procedures requirements. These factors plus system supportability, logistics, and communication needs must be considered as integral elements of the portfolio for a given increment. This process focuses on a coordinated effort to achieve a capability within a given timeframe within development and budgetary constraints.

4.5.1.4 Plan for Contingencies

This activity seeks to anticipate and prepare for possible changes to the portfolio evolution paths. Such changes can be expected in complicated, high risk, and/or high tempo developments. An assessment of the technical, programmatic, and scheduling risks (and mitigation plans) among portfolio programs provides the basis for this process step. The objective is to ensure that risk is assessed for each of the elements and to understand the risk mitigation plans. In addition, a technical assessment is required of the important interrelationships between programs caused by changes to estimate the confidence of continuing on the planned integrated portfolio schedule. The assessment must look at both the first-order effects between the affected programs, as well as the potential for cascading down stream effects to third parties.

This activity must also establish options for resolving portfolio issues and provide the technical and analytical basis for selecting options for consideration by decision makers. The activities will be specific to issues being addressed, but must consider the technical, schedule, and cost implications of each option. In extreme cases, it may be necessary to recede in the process to explore new concept options.

4.5.1.5 Refine Portfolio Plans

This phase refines the interdependencies between portfolio programs, establishes a mutually supportive time-phased integrated portfolio schedule to assure program interdependencies are effectively addressed across the portfolio, and manages the program interdependencies based on engineering activities within the portfolio programs. It is characterized by the application of formal assessments to measure the progress of portfolio programs relative to each other in specific system engineering disciplines. It measures confidence at the portfolio level that the required capability will be achieved and that the portfolio strategy can be executed.

4.5.2 Evaluate Enterprise Impacts

Beyond a single capability portfolio, plans must be assessed for their impacts (technical, capability, and resource) on other capability portfolios and the broader enterprise. Analyses of these factors have typically not been considered until an initiative has transitioned into the acquisition process, but with an earlier discussion of these impacts, decision makers should be better informed regarding the consequences of their decisions and can take appropriate actions. Figure 4.17 outlines the basic enterprise impacts that must be assessed.

4.5.2.1 Identify Technical Impacts

This activity examines the integration and interoperability requirements of the portfolio systems within the larger enterprise. This review assesses the compatibility of the proposed solution concepts with the integration and interoperability standards of the enterprise. The objectives of the analysis are to gain confidence that integration and interoperability objectives can be met, to assess potential concept deficiencies (and the consequences if not addressed), and to assess the overall prospects of accomplishing enterprise integration plans. Problems identified at this stage can then be addressed early in the design process with corrective actions such as reallocating functional and physical interfaces within the portfolio, specifying

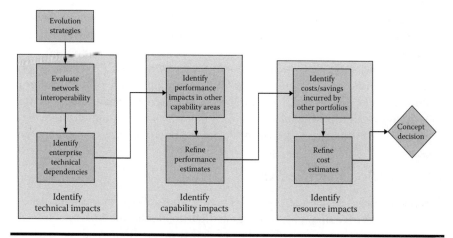

Figure 4.17   Assess enterprise impacts.

integration and interoperability testing, or realigning portfolio program schedules to reduce integration and interoperability risks while retaining synchronized capability fielding plans.

A persistent challenge for ESE is properly evaluating the maturity and stability of proposed technologies and understanding the integration effort, particularly as enterprises engage in software-intensive acquisitions and networked units. An underestimated, but prevalent risk in this effort is the willingness of organizations to change business rules and processes to accommodate improvements in mission performance promised through new technology investments. Such considerations must be part of the technical impact assessment.

4.5.2.2 Identify Capability Impacts

A new solution concept in one capability area may well impact mission performance in other capability areas; this is particularly true for enabling capability areas (e.g., C2 and communications) and their related infrastructure. This step in the analytical process seeks to trace the performance effects of a proposed solution throughout the enterprise and capture the total mission impacts. Such an accounting, though possibly general and rough, is needed to capture the total mission value of the proposed solutions.

It is possible that proposed solution concepts may have negative impacts on other capability areas. For example, assets may be reallocated or reoriented among capability areas, improving some, while deemphasizing others. Such situations must be explicitly examined and evaluated in balancing mission risks and resource allocations across the enterprise.

4.5.2.3 Identify Resource Impacts

Resource decisions are central to all decision processes, and CBEA devotes considerable attention to costing alternatives. At this stage of the analysis, the resource implications of proposed portfolio development plans must be examined with a reasonable level of fidelity. The basic issue is to determine the ability of the enterprise to afford the intended capability improvements. The enterprise implications of investment options (including opportunity costs) should be detailed for decision makers to consider in their evaluations.

As part of the resource impact considerations, divestitures should be explored, focusing on determining when programs (regardless of life-cycle stage) should be terminated because the capability provided is no longer required, or that the residual capability may be adequate at an acceptable degree of risk. In cost-neutral or cost-saving portfolios, tradeoffs must be constantly considered. Using risk guidance from the strategic documents is a way to formulate a ground for divestures. Divesture considerations are increasingly important as topline pressures demand more efficiency in the use of available resources to satisfy enterprise missions.

4.5.3 Concept Decision

This concept decision point represents the key managerial evaluation and selection of capability solution alternatives, constrained by fiscal realities and acceptable degrees of risk (or opportunity). Proposals for the evolution of a capability portfolio will likely impact numerous enterprise stakeholders, and these stakeholders should decide and commit to a concept as early as practical in the process of considering alternatives. A concept decision represents such a decision point based on an analytically derived trade space to determine resource allocation priorities that balance current and future enterprise requirements, while considering associated risks. Such a decision should be based on the cumulative analysis of CBEA that provides a comprehensive perspective of the issues and seeks to effectively balance competing pressures within the enterprise to determine, manage, and resource appropriate COAs.

4.5.3.1 Purposes

The analytical objective for the concept decision is to provide investment decision makers with technically sound, unbiased options to acquire capabilities where the options are beyond the purview of individual programs or resource sponsors. In the simplest and most practical terms, a concept decision would entail the following:

> Defining and broadly communicating the goals and objectives of the capability portfolio
> Establishing and consistently applying a set of criteria that will be used to select among competing projects and initiatives
> Articulating expectations about the type of benefits being sought and the rates of returns to be achieved
> Identifying, minimizing, and diversifying risk—selecting a mix of investments that will avoid undue risk, will not exceed acceptable risk-tolerance levels, and will spread risks across projects and initiatives to minimize adverse impacts; similarly, identifying, maximizing, and diversifying opportunity—selecting a mix of investments that will embrace worthwhile opportunity, will not fall below acceptable opportunity tolerance levels, and will spread opportunities across projects and initiatives to maximize advantageous impacts
> Understanding, accepting, and determining tradeoffs
> Monitoring portfolio performance—understanding the progress that the portfolio is making toward the achievement of the goals and objectives
> Achieving desired objectives—acquiring confidence that the desired outcomes will be achieved given the aggregate of investments that are made

4.5.3.2 Analytical Information

The information and supporting analyses for a concept decision should be tailored and focused on the issues that will most influence the decisions. The requirements for analysis must be balanced with the need for imposing as minimal an impact as practicable on delaying decisions. The analysis must, however, be adequate to characterize a decision trade-space (to include risk and opportunity) for use by senior executives when reviewing alternative COA. The goal is to promote well-informed decisions by carefully examining the objectives, risks and opportunities, and funding of all investments within a capability portfolio and their impact on the larger enterprise.

To support a concept decision, senior decision makers should be provided with an analytical basis characterizing the solution space for addressing capability needs. For material aspects, the concept of life-cycle management is a dominant consideration. Life-cycle management encompasses monitoring programs that are fielded, under development, or being planned to ensure that the resultant capability is responsive to mission needs and is affordable. A summary of enterprise impact analyses (technical, capability, and resource) should provide the most relevant framework for describing and considering the enterprise trade-space and risk and opportunity assessments.

4.5.3.3 Possible Outcomes

Outcomes from a concept decision may include:

1. Approving investment in material and nonmaterial changes to the portfolio
2. Deferring action and accepting the identified levels of risk and opportunity
3. Deferring action until a more robust solution can be provided
4. Providing direction for further maturation of the capability need requirements (to include developing supporting doctrinal and concepts of operations)
5. Recommending the termination or curtailing of programs
6. Addressing acquisition-related issues of special importance affecting the ability to develop and manage the desired capability
7. Deciding resourcing challenges/strategies

4.5.3.4 The Need for Tools

The formulation and execution of analysis to support such portfolio decisions is a highly complex undertaking. Leading portfolio managers utilize a variety of automated tools to help them formulate, manage, and maintain their organizations' portfolios. There are comprehensive investment management tools available, including the federal government's own IT Investment Portfolio System (I-TIPS), several other systems developed by government organizations, and a few commercially available investment management products. These tools vary dramatically in the range of portfolio management and support capabilities they provide. (For a more detailed discussion of the tools that are available and in use across the federal government, the reader can refer to the *Smart Practices in Capital Planning Guide*, prepared by the CIO Council Committee on Capital Planning and IT Management.)

Given the complexity of the endeavor, the portfolio evolution analysis process requires some simplifying assumptions, decisive clarity in its mechanics, and some tolerance for failure:

> "Anyone who imagines that analysts can readily compute the relative worth of an additional fighter aircraft, missile launcher, or company of tanks probably has a simplistic and rigid notion of military operations and a correspondingly simple-minded way of comparing worth ... It is better to adopt the spirit of portfolio analysis and recognize the role of multidimensional tradeoffs and subjective judgments. This view may be heretical to operations researchers, but it is true nonetheless" (Davis, 2002, p. 57).

Capability portfolio analysis requires extensive, ongoing management and maintenance.

4.5.4 Develop Portfolio Roadmap

The results of capabilities-based analyses and decisions should typically be documented in capability roadmaps. Roadmaps are used as fundamental capability planning and management tools to enable the development of solutions that meet enterprise needs. Roadmaps provide integrated, time-phased, and fiscally informed plans that assist in conducting capability assessments, guiding systems development, and defining investment strategies. Roadmaps are also used as a basis for aligning resources and input to strategic guidance documents, program development, and program reviews.

A capability portfolio roadmap integrates systems acquisition plans, technology adoption plans, and other management plans to determine how the capability portfolio and its capabilities will evolve over time, given the level of investment. Figure 4.18 outlines some key activities in developing such a roadmap.

4.5.4.1 Integrate Plans

This module synthesizes programmatic analyses conducted across the capability portfolio to develop an integrated plan aimed at synchronizing the evolution of component structures (DOTMLPF) to support the purposeful evolution of the enterprise. Conducted in the context of partnering relationships discussed

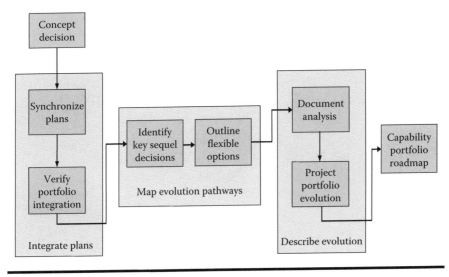

Figure 4.18 Develop capability portfolio roadmap.

previously, this activity must synthesize a common roadmap from the component plans. The basic objectives of this integration include:

1. Aligning resource decisions with capability needs and the interrelated developments of systems that provide the capabilities
2. Establishing the methods and targets for measuring the benefits of investments, initiatives, and programs
3. Integrating projects, services, and assets to provide a coherent technical progression of the enterprise system lifecycles
4. Eliminating unnecessary redundancy in projects and initiatives, consolidating initiatives when it makes sense to do so
5. Improving stakeholder visibility and communication on issues attendant to synchronizing decisions and activities for the evolution of capabilities, promoting understanding across organizational lines

As in all of the CBEA modules, portfolio roadmap efforts are likely situated in a nested management structure, conducted for various levels of compositional and interrelated portfolios. Thus, capability portfolio roadmaps are typically developed and implemented hierarchically, using a spiral approach. Figure 4.19 illustrates the conceptual process. As noted earlier, establishing partnering relationships in a collaborative environment is critical in roadmap development.

4.5.4.2 Map Evolution Pathways

An important element of a portfolio roadmap is to forecast key events or decision points that may significantly alter the portfolio evolution plan. On the basis of the

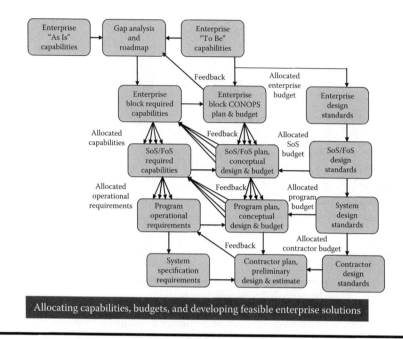

Figure 4.19 A hierarchical approach to roadmaps.

work accomplished in examining alternative evolution strategies (i.e., *in planning for contingencies*), the roadmap should identify significant possible branching points in the portfolio development plan (correlated with sequel decisions to be made) and outline the consequences and possible COAs for each branching point. Such branching points will often correspond with acquisition milestone decisions, or major enterprise financial decisions.

4.5.4.3 Describe Evolution

Documenting the projected portfolio evolution in a capability-based roadmap represents the capstone task of evolutionary planning. Based on the analysis and decisions to this point, the roadmap should

1. Detail the acquisition milestones and significant events which lead to fielding the portfolio of systems over time. It should address how individual systems are developed, evolved, enhanced, or upgraded.
2. Identify the key technologies, their maturation milestones, and fielding plans when the technologies are integrated.
3. Describe the portfolio systems in terms of their initial deployment milestones and the associated upgrade and retirement schedules.

4. Detail training requirements and personnel training profiles per year for each version/evolution of the systems that make up the portfolio.
5. Outline an investment profile to depict how much funding is allocated for each system in the portfolio for the planned and long-term horizon. Funding by category should address research, development, test and evaluation, operations and maintenance, personnel and training, production, fielding, sustainment, and so on.

As part of the roadmap development process, there is also a need to establish the organization and management capacity to continuously monitor portfolio performance and to propose, analyze, and make adjustments in a timely manner. This implies the need to routinely collect and analyze portfolio data and information. Important information must be collected on various programs and projects, including: status of project critical paths, milestone hit rate, deliverables hit rate, actual cost versus estimated cost, actual resources versus planned resources, and high probability, high impact events. Specific criteria and data to be collected and analyzed may include the following:

1. Standard cost management measures
2. Stakeholder impacts, as defined in performance measures
3. Technology impact (as measured by contribution to, or impact on, some form of defined technology architecture)
4. Project schedules
5. Risks, risk avoidance, and risk mitigation
6. Basic project management techniques and measures

Data sources and data collection mechanisms also are important. Many organizations prefer to extract information from existing systems, that is, accounting, financial, and project management systems. In many organizations, mechanisms are in place to enable the creation, participation and "buy-in" of stakeholder coalitions. These mechanisms are essential to ensure that the decision-making process is inclusive and representative. By getting stakeholder buy-in early in the portfolio management process, it is easier to ensure consistent practices and acceptance of decisions across an organization. Stakeholder participation and buy-in can also provide sustainability to portfolio management processes when there are changes in leadership.

The resulting capability-based portfolio roadmap (and its monitoring mechanisms) should provide a living view of development plans for an enterprise capability area. Stakeholders across the enterprise should be able to use the roadmap as a baseline for planning, programming, and acquisition analyses, including periodic iterations of the CBEA process.

4.6 Joint Capabilities Integration and Development System

In recent years, the Joint Chiefs of Staff (JCS) have implemented a capabilities-based approach to identifying improvements to existing capabilities and to develop new warfighting capabilities. JCIDS was established to support the Chairman, JCS and the Joint Requirements Oversight Council (JROC) in advising the Secretary of Defense in identifying, assessing, and prioritizing joint military capability needs (Joint Requirements Oversight Council, 2003, 2005). The approach establishes a collaborative process that utilizes joint concepts and integrated architectures to identify prioritized capability gaps and integrated joint DOTMLPF and policy approaches to resolve those gaps.

Relevant guidance on the approach can be found in Chairman of the Joint Chiefs of Staff Instruction (CJCSI) 3170.01E, JCIDS, and Chairman of the Joint Chiefs of Staff Manual (CJCSM) 3170.01B, *Operation of the JCIDS*.

4.6.1 Top-Down, Concepts-Centric Approach

The JCIDS process is founded on a top-down, concepts-centric approach to capabilities analysis. Joint future concepts are developed from strategic guidance, providing a top-down baseline for identifying future capabilities. The Family of Joint Future Concepts is used to underpin investment decisions leading to the development of new capabilities. New capability requirements, materiel or nonmateriel, must relate directly to capabilities identified through the Family of Joint Future Concepts. The flow from national level and strategic guidance through the concepts is shown in Figure 4.20 from CJCSI 3170.01E.

4.6.2 The Analysis Process

The Family of Joint Future Concepts requires a deliberate analytical process for the examination of needed capabilities through an iterative process of assessment. The purpose of the JCIDS analysis process is to identify capability gaps and redundancies, determine the attributes of a capability or combination of capabilities that would resolve the gaps, identify materiel and/or nonmateriel approaches for implementation, and roughly assess the cost and operational effectiveness of the joint force for each of the identified approaches. Figure 4.21 from CJCSM 3170.01B contains the major elements of a JCIDS approach to CBPA: the Functional Area Analysis (FAA), the Functional Needs Analysis (FNA), and the Functional Solutions Analysis (FSA).

The FAA synthesizes existing guidance to specify the military problem to be studied. The FNA then examines that problem, assesses how well the DoD can address the problem given its current program, and recommends needs that the DoD should address. The FSA takes this assessment as input, and generates recommendations for

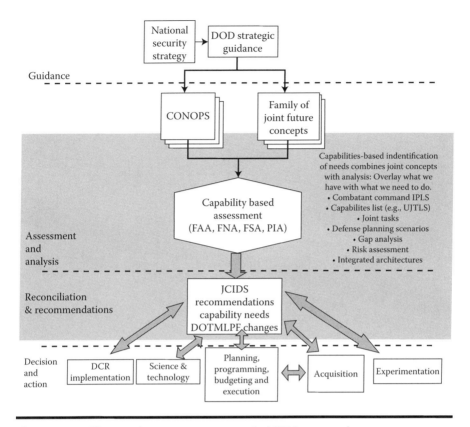

Figure 4.20 The top-down, concepts-centric JCIDS approach.

solutions to the needs. Of course, these simplified inputs and outputs decompose into much more complicated sets of products, and the analyses themselves require much more examination. The point is, however, that JCIDS analysis follows the general logic of most planning problem solving: specify the questions, estimate current and projected abilities, and recommend actions.

4.6.2.1 Functional Area Analysis

The FAA identifies the operational tasks, conditions, and standards needed to achieve military objectives. It uses the national strategies, the Family of Joint Future Concepts, Unified Command Plan assigned missions, CONOPS, joint tasks, the capabilities list (e.g., Universal Joint Task List), the anticipated range of broad capabilities that an adversary might employ, and other sources as input. The FAA identifies the scenarios against which the capabilities and attributes will be assessed. Scenario sources include, but are not limited to, the Defense Planning Scenarios (DPS) published by the Office of the Secretary of Defense (OSD). The FAA produces a prioritized list of capabilities and tasks across all func-

146 ■ *Enterprise Systems Engineering*

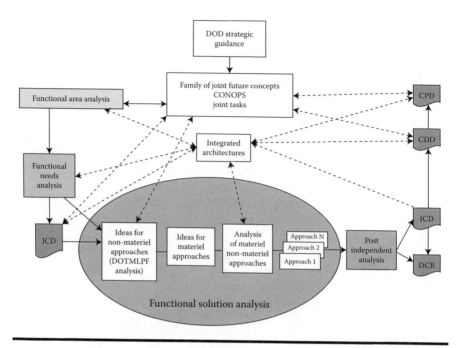

Figure 4.21 The JCIDS analysis process.

tional areas necessary to achieve the military objectives. The capabilities and their attributes should be traceable to the Family of Joint Future Concepts and any other supporting information used to develop the capabilities. These capabilities form the basis for integrated architectures and are reviewed in the follow-on FNA.

4.6.2.2 Functional Needs Analysis

The FNA assesses the ability of the current and programmed warfighting systems to deliver the capabilities the FAA identified under the full range of operating conditions and to the designated MOEs. Using the capabilities and tasks identified in the FAA as primary input, the FNA produces a list of capability gaps that require solutions and indicates the time frame in which those solutions are needed. It may also identify redundancies in capabilities that reflect inefficiencies. The FNA will also provide the relative priority of the gaps identified. The FNA serves to further define and refine the integrated architectures. As an inherent part of defining capability needs, the FNA must assess the entire range of DOTMLPF and policy.

4.6.2.3 Functional Solution Analysis

The FSA is an operationally based assessment of all potential DOTMLPF and policy approaches to solving (or mitigating) one or more of the capability gaps identified in the FNA. It may be appropriate to conduct more than one FSA on the results of an FNA, depending on the scope and variety of capability gaps identified. Applicable integrated architectures are a key component of the FSA to ensure potential approaches to providing the capability are properly linked to existing capabilities and that the relationships are understood. On the basis of the capability needs, potential approaches are identified, including (in order of priority) integrated DOTMLPF and policy changes that leverage existing materiel capabilities; product improvements to existing materiel or facilities; adoption of interagency or foreign materiel solutions; and initiation of new materiel programs. The completed FSA documents the capability gaps and alternative approaches and include integrated architectures linking the approaches to existing systems. Identified capability needs or redundancies (excess to the need) establish the basis for developing nonmateriel and/or materiel approaches are documented in capabilities requirements documents.

4.6.3 Joint Capabilities Integration and Development System Analytical Output

A result of the joint concepts-centric, capabilities-based JCIDS analysis process is a robust cross-component analysis of required capabilities for warfighting. This will ensure the sponsor considers what the warfighter values most in joint force capabilities and the integration of those capabilities early in the process. The development of DOTMLPF and policy solutions must consider appropriate component, cross-component, and interagency expertise; integrated architectures, capability roadmaps, science and technology community initiatives and experimentation results; and joint test results.

Owing to the wide array of issues that are considered in the JCIDS process, the breadth and depth of the analysis must be tailored to suit the issue. For JCIDS analyses performed by the sponsor or a combatant command, Functional Capability Boards (FCBs) provide oversight and assessment as appropriate. The sponsors and combatant commands coordinate with the FCBs throughout the analysis process to reduce redundant analysis, ensure consistency in capability definitions and ensure approaches considered cover the broad range of joint possibilities.

4.6.4 A Critique of Joint Capabilities Integration and Development System

A recent DoD-chartered assessment panel (Kadish, 2006) conveyed criticisms of JCIDS:

> Most of the comments that the Panel received concerning the Joint Capabilities Integration and Development System found it too

complex, with little added value in defining capabilities that require Material Solutions or that establish actionable parameters to guide program definition. The consequence is a widely-held doubt that the Department is acquiring the 'right things' in the 'right quantities' (p. 36).

In addition, key stakeholder groups (especially combatant commanders and industry) reported that they should be more involved in the process. It must be noted, however, that the assessment panel reaffirmed the need for a capabilities-based approach to planning and requirements management, but found the current process inadequate. JCIDS will continue to evolve in response to such criticisms, and the generally negative assessment of this current system does not negate the need for a capabilities-based perspective.

4.7 Air Force Capabilities-Based Planning

The AF currently implements the JCIDS processes in its own comprehensive approach to CBP culminating in its CRRA. The AF applies CBP results to inform the decision makers involved in the Planning, Programming, Budgeting, and Execution (PPBE) cycle, the capabilities requirements process, and the acquisition process. Additionally, agencies refer to CBP for a wide range of activities, such as developing and using appropriate metrics to identify capabilities in DOTMLPF activities, program justifications from a capabilities perspective, directives to the field, and responses to inquiries from the OSD and Congress.

Relevant AF guidance on the approach can be found in the following: Air Force Instruction (AFI) 10-601, *Capabilities Based Requirements Development*; AFI 10-604, *CBP*; Air Force Planning Document (AFPD) 10-28, *AF Concept Development*; and AFI 10-2801, *AF Concept of Operations Development* (Secretary of the Air Force, 2003, 2004, 2005, 2006).

4.7.1 CONOPS-Based Approach

The AF, using a defined set of focused capabilities, describes how operations are conducted to achieve effects using air and space power in its AF CONOPS. These AF CONOPS form the foundation for creating road maps that bridge current AF capabilities with those required for the next 20 years and beyond. Figure 4.22 identifies the AF CONOPS.

The AF assesses the capabilities resident in the AF CONOPS; describes capability objectives, trade-space study areas, and associated risks identified by CBP; proposes recommended guidance across the DOTMLPF spectrum to mitigate risk; and informs the joint community of these results. Presentation of CBP results and recommended guidance to the Air Force Requirements for Operational Capabilities

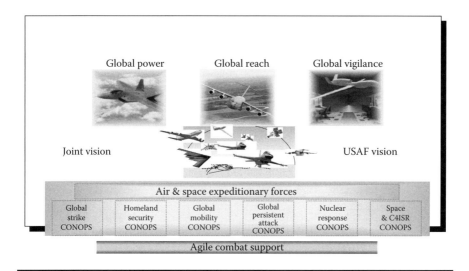

Figure 4.22 AF CONOPS.

Council (AFROCC) serves as the bridge to the capability development process and leads to specific analytically based DOTMLPF solutions and, at times, the development of new capabilities.

The AF CONOPS link the warfighting effects expressed in Joint Operating Concepts (JOC) with the portfolio of AF capabilities that can achieve those effects. As the conceptual foundations for capability analysis, the AF CONOPS, framed by the Master Capabilities Library (MCL), serve as the catalysts of the AF CBP process. AF CBP employs an analytically sound, repeatable, and traceable process to identify, assess, and prioritize AF capability needs and potential trade-space study areas across the DOTMLPF spectrum.

4.7.2 Capability Categories

Defining and analyzing capabilities are core elements of CBP, and analysis is facilitated by the way in which capabilities are viewed. Capabilities can be viewed from functional and/or operational perspectives. Within the AF's MCL, capabilities are categorized in a functional perspective. AF CONOPS and road maps are organized by operational effects and capabilities.

4.7.2.1 Functional

The AF uses a functional perspective to define capabilities, subcapabilities, and tasks that are commonly related. The AF's capabilities are named in the AF's MCL. Functional capability categories are useful because they represent activities or processes that are available, in a specific order with particular relationships, so the AF can achieve its effects. These categories cover warfighting activities (e.g., force application, force projection, communications), direct support and logistics (e.g., security, mission generation, medical), and institutional activities (e.g., acquisition, training).

The MCL is a foundational element of the CBP process. As a repository for all named AF capabilities, the MCL also captures the subcapabilities to the task level. The task level is the first measurable level reached when breaking down top-level capabilities into component subcapabilities. For consistency and ease of analysis, capability statements in all AF CONOPS documents are based on the most current AF MCL to ensure linkage between the AF CONOPS and the capabilities, subcapabilities, and tasks.

4.7.2.2 Operational

The AF uses an operational perspective to conceptualize and analyze desired effects and the capabilities required to achieve them. Operational capability categories reflect the purposes to which capabilities are put, such as large-scale operational and support challenges. The AF CONOPS focus on a general class of operational challenges most pertinent to air and space power. Road maps are developed to display how resources link to operational effects and capabilities, with employment concepts applied to derive sufficiency requirements for those resources. Operational categories provide a more direct link to the combatant commands and the National Defense Strategy than do the functional categories.

4.7.3 Air Force Analytical Process

On the basis of the JCIDS, the AF process identifies AF-wide capability shortfalls, gaps, and trade-space study areas through the application of FAA, FNA, and FSA processes. Senior leaders use these findings and the results of other CBP subprocesses to make comprehensive decisions that yield the best results for the future of the AF and for the joint warfighter. The capabilities review is a structured, repeatable, and traceable analytic process that supports actionable results and feeds the JCIDS process. Figure 4.23 outlines the analytical approach.

The process follows the general top-down approach of CBP, but also uses some distinctive analytical constructs. First, Process Sequence Models (PSMs) are developed to link tasks and capabilities to effects from an operational perspective; the PSMs highlight pivotal capabilities required to achieve a desired effect, in terms of the MCL task structures. Second, the process uses value-focused thinking (VFT) (described in the following sections) to assess and quantify the risk of achieving effects in varying scenarios and conditions. The following steps outline the general capability review process.

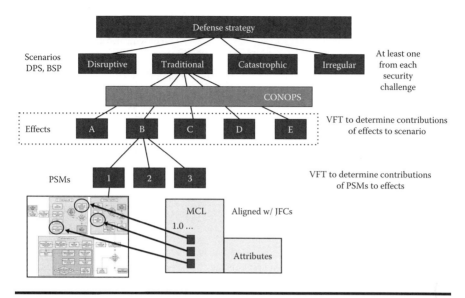

Figure 4.23 **Analytical approach.**

4.7.3.1 Functional Area Analysis

Using strategic guidance, PSMs, results of a data call, scenarios, and leveraging AF and Joint Lessons Learned, SMEs from across the AF collaboratively complete an FAA. In general, an FAA identifies the operational tasks, conditions, and standards needed to achieve military objectives. The military objectives are expressed in the AF CONOPS, and conditions are elaborated through the use of scenarios and standards resident in the AF's MCL. At the HQ USAF level, the FAA results in a preliminary assessment of capability proficiencies and sufficiencies.

4.7.3.2 Functional Needs Analysis

The FNA effort assesses the adequacy and sufficiency of forces to achieve operational capability objectives. SMEs consider contributions of capabilities resident in programs of record in different scenarios, across established timeframes, and existing force structure constraints. Additionally, AF CONOPS Champions and AF CONOPS Flight Leads evaluate risks to refine and prioritize capability objectives and trade-space study areas. Figure 4.24 outlines a general approach to risk assessment using VFT.

Given the existence of a shortfall, AF CONOPS Champions and AF CONOPS Flight Leads determine the consequence to the AF of having a specific amount of capability *and* the likelihood that the shortfall will have an adverse impact on the AF's ability to achieve desired effects for a given time period. As depicted in

152 ■ *Enterprise Systems Engineering*

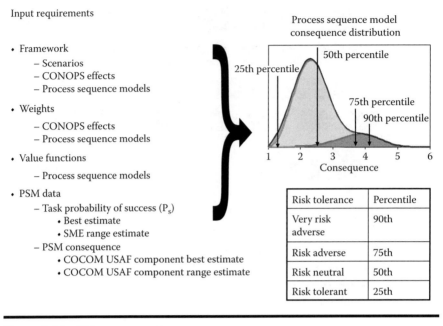

Figure 4.24 Risk-assessment process.

- Process sequence model consequence distributions convey risk as the relative likelihood of the consequence levels

- Risk contributions by tasks are discerned through their correlation and available PSM consequence improvement

- Process sequence model consequence distributions are weighted and rolled up to the effect level

- Effect consequence distributions are weighted and rolled up to the scenario level

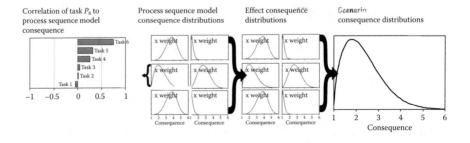

Figure 4.25 Risk-assessment output.

Figure 4.25, the process aggregates the assessments to depict to senior leaders where risks exist in the capability to provide desired effects in the given scenarios.

4.7.3.3 Functional Solution Analysis

Designated sponsors and their teams, using in-depth analysis tools and processes, refine the capability objectives and identify potential COAs. This effort brings the sponsors and the DOTMLPF analytic community together to provide the best recommendations for a comprehensive solution. COAs are developed from across the DOTMLPF spectrum to address the top-priority capability areas.

Work is conducted throughout the AF to develop and test candidate solutions using M&S, architecture models, military utility, and cost analysis to assess capability improvements and validate the proposed COAs. Solutions are derived from multiple COAs that address CBP-identified capability objectives. A COA may encompass any number of DOTMLPF programs. Designated COA offices of primary responsibility (OPR), work with appropriate agencies to integrate COAs into potential solution sets.

4.7.4 Air Force Analytical Output

Several types of key planning documents result from the AF CBP analytical process

1. *Capability Requirements Documents*: Capability requirements documents may be developed to address the capability objectives identified through the subprocesses of CBP, approved by AF senior leadership, and recommended through the AFROCC. The CONOPS Sponsor, assigned by the Executive AFROCC, leads the development and staffing of these capability requirements documents.
2. *Road Maps*: A road map documents projected effects and capabilities over three Future Year Defense Programs (FYDPs). It is based on a long-range vision that implements the national defense strategy and identifies risks incurred when applying projected fiscal constraints. It communicates overarching strategy, force structure, planned recapitalization and modernization, and the timing of key decisions and actions. Road maps incorporate senior leader direction based on the results of the 4-star CRRA, and they are synchronized with the AF programming process to provide timely guidance for Program Objective Memorandum (POM) development.
3. *Annual Planning and Programming Guidance (APPG)*: The APPG contains CBP guidance for planning and programming actions approved through the 4-star CRRA. The APPG provides direction for the AF POM.

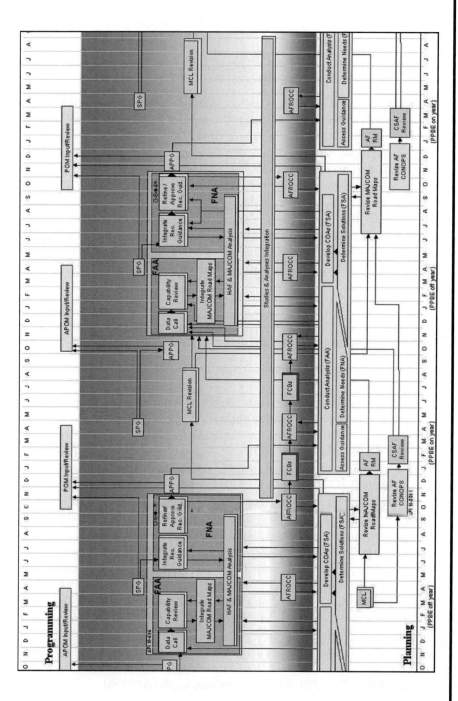

Figure 4.26 AF CBP process flow

4.7.5 Timing

AF CBP events are timed to respond to events in the larger PPBE process they support. A 4-star review is timed to occur every 2 years, late in the PPBE *off-year* (i.e., odd-numbered years). Approved guidance is then transmitted via the APPG and is available in time for the major commands to build their POM inputs for the PPBE *on-year* (i.e., even-numbered years). Figure 4.26 depicts the synchronized cycle.

4.8 Summary and Conclusions

This chapter describes CBEA as a key element of ESE; it is the analytical framework that supports CBP, programming, and acquisition in a systemic approach to the purposeful evolution of the enterprise. The purpose of CBEA is to help enterprise decision makers adjudicate risks through their policy and resource allocation decisions in a highly uncertain, dynamic, and complex environment. In this context, a focus on capabilities provides the guiding principles and approaches for guiding enterprise evolution. To a large degree, capabilities serve as the raison d'être of the enterprise.

This work outlined the key elements of CBEA as a set of modules that can be selected and adapted to provide the engineering analysis processes for a broad set of enterprise activities. The work also defined a set of analytical principles that distinguishes the CBEA approach to analysis. The chapter also describes two recent instantiations of CBP to which CBEA can be applied: the JCIDS and the AF approach to CBP. While these approaches are relatively new, and their details will likely continue to develop, both are founded on the need to focus analysis on end-user capabilities.

This chapter is intended primarily to delimit the basic concepts of CBEA. It describes the concepts and begins to characterize the necessary activities. As experience and data are accumulated on practical applications of these ideas, this work should evolve toward greater insight and detail on the structure and conduct of CBEA.

Bibliography

Ackoff, R. L. and F. E. Emery. 1972. *On Purposeful Systems*. Chicago: Aldine-Atherton.

Adner, R. and C. E. Helfat. 2003. "Corporate effects and dynamic managerial capabilities." *Strategic Management Journal*, 24, 1011–1025.

Afuah, A. N. 2002. "Mapping technological capabilities into product markets and competitive advantage." *Strategic Management Journal*, 23, 2, 171–179.

Aldridge, P. 2003. *Joint Defense Capabilities Study Final Report*. Joint Defense Capabilities Study Team. December 2003.

Anderson, S., M. Coirin, and M. Webb. 2006. "Enterprise Systems Engineering Theory and Practice, Vol. 8: Capabilities-Based Engineering Analysis (CBEA)." MP-5B0000043, MITRE Product, The MITRE Corporation.

Axelsson, B. and G. Easton, eds. 1992. *Industrial Networks: A New View of Reality*. London: Routledge.

Baldwin, C. Y. and K. B. Clark. 2000. *Design Rules: The Power of Modularity*. Cambridge, Massachusetts: MIT Press.

Bankes, S. C. 1993. "Exploratory modeling for policy analysis." *Operations Research*, 41, 3, 435–449.

Barney, J. B. 1986. "Types of competition and the theory of strategy: Toward an integrative framework." *Academy of Management Review*, 11, 4, 791–800.

Barney, J. B. 1991. "Firm resources and sustained competitive advantage." *Journal of Management*, 17, 99–120.

Barney, J. B. and E. J. Zajac. 1994. "Competitive organisational behaviour: Toward an organisationally-based theory of competitive advantage." *Strategic Management Journal*, 15, 5–9.

Beinhocker, E. D. 1999. "Robust adaptive strategies." *Sloan Management Review*, 40, 3, 95–106.

Bolton, D., S. Jones, D. Till, D. Furber, and S. Green. 1994. "Using domain knowledge in requirements capture and formal specification construction." In *Requirements Engineering: Social and Technical Issues*. Jirotka, M. and Goguen, J., eds. New York: Academic Press, pp. 141–162.

Boynton, A. C. 1993. "Achieving dynamic stability through information technology." *California Management Review*, 35, 2, 58–67.

Brown, S. L. and K. Eisenhardt. 1998. *Competing on the Edge: Strategy as Structured Chaos*. Boston: Harvard Business School Press.

Chairman of the Joint Chiefs of Staff Instruction 3170.01E. 2005. 11 May 2005. *Joint Capabilities Integration and Development System*.

Chairman of the Joint Chiefs of Staff Manual 3010.02B. 2005. 1 December 2005. *Joint Operations Concepts*.

Chairman of the Joint Chiefs of Staff Manual 3170.01B. 2005. 11 May 2005. *Operation of the Joint Capabilities Integration and Development System*.

Chandler, Alfred D. Jr. 1990. *Scale and Scope: The Dynamics of Industrial Capitalism*. Cambridge: Harvard University Press.

Chandrasekaran, B. and Josephson, J. R. 1996. "Representing Function as Effect." *AAAI-96 Workshop on Modeling and Reasoning about Function*. Portland, OR.

CIO Council Committee on Capital Planning and IT Management. 2005. *Smart Practices in Capital Planning Guide*.

CJCSM 3170.01E. 2005. "Joint Capabilities Integration and Development System," Chairman of the Joint Chiefs of Staff Manual/Memorandum/Instruction 11 May 2005. Glossary page GL-5.

Cohen, M. and R. Axelrod. 1999. *Harnessing Complexity: Organizational Implications of a Scientific Frontier*. New York: The Free Press.

Collis, D. J. 1994. "Research note: How valuable are organizational capabilities?" *Strategic Management Journal*, 15, 143–152.

Conger, S. 1994. *The New Software Engineering*. London; Boston, MA: International Thomson Computer Press.

Connor, D. R. 1998. *Leading at the Edge of Chaos: How to Create the Nimble Organization*. New York: John Wiley & Sons.

Cotterell, M. and B. Hughes. 1995. *Software Project Management*. London; Boston, MA: International Thomson Computer Press.

Courtney, H. 1997. "Strategy under uncertainty." *Harvard Business Review*, 75, (6), 67–81.

Coyne, K. P. 1986. "Sustainable competitive advantage—What it is and what it isn't." *Business Horizons*, 29, 54–61.

Crissman, LTC Doug. 7 March 2005. The Joint Force Capability Assessment (JFCA) Study and the Development of Joint Capability Areas. Briefing.

Daniel, E. M. and H. N. Wilson. 2003. "The role of dynamic capabilities in e-Business transformation." *European Journal of Information Systems*, 12, (4), 282–296.

Dardenne, A., A. van Lamsweerde, and S. Fickas. 1993. "Goal-directed requirements acquisition." *Science of Computer Programming*, 20, 3–50.

Davis, P. A. 2002. *Analytic Architecture for Capabilities-Based Planning, Mission-System Analysis, and Transformation.* Santa Monica, California: RAND.

Davis, P. K. 2001. *Effects-Based Operations (EBO): A Grand Challenge for the Analytical Community.* Santa Monica, California: RAND, p. xiii.

Davis, P. K. and R. Hillestad. 2001. *Exploratory Analysis for Strategy Problems with Massive Uncertainty.* Santa Monica, California: RAND.

DeMarco, T. and T. Lister. 1987. *Peopleware: Productive Projects and Teams.* New York: Dorset House Publishing.

Department of Defense. 30 September 2001. *Quadrennial Defense Review Report.* Washington, DC.

Department of Defense Architecture Framework Working Group. 30 August 2003. *DoD Architecture Framework, Vol. 1: Definitions and Guidelines.*

Department of Defense Directive 8620.1. 6 December 2002. Data Collection, Development, and Management in Support of Strategic Analysis.

Department of Defense Instruction 8620.2. 21 January 2003. Implementation of Data Collection, Development, and Management for Strategic Analyses.

Dix, A., J. Finlay, G. Abowd, and R. Beale. 1993. *Human–Computer Interaction.* New York: Prentice-Hall.

Donaldson, T. and L. E. Preston. 1995. "The stakeholder theory of the corporation: Concepts, evidence and implications." *Academy of Management Review*, 20, 1, 65–91.

Eason, K. 1987. *Information Technology and Organisational Change.* London: Taylor & Francis.

Eisenhardt, K. and J. Martin. 2000. "Dynamic capabilities: What are they?" *Strategic Management Journal*, 21, 1105–1121.

Epstein, J. 1998. "Scenario planning: An introduction." *Futurist*, 32, 6, 50–52.

Federal CIO Council 2002. A Summary of First Practices and Lessons Learned in Information Technology Portfolio Management.

Foss, N. J. 1993. "Theories of the firm: Contractual and competence perspectives." *Journal of Evolutionary Economics*, 3, 127–144.

Freeman, R. E. 1984. *Strategic Management: A Stakeholder Approach.* Boston: Pitman.

Galal, G. H. and A. Finkelstein. 1999. "*Thoughts on Identifying and Managing Stakeholders.*" Department of Computer Science Research Note RN/99/16. University College London, University of London.

Galal, G. H. and R. J. Paul. 1999. "A qualitative scenario approach to managing evolving requirements." *Requirements Engineering Journal*, 4, 2, 92–102.

Gharajedaghi, J. 1999. *Systems Thinking.* Boston: Butterworth-Heinemann.

Ghemawat, P. 1986. "Sustainable Advantage." *Harvard Business Review*, 64, 53–58.

Gotel, O. C. Z. and A. C. W. Finkelstein. 1995. "Contribution structures." In *Proceedings of the 2nd International Symposium on Requirements Engineering, March 27–29, 1995, York, England (RE 95).* IEEE Computer Society Press, pp. 100–107.

Grant, R. M. 1991. "The resource-based theory of competitive advantage: Implications for strategy formulation." *California Management Review*, 33, 114–135.

Grant, R. M. 1996. "Towards a knowledge-based theory of the firm." *Strategic Management Journal*, 17 (Special Issue), 109–122.

Haeckel, S. 1999. *Adaptive Enterprise*. Boston, Massachusetts: Harvard Business School Press.

Hall, R. 1993. "A framework linking intangible resources and capabilities to sustainable competitive advantage." *Strategic Management Journal*, 14, 607–618.

Helfat, C. E. 1997. "Know-how and asset complementarity and dynamic capability accumulation." *Strategic Management Journal*, 18, 5, 339–360.

Helfat, C. E. and M. A. Petered. 2003. "The dynamic resource-based view." *Strategic Management Journal*, 24, 997–1010.

Holtzblatt, K. and S. Jones. 1993. "Contextual inquiry: A participatory technique for systems design." In *Participatory Design: Principles and Practice*. Schuler, D. and Namioka, A., eds. New Jersey: Lawrence Erlbaum, p. 210.

Jacobson, I., M. Christerson, P. Jonsson, and G. Overgaard. 1992. *Object-oriented Software Engineering: A Use Case Driven Approach*. Reading, Massachusetts: Addison-Wesley.

Joint Chiefs of Staff (JCS) J-8. 2004a. July 2004. *SPG-Directed Planning Task: Integrated Architectures*. Briefing.

Joint Chiefs of Staff (JCS). 2004b. Joint Concept Development and Revision Plan (JCDRP).

Joint Requirements Oversight Council. 2005. Modifications to the Operation of the Joint Capabilities Integration and Development System. JROC Memorandum.

Joint Requirements Oversight Council, JROCM 199-03. 20 October 2003. *Joint Forcible Entry Operations Study (PDM II)*.

Jumper, J. (Gen.), Chief of Staff (CSAF), United States Air Force. 2002. *CSAF Sight Picture: Capabilities Review and Risk Assessment*. http://afconops.hq.af.mil/support/index3.htm:1.

Kadish, R. 2006. Defense Acquisition Performance Assessment Report.

Kelly, S. and M. A. Allison. 1999. *The Complexity Advantage: How the Science of Complexity Can Help Your Business Achieve Peak Performance*. New York: McGraw-Hill.

Kent, G. A. 1989. *A Framework for Defense Planning*. Santa Monica, California: RAND.

King, A. A. and C. L. Tucci 2002. "Incumbent entry into new market Niches: The role of experience and managerial choice in the creation of dynamic capabilities." *Management Sciences*, 48, 2, 171–186.

Kirkwood, C. W. 1997. *Strategic Decision Making: Multiobjective Decision Analysis with Spreadsheets*. Belmont, CA: Duxbury Press.

Kotonya, G., and I. Sommerville. 1998. *Requirements Engineering: Processes and Techniques*. New York: John Wiley.

Langlois, R. N. 1988. "Economic change and the boundaries of the firm." *Journal of Institutional and Theoretical Economics*, 144, 635–657.

Langlois, R. N. 1992. "Transaction-cost economics in real time." *Industrial and Corporate Change*, 1, 99–127.

Langlois, R. N. and N. J. Foss. 1997. "Capabilities and governance: The rebirth of production in the theory of economic organization." *DRUID Working Papers 97-2*, DRUID. Copenhagen Business School, Department of Industrial Economics and Strategy/Aalborg University, Department of Business Studies.

Langlois, R. N. and P. L. Robertson. 1993. "Business organization as a coordination problem: Toward a dynamic theory of the boundaries of the firm." *Business and Economic History*, 22, 31–41.

Lempert, R., E. Schlesinger, and S. C. Bankes. 1996. "When we don't know the costs or the benefits: Adaptive strategies for abating climate change." *Climatic Change*, 33, 2, 235–274.

Leonard-Barton, D. 1992. "Core capabilities and core rigidities: A paradox in managing new product development." *Strategic Management Journal*, 13, Summer Special Issue, 111–125.
Lissack, M. and J. Roos 1999. *The Next Common Sense: Mastering Corporate Complexity through Coherence*. London: Nicholas Brealey Publishing.
Lyytinen, K. and R. Hirschheim. 1987. "Information systems failures—A survey and classification of the empirical literature." *Oxford Surveys in Information Technology*, 4, 257–309.
MacCaulay, L. 1994. "Cooperative requirements capture: Control Room 2000." In *Requirements Engineering: Social and Technical Issues*. Jirotka, M., and Goguen, J., eds. London: Academic Press, pp. 67–85.
MacIntosh, R. and D. MacLean, 1999. "Conditioned emergence: A dissipative structures approach to transformation." *Strategic Management Journal*, 20, 4, 297–316.
Mahoney, J. T. 1995. "The management of resources and the resource of management." *Journal of Business Research*, 33, 2, 91–101.
Makadok, R. 2001. "Toward a synthesis of the resource-based and dynamic-capability views of rent creation." *Strategic Management Journal*, 22, 5, 387–402.
McGrath, G. R., I. C. MacMillan, and S. Venkataraman. 1995. "Defining and developing competence: A strategic process paradigm." *Strategic Management Journal*, 16, 251–275.
Nelson, R. R. and S. G. Winter. 1982. *An Evolutionary Theory of Economic Change*. Cambridge, Massachusetts: Belknap Press.
Newman, W. M. and M. G. Lamming. 1995. *Interactive System Design*. Reading, Massachusetts: Addison-Wesley.
Office of Aerospace Studies (OAS/DR). 2004. *Analysis Handbook: A Guide for Performing Analysis Studies for Analyses of Alternatives or Functional Solution Analyses*. Air Force Materiel Command.
Pascale, R. T. 1999. "Surfing the edge of chaos." *Sloan Management Review*, 40, 3, 83–95.
Pirnie, B., and S. Gardiner. 1996. *An Objectives-Based Approach to Military Campaign Analysis*. Santa Monica, CA: RAND National Research Defense Institute.
Pouloudi, A. 1997. "Stakeholder analysis as a front-end to knowledge elicitation." *AI & Society*, 11, 122–137.
Pouloudi, A. and E. A. Whitley. 1997. "Stakeholder identification in inter-organizational systems." *European Journal of Information Systems*, 6, 1, 1–14.
Prahalad, C. K. and G. Hamel. 1990. "The core competence of the corporation." *Harvard Business Review*, 68, 3, 79–91.
Priem, R. L. and J. E. Butler. 2001. "Is the resource-based 'View' a useful perspective for strategic management research?" *Academy of Management Review*, 26, 1, 22–40.
Reed, R. and R. J. DeFillippi. 1990. "Causal ambiguity, barriers to imitation and sustainable competitive advantage." *Academy of Management Review*, 15, 88–102.
Ryan, P. B. 1985. *The Iranian Rescue Mission: Why It Failed*. Annapolis, MD: Naval Institute Press.
Sanchez, R. and J. T. Mahoney. 1996. "Modularity, flexibility, and knowledge management in product and organization design." *Strategic Management Journal*, 17, 63–76.
Sanders, T. I. 1998. *Strategic Thinking and the New Science: Planning in the Midst of Chaos, Complexity, and Changes*. New York: The Free Press.
Schaefer, S. 1999. "Product design partitions with complementary components," *Journal of Economic Behavior & Organization*, 38, 3, 311–330.
Schilling, M. 2000. "Toward a general modular systems theory and its application to inter-firm product modularity." *Academy of Management Review*, 25, 2, 312–334.

Schoemaker, P. 1997. "Disciplined imagination: From scenarios to strategic options." *International Studies of Management and Organization*, 27, 2, 43–70.
Scott, J. 1991. *Social Network Analysis*. London: Sage.
Secretary of the Air Force. 2003. Air Force Planning Document (AFPD) 10-28. *Air Force Concept Development*.
Secretary of the Air Force. 2004. Air Force Instruction (AFI) 10-601. *Capabilities Based Requirements Development*.
Secretary of the Air Force. 2005. AFI 10-2801. *Air Force Concept of Operations Development*.
Secretary of the Air Force. 2006. AFI 10-604. *Capabilities-Based Planning*.
Secretary of Defense. 18 March 2002. "Requirements System." Memorandum.
Secretary of Defense. March 2005. National Defense Strategy of the United States of American.
Secretary of Defense. 6 May 2005. "Operational Availability (OA)-05/Joint Capability Areas." Memorandum.
Simon, H. A. 1962. "The architecture of complexity." *Proceedings of the American Philosophical Society*, 106, 467–482. Reprinted in: Simon, H. A. 1981. *The Sciences of the Artificial*, 2nd edition. Cambridge, Massachusetts: MIT Press.
Spewak, S. 1992. *Enterprise Architecture Planning*. New York: John Wiley & Sons.
Stalk, G., Jr., P. Evans, and L. E. Schulman. 1992. "Competing on capabilities: The new rules of corporate strategy." *Harvard Business Review*, 70, 2, 57–70.
Teece, D. J. 1986. "Firm boundaries, technological innovation and strategic management." In *The Economics of Strategic Planning*. Thomas, L. G., ed. Lexington, Massachusetts: Lexington Books, pp. 187–199.
Teece, D. J., G. Pisano, and A. Shuen. 1997. "Dynamic capabilities and strategic management." *Strategic Management Journal*, 18, 509–533.
Troxell, J. F. 1997. *Force Planning in an Era of Uncertainty, Two MRCs as a Force Sizing Construct*. Carlisle, PA: U.S. Army War College, pp. 14–15.
USAF Scientific Advisory Board. 26 November 1996. Vision of Aerospace Command and Control for the 21st Century, Executive Summary. SAF/PAS 96-1117. 13.
Wade, M. and J. Hulland. 2004. "Review: The resource-based view and information systems research: Review, extension, and suggestions for future research." *MIS Quarterly*, 28, 1, 107–142.
Waldrop, M. 1992. *Complexity: The Emerging Science at the Edge of Chaos*. New York: Touchstone Publishing.
Washburn, A. R. 2001. "Bits, bangs, or bucks?. The coming information crisis." *PHALANX*, 34, 3 , 6–7 (Part 1); 34, 4, 10–11, 31–32 (Part 2).
White, B. E. 2006. "Enterprise opportunity and risk." *INCOSE Symposium*, Orlando, FL, 9–13 July 2006.
Williams, J. R. 1992. "How sustainable is your competitive advantage." *California Management Review*, 34, 29–51.
Winter, S. 2003. "Understanding dynamic capabilities." *Strategic Management Journal*, 24, 10, 991–995.
Winter, S. G. 2000. "The satisficing principle in capability learning." *Strategic Management Journal*, 21, Special Issue, 981–996.
Zahra, S. A. and G. Georges. 2002. "The net-enabled business innovation cycle and the evolution of dynamic capabilities." *Information Systems Research*, 13, 2, 147–150.

Chapter 5

Enterprise Opportunity and Risk*

Brian E. White

Contents

5.1 Introduction 162
5.2 Traditional View of Risk and Opportunity 165
    5.2.1 What Is Risk? 165
    5.2.2 Quantifying Risk..167
    5.2.3 What Is Opportunity? 168
5.3 System and Systems of Systems Risk Management..................................173
    5.3.1 Traditional Systems Engineering..173
5.4 System, System of Systems, and Enterprise Opportunity Management....173
    5.4.1 The Regimen..173
        5.4.1.1 Opportunities and Risks in Establish Rewards 174
        5.4.1.2 Opportunities in Characterize Continuously..................174
5.5 Further Thoughts on Opportunity/Risk Concerning TSE, SoS 175
    5.5.1 Engineering and ESE...175
5.6 Qualitative Distinctions between TSE, and SoS and ESE Opportunity...177
    5.6.1 Management...177
5.7 Concluding Remarks 178
5.8 Suggestions for Future Work...179
References 180

---

* This chapter is based on a paper and presentation originally published at the *International Council on Systems Engineering (INCOSE) 2006 Symposium* (White, 2006).

161

5.1 Introduction

In a recent book on opportunity management, Hillson (2004) makes a rather convincing case that opportunities get "short shrift" in most programs. See, for example, Hillson's comments (on pp. iii, vii, and xvi): "There is … a systemic weakness in risk management as undertaken on most projects. The standard risk process is limited to dealing only with uncertainties that might have negative impact (threats). This means that risk management as currently practiced is failing to address around half of the potential uncertainties—the ones with positive impact (opportunities)." Furthermore, anecdotally, the author of this chapter has noticed that many—if not most—risk and risk management documents, tools, processes and websites, courses, and so on, do not even mention or discuss opportunities or opportunity management (e.g., GAO, 2005; Martin, 2005; ODU Advertisement, 2005; Roman, 2005; Stein, 2005). Therefore, it seems to make sense to appreciate opportunity at the system scale. In fairness, some publications, for example, Forsberg and Mooz (1996) treat both risk and opportunity.

Hillson views opportunity as the opposite of threat and adopts the position that these two factors together constitute risk. But, he also provides extensive evidence of the viewpoint that treats opportunity as the opposite of risk. In the present author's opinion, the latter view is more traditional and straightforward; though Hillson makes the case that the former is becoming more prevalent in academic and professional circles. Nonetheless, in this theoretical chapter, opportunity is viewed as the opposite of risk. The author hypothesizes that in systems engineering (SE) at an enterprise scale the focus should be on opportunity, as depicted in Figure 5.1; enterprise risk should be viewed more as threatening the pursuit of enterprise opportunities.

Before proceeding some clarification may be appropriate. First, enterprises often are thought of as social organizations that develop, test, deploy, use, and upgrade systems, system of systems (SoS), or families of systems, and are usually not thought of as systems in themselves. The U.S. Air Force and Ford Motor Company are examples of enterprises. However, in a broad definition of system used here (White, 2005), that is,

> *System*: An interacting mix of elements forming a whole greater than the sum of its parts.
> *Features:* These elements may include people, cultures, organizations, policies, services, techniques, technologies, information/data, facilities, products, procedures, processes, and other human-made or natural entities. The whole is sufficiently cohesive to have an identity distinct from its environment.
> *Note:* In this definition a system does not necessarily have to be fully understood, have a defined goal/objective, or have to be designed or orchestrated to perform an activity.

An enterprise can be thought of as a system. Second, Figure 5.1 is not intended to suggest a hierarchy, of management levels, or otherwise, but rather three different

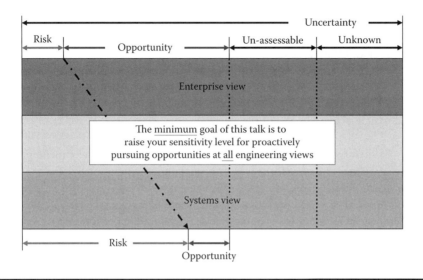

Figure 5.1 Relative importance of opportunity and risk at distinct engineering scales, acknowledging other kinds of uncertainty.

engineering scales that should be continually examined collectively to observe patterns and emerging behaviors that help inform the observer, practitioner, or intervener of the underlying reality that often is elusive or even impossible to comprehend.

Figure 5.1 is meant to suggest that the importance of opportunity management should increase qualitatively as one proceeds from system, to SoS, to enterprise scales. This is partially based on the premise, supported by historical fact and *ad hoc* observations that risk management tends to dominate at a system scale. At an enterprise scale, the author tries (throughout this chapter) to develop the rationale for paying much more attention to opportunity management than risk management. It might then follow that opportunity management and risk management would be roughly coequal at an SoS scale. These statements can be viewed as both descriptive and prescriptive. Nevertheless, further testing of hypotheses concerning the greater importance of opportunity management at SoS and enterprise scales is appropriate as part of future work. Further, the relative impact of the opportunities and risks themselves at any engineering scale is a topic distinct from the relative importance of opportunity versus risk management; for example, the potential impacts of risks at the enterprise scale well may be larger than at a single system scale.

Mike Kuras (M. L. Kuras, pers. comm., 2005) observed that risk and opportunity can be thought of as assessable uncertainties. Clearly there exist unassessable uncertainties and unknown uncertainties. So, the topic of uncertainty management is more general than what is treated in this chapter. This idea is merely acknowledged in Figure 5.1, where there is no attempt to depict relative importance of these other uncertainties at any of the three scales shown. A good framework for the treatment of uncertainties is provided in Hastings and McManus (2004). Another excellent work on uncertainty management that emphasizes opportunities is that of de Neufville et al. (2004).

A previous paper (Kuras and White, 2005) offered some definitions of system, SoS, and enterprise in the context of complex-system engineering (CSE). Three additional definitions of SE, enterprise systems engineering (ESE), and CSE are offered to help clarify the use of these terms:

> *SE:* An iterative and interdisciplinary management and development process that defines and transforms requirements into an operational system.
> *Features:* Typically this process involves environmental, economical, political, and social aspects. Activities include conceiving, researching, architecting, utilizing, designing, developing, fabricating, producing, integrating, testing, deploying, operating, sustaining, and retiring system elements.
> *ESE:* A regimen for engineering (methodically conceiving and implementing solutions to real problems, with something that is meant to work) successful enterprises.
> *Features:* ESE is SE but with an emphasis on that body of knowledge, tenets, principles, and precepts having to do with the analysis, design, implementation, operation, and performance of an enterprise. The enterprise systems engineer concentrates on the whole as distinct from the parts, and its design, application, and interaction with its environment. Some potentially detrimental aspects of traditional systems engineering (TSE) are given up, that is, not applied, in ESE. (See the Traditional Systems Engineering subsection for more on a TSE definition.)
> *CSE:* ESE but with additional conscious attempts to further open the enterprise to create a less stable equilibrium among many interdependent component systems.
> *Features:* In CSE, special attention is paid to emergent behavior, especially due to the openness quality, which can either be desirable or undesirable. One tries to instill the deliberate and accelerated management of the natural processes that shape the development of complex systems.

Again, before proceeding some further clarification may be appropriate. By further open the enterprise and openness quality, the author indicates either the acceptance of new inputs from an existing context or perhaps an expanding context. Five principles should be kept in mind:

1. Any system or enterprise tends to be nested, that is, part of a larger system or enterprise, respectively, and composed of smaller systems or enterprises.
2. The precise boundary of either a system or an enterprise can be elusive; usually it is beneficial to everyone's understanding to have discussions as to where these boundaries, albeit fuzzy ones, might lie.

3. One should always ask how a system function helps manifest a capability in the enterprise in which the system is a component.
4. High risk/opportunity for one organization is often high opportunity/risk for another; for example, think of organizations in government versus private industry.
5. In discussing risk/opportunity, there may be consequences beyond the mission that might be included in the unassessable or unknowable category; it is these very consequences to which the openness quality may be referring.

5.2 Traditional View of Risk and Opportunity

Before treating opportunity, the more familiar risk-related definitions and ideas are introduced:

$$P_o = \text{Probability of occurrence}$$
$$C_f = \text{Consequence of failure}$$
$$A_r = \{P_o, C_f\} = \text{Risk assessment}$$

Consider an example of buying a new personal computer (PC) and worrying about a successful virus attack occurring with probability, P_o, estimated from anecdotal evidence or some source of published statistics. The consequence of disruption, C_f, from such an attack is evaluated by contemplating potential loss of critical files or work time, say. P_o can be reduced with risk mitigation, for example, by obtaining and installing virus protection software. Backing up files independently can reduce the impact of a successful attack. The difference between the before and after combinations $\{P_o, C_f\}$ determines the improvement in risk assessment. The foregoing is risk management. (Buying a bigger central processing unit than needed initially with expansion slots for future—yet unknown—use is opportunity management.) This model of risk assessment being the pair $\{P_o, C_f\}$ should be kept in mind when evaluating program/project risks.

5.2.1 What Is Risk?

"A *risk* is a potential event that, if it occurs, will adversely affect the ability of the system to perform its mission. Thus risk is a probabilistic event. In contrast, an *issue* is an adverse event that has already occurred or will occur with certainty." [According to this definition (Garvey, 2000) an earthquake that has just occurred is an issue; a future earthquake, although certain to occur sometime, somewhere, in the indefinite future, still would be treated as a risk.] Issues need to be tracked, even though, technically, they are not risks. The limited resources (in most cases) for mitigation

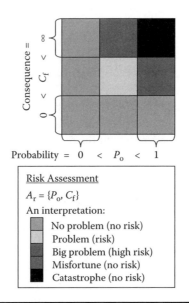

Figure 5.2 Perspectives on risk assessment.

need to be balanced to address both issues and risks with a proper holistic allocation, considering both at the same time. There is no risk if something (see Figure 5.2)

- Never happens, no matter what the consequence
- Happens with no consequence
- Surely happens (this is subtler, implying an issue and maybe a consequence)

How can something undesirable that will surely happen not be a risk? Program planning must eliminate such an eventual occurrence from being relevant. If not, that part of the program should be halted, replanned, and restarted. In other words, undesirable certainties are put outside the box. This is a situation when it is good to stay inside the box or redefine one's box to be smaller to eliminate the unfortunate certainty. Recognize, however, that what might happen with this program redefinition is the creation of a new risk (with a probability between 0 and 1). Also, do not be too confused by Figure 5.2; as Paul Garvey pointed out, the range of the consequence variable, C_f, is often taken to be [0, 1]; in such as case, $C_f = \infty$ is moot. In qualitative risk management, probabilities typically are estimated with only 5–10% accuracy, and consequences are rated with only 3–5 categories. Many organizations use quantitative risk management. In commercial engineering, it is not uncommon to know the probabilities of failure to a high level of accuracy based on tests or models, and to know the impact in dollars to two-place accuracy. Furthermore, impact is often defined on a continuous scale rather than just heaping effects into 3–5 categories.

5.2.2 Quantifying Risk

What simple mathematical expression can express risk?

- Suppose risk, R, were defined as $R = P_o \times C_f$
  Then R would not match our definition of risk and assessment interpretation of Figure 5.2, that is, when $P_o = 1$, there would be no risk, but $R = C_f$ could be >0. More importantly, a formula can be quantitative only if C_f as well as P_o, assumes a cardinal (quantitative), not ordinal (qualitative) value. Instead of quantifying risk, an event, why not express the estimated disruption, D_e, one would face considering risk?
- Let $D_e = P_o \times C_f$. (note: this is the same as the expected consequence)
  - Estimated comes from uncertainty, that is, P_o
  - Disruption measures aggravation to the program or mission
  - This definition is consistent with our definition of risk and our interpretation of assessment, for example,

  $D_e = 0$, when there is no problem
  $D_e = \infty$, when there is a big problem or catastrophe
  $D_e > 0$ (and $< \infty$), when there is a problem or misfortune

When this topic is treated in more depth, the mathematical treatment and the formula(s) used in risk models are slightly more sophisticated. In general, one can plot C_f versus P_o to express a nonlinear relationship, see Figure 5.3 (P. R. Garvey,

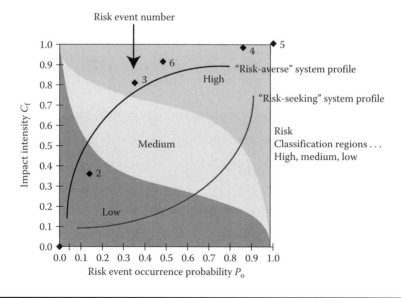

Figure 5.3 Risk classification example.

pers. comm., 2005). Mathematical modeling and correctness are nice, but the more important point is to understand the concept of how the two factors, P_o and C_f, work together to impact one's risk management plans and operation. Suppose C_f is defined as a real number in the range [0, 1] and is called impact intensity, for example. Then low, medium, and high risks, respectively, could be defined as in the dark gray, light gray, and medium gray areas. The boundaries are drawn arbitrarily to depict hypothetical value judgments of the risk manager based upon his/her assessments of the risk probability and consequence pairings $A_r = \{P_o, C_f\}$. Here the probability of risk event occurrence is identical to P_o. One could draw any single-valued curve in this space as a given instance of the risk assessment (relationship) between occurrence probability and impact intensity, $A_r = \{P_o, C_f\}$. A "risk-averse" system would be represented by a convex-up curve, as shown, because the larger the increase in probability, the lesser the increase in consequence.

As an example of a risk-averse system profile, consider again buying a PC and protecting against virus attacks. As the vulnerability to the PC gradually increases with increasing sophistication of viruses, the probability, P_o, of a successful attack increases. One would first try to reduce this probability by acquiring better virus protection software. Then, compared to doing little to mitigate the significant consequences of an attack when $P_o < 0.2$, for example, for $P_o > 0.3$, say, one might continually escalate efforts to protect themselves by more regular independent backups. The extra backups would help, but they might not cover fully a greater loss in expected work effectiveness. C_f may still increase with the associated increase in P_o, but in a risk-averse system, the additional impact on the PC user is not as great.

A "risk-seeking" system would be represented by a convex-down curve, as shown, because the larger the increase in probability the greater the increase in consequence.

A straight line, $C_f = P_o$, in the square of Figure 5.3, from [0, 0] to [1, 1], would be labeled "risk neutral." In this case, one could prioritize risk by the product, $D_e = P_o \times C_f = P_o^2 = C_f^2$.

Contact the author if the risk-averse/-seeking system profile shapes of Figure 5.3 seem counterintuitive.

5.2.3 What Is Opportunity?

- Opportunities are events or occurrences that assist a program in achieving its cost, schedule, or technical performance objectives.
- In the larger sense, explored opportunities can enhance or accomplish the entire mission.
- Opportunity is also associated with uncertainty and impact.
- There is a duality or parallelism to risk that can be applied.
- For an opportunity, let Q_o be the probability of occurrence, and B_s, the benefit of success.

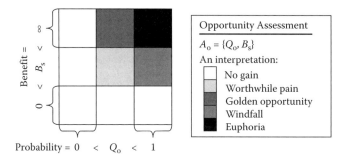

Figure 5.4 Perspectives on opportunity assessment.

- Similarly, to estimated disruption, we can pose the simple formula $E_e = Q_o \times B_s$, the estimated enhancement or expected benefit.

Figure 5.4 is the dual of Figure 5.2. Again, do not be confused by Figure 5.4. As Paul Garvey (P. R. Garvey, pers. comm., 2005) pointed out, the range of the benefit variable, B_s, could be taken to be [0, 1]; in such a case, $B_s = \infty$ is moot.

In this author's opinion, based on personal observation of several program managers (PMs), a natural tendency for military officers in the acquisition field is to be aggressive in minimizing downside risk and conservative in maximizing upside opportunities. However, others may feel that colonels would tend to be aggressive with opportunities and conservative with risks, just the opposite of what a typical systems engineer might do. One needs to consider incentives and incentive systems that promote one behavior versus another before deciding. Are colonels aggressive because they have to do something big to be promoted to general? Do systems engineers fear that they will be punished if the systems they deploy fail during use? These are interesting questions worthy of further study.

As with the discussion on risk, remember that opportunity has two components: both the probability of occurrence, Q_o, and the benefit of success, B_s. One can get excited about all kinds of eventualities, but if Q_o is very small, one should not waste resources pursuing this opportunity event unless B_s is very good. If Q_o is very large and the B_s is significant, one should seriously consider restructuring a program/project to help ensure the possibility of enhancing the overall mission. As before, this formula for estimated enhancement is quantitative only if B_s assumes cardinal (quantitative) values.

Again, when this topic is treated in more depth, the mathematical treatment and the formula(s) used in the opportunity models would be more sophisticated. In general, one can plot B_s versus Q_o to express a nonlinear relationship (see Figure 5.5). Mathematical modeling and correctness are nice, but it is more important to understand how Q_o and B_s work together to impact one's opportunity management plans and operation. Suppose B_s is a real number in the range [0, 1] and is called impact

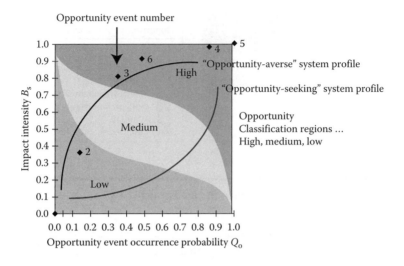

Figure 5.5 Opportunity classification example.

intensity, then low, medium, and high opportunities, respectively, could be defined as in the medium gray, light gray, and dark gray areas. The boundaries are drawn arbitrarily to depict hypothetical value judgments of the opportunity manager based on his/her assessments of the opportunity probability and benefit pairings $A_o = \{Q_o, B_s\}$. Here the probability of opportunity event occurrence is identical to Q_o. One could draw any single-valued curve in this space as a given instance of the opportunity assessment (relationship) between probability of occurrence and impact intensity, $A_o = \{Q_o, B_s\}$. An "opportunity-averse" system would be represented by a convex-up curve, as shown, because the larger the increase in probability, the lesser the increase in benefit.

As an example of an opportunity-seeking system profile, suppose the quality of a joint software upgrade program gradually degrades with time because it becomes increasingly difficult to integrate the results from the existing set of contractors into the operational system, the probability of failure, P_o, increases, but the opportunity to do something about this, that is, Q_o, increases. MITRE's Kenneth Brayer suggests that Brooks' Law (Brooks, 1995) may apply. To paraphrase: adding more people to a late software project usually implies an initial penalty but might pay off in the long run. Compared to doing little to encourage the marginal benefits of integration when $Q_o < 0.2$, for example, for $Q_o > 0.3$, suppose program office policy is to add other contractors to the mix while permitting additional opportunities to learn what works best by engineering the operational system. B_s may increase with the associated increase in Q_o—but by even more, say, than at smaller values of Q_o. The extra involvement with the users helps focus the integration effort while effectively shaping the composition of the productive contractors.

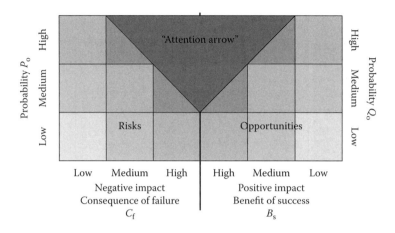

Figure 5.6 Risk/opportunity representation on probability/impact grid. (After Hillson, D. 2004. *Effective Opportunity Management for Projects*. New York: Marcel Dekker, Inc. p. 126.)

An "opportunity-seeking" system would be represented by a convex-down curve, as shown, because the larger the increase in probability, the greater the increase in benefit.

Again, a straight line, $B_s = Q_o$, from [0, 0] to [1, 1] in Figure 5.5 would be labeled "opportunity neutral." Here one could prioritize opportunity by the product, $E_e = Q_o \cdot B_s = Q_o^2 = B_s^2$.

Contact the author if the opportunity-averse/-seeking profiles of Figure 5.5 seem counterintuitive.

Figure 5.6 is another way to evaluate where to pay attention (e.g., concentrate on the triangular shaped "attention arrow" area), especially with limited resources. Obviously, emphasis should be placed on the higher probability, higher impact squares. Inspired by a figure in Hillson's book, the diagram has been modified to be consistent with the present author's choice of treating risk and opportunity as dual entities, as well as the terminology used in this chapter.

A related point of MITRE's Joe DeRosa: Another aspect of duality between risk and opportunity is the *number* of risks and opportunities versus the analysis of any single risk or opportunity. It is advisable to have few risks in a program and to perform risk analysis and mitigation. In our opinion it is also desirable to have many opportunities. Perhaps only a small subset of them will potentially materialize, and there is a risk of dilution in trying to track too many opportunities at once. Clearly, subject to the resources available, only the most promising opportunities should be tracked, and prioritized just as one especially tracks the most threatening downside risks. Nevertheless, good ESE (and CSE) should create an environment that decreases the number of risks and increases the number of opportunities.

172 ■ *Enterprise Systems Engineering*

	G	Y	R
Process	In a SoS, this action step requires a <u>similar</u> effort and scope as compared to a TSE environment	In a SoS, this action step requires a <u>modest increase</u> effort and scope as compared to a TSE environment	In a SoS, this action step requires a <u>significant increase</u> effort and scope as compared to a TSE environment
	or	or	or
Tools/ constructs	In a SoS environment, tools and/or analytical constructs <u>similar</u> to those applied in TSE environments can be used with few (if any) modification as they are applied in a TSE environment	In a SoS environment, tools and/or analytical constructs similar to those applied in TSE environments can be used; however, they require <u>modest changes or modest extensions</u> to their underlying logic designs <u>or modest extensions</u> to be properly applied in an SoS environment	In a SoS environment, tools and/or analytical constructs similar to those applied in TSE environments can be used; however, they require <u>significant changes</u> to their designs or <u>significant extensions</u> to their underlying logic to be properly applied in an SoS environment. In some areas, new tools and/or analytical constructs may also be needed

Figure 5.7 Qualitative distinctions between TSE and SoS risk management.

5.3 System and Systems of Systems Risk Management

Garvey's color scheme (Figure 5.7) (P. R. Garvey, pers. comm., 2005) can be used for either risk or opportunity management, although he neither considered opportunity management nor enterprise-scale risk management that this chapter addresses. Garvey assessed risk management for SoS compared to TSE. The present work attempts to break new ground in the corresponding distinctions for opportunity, not only between TSE and SoS, but also between SoS and ESE. These will be treated after characterizing TSE.

5.3.1 Traditional Systems Engineering

Arguably, TSE's underpinnings derive from classical linear system analysis, where much can be formulated, predicted, repeated, and explained. The central concept of superposition, that is, the well-behaved output summation of separate inputs, works well and leads to the notions of system decomposition and hierarchical composition of separately engineered subsystems. Unfortunately, this approach does not always work well in nonlinear—let alone complex—system environments.

TSE addresses a solution's form, fit, and function in two steps: functionality and implementation. It typically starts with specifications (predictions engineered to come true). Predictions (as plans, roadmaps, etc.) carry much weight; systems are built to stand alone. Developmental tests often disregard implementation. If there's divergence from the plan during development, a big effort is made to get back on track to restore the predictions' validity. But, in complex environments, the approach may miss potentially superior solutions.

5.4 System, System of Systems, and Enterprise Opportunity Management

5.4.1 The Regimen

MITRE's Doug Norman and Mike Kuras have developed a regimen for CSE (Kuras and White, 2005) that consists of eight interrelated activities (whose names have been updated)

- Analyze and shape the environment.
- Tailor developmental methods to specific regimes and scales.
- Identify and define targeted outcome spaces.
- Establish rewards (and penalties).

- Judge actual results and allocate rewards.
- Formulate and apply developmental stimulants.
- Characterize continuously.
- Formulate and enforce fitness regulations (policing).

To some extent, all these activities relate to opportunities and risks, but two (discussed next) relate explicitly: establishing rewards and characterize continuously.

5.4.1.1 Opportunities and Risks in Establish Rewards

The venture capital system of rewarding for results, and an entrepreneur's incentives for working within this outcome space, is a specific example of what is indicated here. Also, see Courtney (2001).

Imagine that a suitable outcome space has been identified and (although not necessary) communicated to a complex system's autonomous agents who must develop specific outcomes that fit in this space. In doing so, they take advantage of any *opportunities* that appear or that they can uncover. At the same time, they face the *risks* of developing products that may not be classified as outcomes, or that are less desirable outcomes. These *risks* are either not rewarded or are rewarded less than more desirable outcomes.

Because rewards may be granted to many outcomes fitting the space, agents would likely pursue *opportunities* to achieve outcomes more aggressively rather than *risking* not achieving an outcome. *Risk* mitigation might be relegated to ordering outcomes according to their perceived rewards. Presumably this ordering would be pursued with other autonomous agents because only targeted populations of agents, not individual agents, are offered rewards. This would be an example of cooperation among autonomous agents.

Despite the apparent logic of this hypothesis—that *opportunities* would be treated more aggressively than *risks*—validation from actual case studies should be sought.

5.4.1.2 Opportunities in Characterize Continuously

This CSE activity is the continual generation and refinement of complex-system characterizations. Characterization is crucial for autonomous agents so they can independently develop metrics to guide their local decision making to be congruent. The specific outcomes used to judge the actual results should be characterized, as should the rationale that eventually explains the subsequent judging decisions.

Rewards (and perhaps outcome spaces) initially should be characterized with succinct bumper sticker labels (e.g., the U.S. Army's visionary slogan, "Own the Night"). Pithiness encourages a greater variety of *opportunities* for good and bad outcomes. Initially, at least, it is probably better to have a very open outcome space with the potential for great rewards. But this implies a greater inconsistency in how rewards (and outcome spaces) may be interpreted.

To the extent that consistency matters, a complex system benefits from continually developing and espousing more detailed and complete characterizations. Still, in complex-system evolution, characterizations can become too refined. In many situations, new outcome spaces may need to be added and characterized to encourage new possibilities that otherwise might not be explored.

5.5 Further Thoughts on Opportunity/Risk Concerning TSE, SoS

5.5.1 Engineering and ESE

Moving from TSE to SoS engineering, and ESE (and CSE), one should consider opportunity/risk with respect to a complex system's environment as well as (creating potential program contingencies in) the system. There may be many more opportunities to pursue in a system's environment compared to the system itself due to: the environment's larger scope (which contains the system's scope) and the relative freedom from apparent (from the system's view) constraints, compared to what the system experiences as stress from its own environment. Vigorously pursuing opportunities in the system's environment can reduce the stress and may lead to system solutions that are not only better, but even easier to implement. Because of this strong upside, environmental risks seem less important than opportunities.

As more open systems are fielded, opportunity management will become more main-stream. Today, as MITRE's Mike Bloom points out, a good deal of opportunity management is going on at the enterprise scale in the U.S. Department of Defense (DoD). However, these activities are mainly associated with competing for huge fixed pots of money; there is lots of inertia in big systems' multiyear funding profiles. One could view opportunity (and its associated risk) philosophically ("nothing ventured, nothing gained"). Downside risk would be more concerned with not incurring damage to the system's environment that might stifle the aforementioned opportunities that would, in effect, hinder potential progress in obtaining mission-critical capabilities from the environment's developing systems.

As a property of complex systems and thus of enterprises, openness suggests a predisposition for opportunities (and risks). In creating new, potentially attractive solutions (desired outcomes in targeted outcome spaces), one should try to open the system further to create more opportunities for emergent behavior (alternative outcomes) while paying close attention to emerging risks. One has a license to more aggressively identify, explore, and develop opportunities in an enterprise than in TSE.

Enterprise risks can be mitigated by creating a management process with the adaptive built-in abilities to: quickly assess whether the emergent behavior is desirable; encourage desirable behavior; eventually eliminate (or mitigate) undesirable behavior; and be more accepting of risks to prevent returning to preconceived notions that might lock out ultimately desirable emerging capabilities.

The "messy frontier" (Stevens, 2008) is characterized by political engineering with the promised personal benefits of power and control. There are both potentially high risks and correspondingly high rewards. Cooperative behavior must be fostered to counterbalance continual competition. At the SoS and enterprise scales, several PMs usually need to be taken into account—not just one. Thus a risk and opportunity management integrated product team (IPT), for example, would have to interface with several PMs and their supporting functional areas. Remember that what may be an opportunity for the enterprise might be a risk for a particular program/PM.

Mike Kuras has suggested that research should be performed on what economists do about opportunity and risk at multiscales of analysis, that is, macroeconomics and microeconomics. Perhaps they have done more than anyone in the enterprise opportunity/risk arena.

Consider Figure 5.8, in which the *x*-axis and *y*-axis labels of resolution and field of view of an earlier chart (Kuras and White, 2005) have been replaced by risk avoidance and opportunity pursuit, respectively. As before, the axes represent arbitrary scales indicating low to high values of the two variables. The accessible region comprises combinations of the risk avoidance and opportunity pursuit pair, where humans can comprehend a vision of the underlying reality. Humans are incapable of attaining any visualization in the inaccessible region. The boundary between these

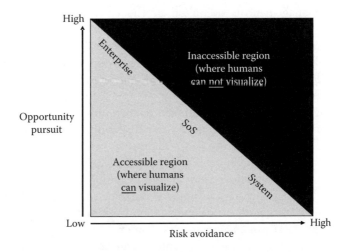

Figure 5.8 Multiscale hypotheses for opportunity and risk.

two complementary regions exists but is not necessarily known or quantifiable. The preceding discussion suggests these *hypotheses* (also see Figure 5.1):

- At a given *System* scale, one typically focuses more on avoiding risk than on pursuing opportunity.
- At an *Enterprise* scale, the emphasis should be on opportunity, not risk.
- At an *SoS* scale, about equal attention is paid to opportunities risk.
- Further evidence to support or refute these hypotheses is needed.

Thus, the opportunities for intervening in an enterprise environment to accelerate the natural processes of evolution are great. The greatest enterprise risk may be in allowing this process to atrophy.

5.6 Qualitative Distinctions between TSE, and SoS and ESE Opportunity

5.6.1 Management

For now this author has decided not to change the color descriptions when contemplating enterprise opportunity (or risk) management. This decision is tentative because enterprises are generally more complex than SoSs and require a different scale of analysis. The discussion is now concluded with summary red(R)-yellow(Y)-green(G) characterizations of opportunity management for SoS and ESE. Unfortunately, because of space limitations, only a modicum of elaboration to justify the choice of colors is offered here. Figure 5.9 shows some preliminary results. Compared to TSE environments, resources in SoS and (especially) enterprise environments are deemed harder to control because many independent organizations are usually involved. This may be particularly true for opportunities as they seem to receive less focus than risks in traditional environments. Similarly, it would be more difficult to assemble a high-performance team in SoS and enterprise environments.

Paradoxically, understanding the mission at the enterprise scale may be easier than at the SoS or program scale because the vision and mission statements for the enterprise scale must be kept rather general (and thus easier to understand) compared to those for the SoS or program scale, which might become more involved (and thus harder to understand). With the regimen for CSE mind-set, it may be easier to think of and try to apply opportunities for managing the enterprise environment at the enterprise scale. The SoS scale may present more constraints for the communities of interest (COIs) that have worked longer at this scale than at the enterprise scale. This chart's colors are selected according to a rationale similar to that of Figure 5.7. The colors tend to complement those of Garvey's comparative risk colors. Note that these initial qualitative assessments of opportunity management could change as more is learned about SoS engineering, ESE, and CSE.

Comparing TSE, and SoS and ESE opportunity management

SoS opportunity management		Enterprise opportunity management	
Assessment	Action steps and substeps	Assessment	Action steps and substeps
Yellow	Step 1 Prepare	Red	Step 1 Prepare
Y: Action 1	Commit resources	R: Action 1	Commit resources
Y: Action 2	Form the team	R: Action 2	Form the team
Y: Action 3	Know the mission	R: Action 3	Know the mission
R: Action 4	Think opportunities	Y: Action 4	Think opportunities
Yellow	Step 2 Identify the opportunities	Yellow	Step 2 Identify the opportunities
Y: Action 1	Establish team	R: Action 1	Establish team
Y: Action 2	Develop understanding	R: Action 2	Develop understanding
Y: Action 3	Identify opportunities	Y: Action 3	Identify opportunities
G: Action 4	Classify opportunities	G: Action 4	Classify opportunities
G: Action 5	Write opportunity statements	G: Action 5	Write opportunity statements
R: Action 6	Correlate related opportunities	Y: Action 6	Correlate related opportunities
Yellow	Step 3 Assess and prioritize opportunities	Red	Step 3 Assess and prioritize opportunities
Y: Action 1	Impact assessment	R: Action 1	Impact assessment
G: Action 2	Probability assessment	Y: Action 2	Probability assessment
R: Action 3	Time frame assessment	R: Action 3	Time frame assessment
Y: Action 4	Reassess opportunities	R: Action 4	Reassess opportunities
Y: Action 5	Rank opportunities	R: Action 5	Rank opportunities
G: Action 6	Coarse sort; identify handling bands	Y: Action 6	Coarse sort; identify handling bands
Green	Step 4 Decide on handling options	Yellow	Step 4 Decide on handling options
G: Action 1	Identify options within each opportunity band	Y: Action 1	Identify options within each opportunity band
G: Action 2	Easy opportunities	Y: Action 2	Easy opportunities
Y: Action 3	Hard opportunities	R: Action 3	Hard opportunities
Y: Action 4	Assign OPRsG	R: Action 4	Assign OPRsG
G: Action 5	Update opportunity database	G: Action 5	Update opportunity database
Yellow	Step 5 Establish handling plans	Red	Step 5 Establish handling plans
Y: Action 1	Develop plans and estimates	R: Action 1	Develop plans and estimates
R: Action 2	Review and approve	R: Action 2	Review and approve
Y: Action 3	Fund, direct, integrate	R: Action 3	Fund, direct, integrate
Yellow	Step 6 Implement opportunity handling	Red	Step 6 Implement opportunity handling
Y: Action 1	Finalize opportunity management plan	R: Action 1	Finalize opportunity management plan
Y: Action 2	Provide mechanisms to monitor	Y: Action 2	Provide mechanisms to monitor
Y: Action 3	Implement handling plans	R: Action 3	Implement handling plans
Y: Action 4	Monitor progress	Y: Action 4	Monitor progress
Green	Step 7 Monitor handling plans	Yellow	Step 7 Monitor handling plans
G: Action 1	Periodically review handling plans	Y: Action 1	Periodically review handling plans
Y: Action 2	Modify or stop, if required	R: Action 2	Modify or stop, if required
G: Action 3	Retire opportunities	Y: Action 3	Retire opportunities

G = green   Y = yellow   R = red

Figure 5.9 ESE versus SoS opportunity management assessment summary.

5.7 Concluding Remarks

The greatest enterprise risk may be in not pursuing enterprise opportunities. Enterprise risk, *per se*, is mainly concerned with keeping an enterprise healthy, that is, open to future possibilities for change. There is duality in treating risks and opportunities among systems, SoS, and enterprises. As the scales and associated emerging patterns change from systems to enterprises, it is asserted that opportunities become more important (see Figure 5.1).

Opportunity and risk management are team sports. Everyone should be sensitive to what is happening and participate in altering course, as appropriate. This is more challenging in enterprise environments. ESE is the big leagues in opportunity management. Keep in mind there are unknowns and unknowables. Be rather humble when confronting the complexities of an enterprise. Opportunities in ESE are abound. Be open to them in creative ways, while being mindful of the need to monitor results carefully to protect against stagnation or even chaos. Qualitative assessments of opportunity management tend to be more difficult for enterprises than for SoS or systems. These assessments might easily change when more is understood about ESE.

Proposed rule of thumb: In ESE, be aggressive with opportunity and accept risk. This is just the opposite of what one might think of as a corresponding TSE tenet. Nevertheless, additional hypothetical examples and—better yet—validation from actual case studies should be sought.

5.8 Suggestions for Future Work

Future work could include the exploration of "Real options to give weight to upside opportunities associated with uncertainty, in addition to the traditional concern with downside losses and risks" (Haberfellner and de Weck, 2005). Other suggestions include:

- Provide more examples to illustrate benefits of pursuing (especially enterprise) opportunities to strengthen validation of this thesis of emphasizing opportunities at enterprise engineering scales.
- Further explore and evaluate appropriateness of enterprise opportunity (versus risk) emphasis in example enterprise environments (suggested by Prof. Olivier de Weck of MIT-ESD during the author's talk on the CSE Regimen at the October 5, 2005 meeting of INCOSE's New England Chapter).
- Do most military PMs (colonels) focus on risks or on opportunities?
- Update Figure 5.9 based on results of the above work.
- Create business case for pursuit of (especially enterprise) opportunities.
- Questions paraphrased from suggestions by MITRE's Michele Steinbach:
  - To what extent does preoccupation with risk lead to a shorter list of options?
  - How can lost opportunities be quantified and included in risk management cost?
  - What is easier, recognizing risks or recognizing opportunities?
  - How does one measure consequence?
  - How much correlation is there between risk-averse (seeking) and opportunity-averse (seeking) individuals or organizations?
  - How does one determine where an organization's culture is between risk adverse and opportunity seeking? Under what circumstances does that motivate change, and what changes should be made, and how should they be accomplished?

- Further investigate risks and opportunities from broader viewpoint to gain more insight from interdependencies. Use parallel, multiscale analysis in considering system and enterprise hierarchies.

References

Brooks, F. P. 1995. "The mythical man-month: Essays on software engineering." In *20th Anniversary Edition (Paperback)*. Reading, MA: Addison Wesley. 2nd (anniversary) expanded edition. http://www.amazon.com/gp/product/customerreviews/0201835959/ref=cm_crdp_pt/002-1403359-6272017?%5Fencoding=UTF8&n=283155&s=books

Courtney, H. 2001. "Making the most of uncertainty." *The McKinsey Quarterly*, Number 4. Stephen Giles. MemberEdition. February 4, 2005.

de Neufville, R. et al. 2004. "Uncertainty management for engineering systems planning and design." *Engineering Systems Monograph. 2004 Engineering Systems Symposium*. Engineering Systems Division. March 29–31, 2004, Cambridge, MA: MIT.

Forsberg, K. and Mooz, H. 1996. "Risk and opportunity management. Center for systems management." *INCOSE 1996 Symposium*, Cupertino, CA.

Garvey, P. R. 2000. *Probability Methods for Cost Uncertainty Analysis—A Systems Engineering Perspective*. New York: Marcel Dekker, Inc.

Government Accountability Office. 2005. *Defense Management: Additional Actions Needed to Enhance DOD's Risk-Based Approach for Making Resource Decisions*. GAO-06-13. http://www.gao.gov/cgi-gin/getrpt?GAO-06-13.

Haberfellner, R. and Olivier de, W. 2005. "Agile systems-engineering versus agile-systems engineering." *INCOSE 2005 Symposium*, July 10–15, 2005, Rochester, NY.

Hastings, D. and McManus, H. 2004. "A framework for understanding uncertainty and its mitigation and exploitation in complex systems." Engineering Systems Division, *2004 Engineering Systems Symposium*, March 29–31, 2004, Cambridge, MA: MIT.

Hillson, D. 2004. *Effective Opportunity Management for Projects*. New York: Marcel Dekker, Inc.

Kuras, M. L. and White, B. E. 2005. "Engineering enterprises using complex-system engineering." *INCOSE 2005 Symposium*, July 10–15, 2005, Rochester, NY.

Martin, V. L. 2005. "Tolerance for change decision map: Processes for releasing flexible technology." 9 December 2005; also, 8 February 2006 abstract and 21 February 2006 briefing.

ODU Advertisement. 2005. "Professional certificate in systems engineering—Risk analysis & management (Elective)." Old Dominion University. *INCOSE INSIGHT*, 8, 1, 53 (overleaf).

Roman, C. S. 2005. "The cultural sources of acquisition risk." *Risk Management, Defense AT&L*: Part II—November–December 15–17, 2005; Part I—September–October 17–19, 2005. chris.roman@dau.mil.

Stein, J. 2005. "Risk management working group news." *INCOSE INSIGHT*, 8, 1, 38.

Stevens, R. 2008. "Profiling complex systems." *IEEE Systems Conference*, April 7–10, 2008, Montreal, Quebec, Canada.

White, B. E. 2005. "A complementary approach to enterprise systems engineering." *National Defense Industrial Association 8th Annual Systems Engineering Conference*, October 24–27, 2005, San Diego, CA.

White, B. E. 2006. "Enterprise opportunity and risk." *INCOSE 2006 Symposium*, July 4–13, 2006, Orlando, FL.

Chapter 6

Architectures for Enterprise Systems Engineering

Carlos Troche, Joseph K. DeRosa, J. Howard Bigelow Jr., Terence J. Blevins, Julio C. Fonseca, Michael J. Webb, Michael R. Coirin, Herman L. Karhoff, Jay M. Vittori, Bhupender (Sam) Singh, Richard W. Fox, Jeffery S. Cook, Cindy A. Plainte, Mark D. Baehre, and Michael R. McFarren

Contents

6.1 Introduction 182
    6.1.1 Read on if You 184
    6.1.2 What Is at Stake? 184
    6.1.3 Why Do Architectures Have Such a Negative Reputation? 187
6.2 Onward: Understanding Enterprises and Architectures 190
6.3 Architectures and the Enterprise 192
6.4 Case Studies of Architecture Use 197
    6.4.1 Architectures for Procurement Transformation 197
    6.4.2 Architectures for Air Force Command and Control—
           A Practical Illustration of Scale and Architectures 197
    6.4.3 Architectures for Multiyear Budgeting 198

6.4.4 Architectures for the Royal Saudi Air Force (RSAF)—
Applying Architectures to Build a SoS 198
6.4.5 Architectures to Develop Combat Capabilities—The JJCIDS 199
6.4.6 Architectures for Combat Capability Development—
The Deliberate and Crisis Action Planning and Execution
Segment (DCAPES) 199
6.4.7 Architectures for E-10 Program Design Activities—
A TSE Perspective..203
6.5 Conclusion 203
References 204

6.1 Introduction

This chapter is about a *tool* that enables you to visualize and analyze the *whole*, each *part*, and their interactions. We call this tool *Architecture* and believe it would have made our six wise men in "The Blind Men and the Elephant" story realize the singular nature of what they described in six different ways. We write this chapter firmly believing that traditional systems engineering (TSE) unintentionally promoted attitudes very similar to our friends and their elephant; and that simply lacing the pieces together was not sufficient as it did not account for emergence and the unique effects of scale; properties that architectures attempt to include and illustrate. In short, we believe this tool has great promise to assist in enterprise systems engineering (ESE) by enabling the rigorous and repeatable description of the Enterprise and its impacts.

Proceed further with this chapter if you are interested in learning more about this tool and how it is used in ESE. Go on to other chapters in this body of knowledge if you seek in-depth information on aspects of ESE that uses architectures.

THE BLIND MEN AND THE ELEPHANT

John Godfrey Saxe

It was six men of Indostan
To learning much inclined,
Who went to see the Elephant
(Though all of them were blind),
That each by observation
Might satisfy his mind

The First approached the Elephant,
And happening to fall
Against his broad and sturdy side,
At once began to bawl:

God bless me! but the Elephant
Is very like a wall!

The Second, feeling of the tusk,
Cried, Ho! what have we here
So very round and smooth and sharp?
To me tis mighty clear
This wonder of an Elephant
Is very like a spear!

The Third approached the animal,
And happening to take
The squirming trunk within his hands,
Thus boldly up and spake:
I see, quoth he, the Elephant
Is very like a snake!

The Fourth reached out an eager hand,
And felt about the knee.
What most this wondrous beast is like
Is mighty plain, quoth he;
'Tis clear enough the Elephant
Is very like a tree!

The Fifth, who chanced to touch the ear,
Said: Even the blindest man
Can tell what this resembles most;
Deny the fact who can
This marvel of an Elephant
Is very like a fan!?

The Sixth no sooner had begun
About the beast to grope,
Than, seizing on the swinging tail
That fell within his scope,
I see, quoth he, the Elephant
Is very like a rope!

And so these men of Indostan
Disputed loud and long,
Each in his own opinion
Exceeding stiff and strong,
Though each was partly in the right,
And all were in the wrong!

Understand that this is an evolving body of work. We are deliberately provocative and instructive, the former a prerequisite for the latter. We aim to describe architectures and their use in the enterprise by including seven examples from the contributors. However, we will not be fully representative in all instances. Through this chapter, we will achieve our goal if the systems engineering practitioner learns about architectures and their uses, and his or her manager gains a new appreciation for the tool and its place in the portfolio of tools that staff engineers should use.

6.1.1 Read on if You ...

- Are involved in determining and trading off business needs, service offerings, and system or capability requirements often in spite of short-term interests and personalities.
- Have been instructed to learn about or develop architectures (see Case Studies Section 6.4).
- Are a practicing engineer who wants to learn more about the subject.
- Are a manager of engineers, concerned about the time a few of your people spend chasing abstractions like architectures instead of "real engineering."
- Are a business leader for whom the benefits of architectures are intended.

6.1.2 What Is at Stake?

Before going much further, you should have an inkling of why ESE requires a tool like architectures that enables you—the practitioner—to visualize the whole, its parts, and their interactions.

Detailed Corporate Memory is Lost or Abandoned without an Adequate Description of the Enterprise: Without a tool to describe the whole, its parts, and their interactions, succeeding generations using the enterprise will progressively evolve and degenerate it, often with tragic consequences. This degeneration has been measured, and the consequences go beyond wasted resources. A review of different bridge designs over 150 years showed that each design degenerated into a bridge collapse incident within roughly 30 years of it first being instituted [1].* It seems that the lack of a complete appreciation of the whole system, its components, and their interactions (including failures) result in the progressive elimination of safety elements of the design until it precipitates a complete failure sometimes with tragic

---

* Petroski goes on to postulate that absent the systematic study of failure (which, in turn, relies on comprehensive documentation of the system and its components), this pattern will continue. Perhaps coincidentally, this generation-length interval happens to separate the conception and beginning of Operation Iraqi Freedom (2002–3) and the end of the Vietnam conflict (1973–75), a close homolog (according to its detractors).

results. Now extrapolate for a moment and consider the consequences of this phenomenon in enterprises such as the military, where a generation may be less than 30 years and the effective "corporate memory," that is, one intact enough to support individual accountability, is thrice reduced by the 36-month rotation of personnel. In cases like this one, a common consequence of this lack of corporate memory (in turn, because of the lack of a comprehensive tool) is the reevaluation of failed concepts every 6–8 years in complete disregard of the previous track record, wasting valuable resources in the process or contributing to serious accidents.

Current Methods for Analyzing the Enterprise are Startlingly Inadequate: We—managers and engineers alike—have become increasingly accustomed to tools and methodologies that decidedly "dumb down" the audience, sometimes with tragic results. The illustration in the next page is a copy of a section of the Space Shuttle Columbia Accident Investigation Board report. It relates how NASA's penchant for using business graphics tools to present and discuss detailed engineering data led to an oversimplification of the problems raised by the engineering staff; so, what were critical technical concerns quickly became management toss-ups as they floated up the chain. What is perhaps more tragic is that this same problem was noted in the Space Shuttle Challenger Accident Investigation Board report 20 years ago (back to fading corporate memories).

ENGINEERING BY VIEWGRAPHS

The Debris Assessment Team presented its analysis in a formal briefing to the Mission Evaluation Room that relied on PowerPoint slides from Boeing. When engineering analyses and risk assessments are condensed to fit on a standard form or overhead slide, information is inevitably lost. In this process, the priority assigned to information can be easily misrepresented by its placement on a chart and the language that is used. Dr. Edward Tufte of Yale University, an expert in information presentation who also researched communications failures in the *Challenger* accident, studied how the slides used by the Debris Assessment Team in their briefing to the Mission Evaluation Room misrepresented key information.[38]

The slide created six levels of hierarchy, signified by the title and the symbols to the left of each line. These levels prioritized information that was already contained in 11 simple sentences. Tufte also notes that the title is confusing. "Review of Test Data Indicates Conservatism" refers not to the predicted tile damage, *but to the choice of test models used to predict the damage.*

Only at the bottom of the slide do engineers state a key piece of information: that one estimate of the debris that struck *Columbia* was 640 times larger than the data used to calibrate the model on which engineers based

continued

their damage assessments. (Later analysis showed that the debris object was actually 400 times larger.) This difference led Tufte to suggest that a more appropriate headline would be "Review of Test Data Indicates Irrelevance of Two Models."[39]

Tufte also criticized the sloppy language on the slide. "The vaguely quantitative words 'significant' and 'significantly' are used 5 times on this slide," he notes, "with de facto meanings ranging from 'detectable in largely irrelevant calibration case study' to 'an amount of damage so that everyone dies' to 'a difference of 640-fold.'"[40] Another example of sloppiness is that "cubic inches" is written inconsistently: "3cu. In," "1920cu in," and "3 cu in." While such inconsistencies might seem minor, in highly technical fields like aerospace engineering a misplaced decimal point or mistaken unit of measurement can easily engender inconsistencies and inaccuracies. In another phrase "Test results do show that it is possible at sufficient mass and velocity"; the word "it" actually refers to "damage to the protective tiles."

As information gets passed up an organization hierarchy, from people who do analysis, mid-level managers to high-level leadership, key explanations and supporting information are filtered out. In this context, it is easy to understand how a senior manager might read this PowerPoint slide and not realize that it addresses a life-threatening situation.

At many points during its investigation, the Board was surprised to receive similar presentation slides from NASA officials in place of technical reports. The Board views the endemic use of PowerPoint briefing slides instead of technical papers as an illustration of the problematic methods of technical communication at NASA.

Review of test data indicates conservatism for tile penetration

- The existing SOFI on tile test data used to create crater was reviewed along with STS-107 Southwest Research Data
    - Crater overprotected penetration of tile coating significantly
        - Initial penetration to be described by normal velocity
            - Varies with volume/mass of projectile (e.g., 200 ft/s for 3cu. In)
        - Significant energy is required for the softer SOFI particle to penetrate the relatively hard tile coating
            - Test results do show that it is possible at sufficient mass and velocity
        - Conversely, once tile is penetrated SOFI can cause significant damage
            - Minor variations in total energy (above penetration level) can cause significant tile damage
    - Flight condition is significantly outside of test database
        - Volume of ramp is 1920cu in vs 3cu in for test

6.1.3 Why Do Architectures Have Such a Negative Reputation?

We close this introduction by addressing a theme of interest to *managers*, especially those too busy to ponder the use of architectures in their programs.

Too Much Detail: Architectures have generally developed a poor reputation among managers because they emphasize the detailed and disciplined use of information and its relationships based on widely accepted and credible references. Unlike our friends from "Indostan"
>And so these men of Indostan
>Disputed loud and long,
>Each in his own opinion
>Exceeding stiff and strong,
>Though each was partly in the right,
>And all were in the wrong!

or the tragedy of Columbia's engineers and their viewgraphs, the architecture discipline requires an attention to detail in both sources and exposition that many find daunting and unnecessary. The art in architecture is to provide the right level of detail for the purpose at hand—too little, the architecture is superfluous; too much, it is prohibitively expensive. Sometimes the architect is at fault for becoming more focused on the architecture than on its utility, inundating the manager with unnecessary details, especially in those cases where the manager clearly understands when an architecture is well enough to specify what he or she needs. Oftentimes, the architecture tends to also threaten the power structure by highlighting the potential folly of ill-conceived decisions, flagging obfuscation, petty interests, or highlighting the nonexecutability of programs.

Different Personal or "Gut" Experience: This attention to detail, disciplined referential integrity, and focus on exposition and visualization tend to also "disenfranchise" many experts because the level of detail and the referential integrity enforced by the architecture discipline are such that no one single person can claim adequate command of all of the details. What many consultants bring to any issue they are working on is credibility borne of "been there, done that" personal experience. Of necessity, any one's experience will be a limited set of examples with a limited focus, often insufficient to draw broad generalizations on which to predict future courses of action. The minute the consultant is required to map those experiences against a broadly accepted set of principles as needed in architectures, often that personal experience falls short of the task and unavoidably looks like an invalid representation of the principles. In short, the architecture discipline tends to devalue that which most consider "invaluable": personal experience. This tendency is most noticeable in work groups introduced to the architecture concept and its requirement for referential integrity; that is, to go beyond personal experience to widely and publicly accepted renditions of the processes being examined so as to rise above personal limitations. In such cases, many of the "experts" find themselves silenced as they seek to make their personal experiences relevant to the more widely accepted whole.

Too counterintuitive: It is, however, this multidimensional nature of architecture—whole, parts, and interactions—that repels its detractors most. Culturally, we have been taught that the way to understand the world around us is through its repeated decomposition of what we observe into its elemental components, the thorough examination of each component, and the enumeration of the behaviors and characteristics of all components to make the whole. Architectures, on the other hand, visualize the whole, so that one can draw understanding from the totality of the system and visualize the component as well, often reflecting the complexity of the system where the understanding of the component does not necessarily scale up to a system-wide conclusion [2]. Consider an illustrative analogy. Imagine yourself at a medical school. As part of your medical education, you would dissect cadavers; evaluate human anatomy, repeatedly studying the three-dimensional location, name, and function of organs, tissues, bones, sinews, and other structures. Your anatomical studies would enable you to appreciate the structure of a biped creature mostly made of fluid, looking rather shriveled when the fluid is removed. In a culturally reductionist way, your command of these details could equip you to trace the movement of a corpse's finger from brain through nerve to muscle, bone, and ligament, "adding" each component's function until the whole produced the movement predicted. Culturally, this is the point at which most managers and engineers alike think they need to know no more. However, as a future doctor in medicine, you would find this knowledge of anatomy essential but insufficient to your task of "first, do no harm," as it negates the understanding of the interactions of all the organs at the macrolevel, at the level of the whole being. A phenomenon such as homeostasis (equilibrium of all body systems) would escape the limited—component-focus—understanding that anatomy proffers, as there is no single organ, tissue, sinew, or *combination* responsible for it. The interaction of medical remedies, and the whole prediction and management of side effects (positive and negative) would escape those limited to the reductionist view of the anatomist only because these often do not evidence themselves, except as system-wide effects. You would need to complement your understanding of anatomy with that of physiology, pharmacology, and many other disciplines that look at the human being as a whole in much the same way that architectures evaluate the enterprise [3].*

---

* Ackoff likens the treating of a headache by adding a chemical to the stomach to the holistic solution of an enterprise problem in sales by changing something in manufacturing.

Need a tool that tells me "What To Do?" instead of "How to Do It?": Architectures, such as anatomy and physiology, reveal and explain both at the component and at the macro (or enterprise) level. Culturally, this makes them somewhat counterintuitive, more closely aligned with natural processes than with the reductionist approaches that gave us scientific management and division of labor. For managers and staff weaned on these principles, architecture represents an extremely discomfiting development, often reflected in the excuses employed to reject them. A common one echoed in the Department of Defense (DOD) is that architectures are not useful to determine "*what to do*" but are useful in determining "*how to do it*."* Note that such a limitation, if it were so, could be assigned to any methodology as—by itself—no single tool (architectures or otherwise) will tell its users what to do and, if it did, it probably is not a tool that should be used because it would contain an inherent bias that does not permit the dispassionate comparison of alternatives. Indeed, if as a manager or staff, you need a tool to tell you "*what to do*," we strongly wish to dissuade you from using architectures or any other tool that provides a detailed comparison of alternatives and nuanced distinctions. We can only help but think a tool that told the user "what to do" would also make its user superfluous, since concluding "what to do" presumes the judgment normally reserved for those employing it, yet another capability (judgment) that—in our view—no tool should ever have. If, on the other hand, you seek a tool that provides a means to view the whole, its parts, and the *behavior* of each at their scale (e.g., homeostasis) in order to compare alternative trade-offs among them, then architectures can fill the need in a way that many other tools cannot without prejudging outcomes.

Yet, managers endorse others' methods like architectures: Contrastingly, in the face of all these complaints, businesses invent methodologies for analyzing their operations that contain all the same element managers find so distasteful about architectures. Six Sigma and Lean Manufacturing are but two of several techniques that analyze—sometimes in more excruciating multidimensional detail than architectures—the different activities that add value (read produce revenue) to their enterprises. It is as if, after all is said and done, managers begrudgingly accept their need to understand those details sufficiently to make intelligent decisions that the current PowerPoint "engineering" methodology is at best insufficient, at worse dangerous, and that the "art" of management—the oft-cited rationale for refusing systematic study—needs to be wholly or partially captured in some repeatable, reputable form to be fully exploited and exported into other areas of the enterprise.

---

* Ascribed to a Joint Staff assessment of architectures July 2004 as recited in a 2005 Joint Capabilities Assessment on the Joint Capabilities Integration and Development System.

6.2 Onward: Understanding Enterprises and Architectures

As this is a chapter on architectures for ESE, we will first discuss what an enterprise is, what architectures are, and how they relate. Within the context of this chapter, an **enterprise**[*] is any *purposeful* collection of activities of interest that exhibits multiple states across time, and may exhibit emergent behavior *potential* that can be identified or suspected beforehand, often at different scales. The notion of emergent *potential* has been included in the definition to address the concern that often emergence can only be identified after it happens; hence, there is little that can be done to use it or—in some cases—harness it. Within the context of this definition, our interest is in recognizing the potential for emergence by identifying the conditions that promote it in a useful form. The construct of System of Systems (SoS) proposed by the Air Force Scientific Advisory Board is a specific instance of the general notion of enterprise we are defining here. According to that study, emergence is one of five main characteristics that distinguish SoS, that is, "the SoS has *emergent* capabilities and properties that do not reside in the component systems" [4].

A *state* is a description of the collection of activities and structures at a point in time existing according to a preagreed set of rules. For instance, if one were to discuss a "hunting" enterprise, one state of the enterprise will be when the hunter's weapon safety is on, while another will be when the safety is off as the hunter gets ready to take the shot. Safety rules would define what actions a hunter would take to inform those around him he was about to take a shot. Transitions between states add dynamism to the enterprise.

Emergent behavior refers to those behaviors that cannot be observed by just looking at the components of the system individually. For instance, no evaluation of a *single* element of the flight envelope (e.g., airspeed, lift, drag, surface area, air density, etc.) will enable an observer to discover the phenomenon of stalling. You need to consider all the factors and the variability within each factor to predict or observe stalling in flight. Many consider *emergence* a "red herring" because it is not unique enough to be distinguishable, cannot be identified until it happens, and it therefore cannot be harnessed. And, the reasoning goes, if emergence is a "red herring," so is ESE, as it is nothing more than TSE at a different scale. We differ from these views. Within the confines of this chapter, we believe emergence is a real property that can be suspected and promoted before it is fully observed, within the context of the collection of purposeful activities of *interest* (i.e., the enterprise), in much the same way as a gardener's attention to the details of soil composition promotes growth of certain varieties over others. Hence, our emphasis on emergence is on its potential and the use of architectures as a way of "teasing" that potential from what we believe are enterprises working at different scales.

---

[*] Synthesized from the Government Accountability Office, contributors to the Write the Book effort, and the New England Complex Systems Institute.

Architecture: The structure of components, their relationships, and the principles and guidelines governing their design evolution over time [*IEEE Std 610.12, December 10, 1990* & DOD *Architecture Framework v 2.0, April 23, 2007.*]

A widely accepted definition of **architectures** is quoted above. A more practical definition is that an architecture is a **model** (but not necessarily a single model) that details *what* a system* does (*activities*), the relationships among those activities, *how* a system performs those activities (*processes*), the *functions* of those tools that are used to perform the activities in total or in part (*system functions*), the actual *tools* themselves that *perform* those system functions (*systems*), the skills and experience of those performing the activities (*roles*), all *related* through a documented, repeatable, and agreed-upon structure (*taxonomy*), and rendered through an equally approved documented and agreed-upon rule set (*framework*). Each of these activities, processes, system functions, systems, and roles—in turn—is described in terms of *attributes*. Depending on the scale, state, and emergent properties of the model, the model may turn out to be a collection of models, some related, others nested. For the sake of simplicity and to overcome human cognitive limitations, architectural models are often rendered as a set of different perspectives or *views*, allowing a more focused look into one or more aspects of interest, in much the same way that models of the human body often have separate views of the circulatory, skeletal, muscular system, and so on. The key, however, is that all of these views are integrated; for example, the rendition of the circulatory system must acknowledge the shape, curves, and special structures the skeleton gives the cranium to accurately show blood flow to the brain. Let us apply these two definitions to the more specific cases of TSE and ESE.

Architectures and TSE: In TSE, architectures decompose, dissect the system into its components, and evaluate the role of each component within the integrated whole of the system. In TSE, architectures serve as the equivalent of the anatomy text, carefully identifying the organs, tissues, bones, sinews, and so on, their functions, and relationships. For those supporting the defense establishment, the use of architectures to define a new capability under the Joint Capabilities Integrated Development System (JCIDS) is a very good example of architectures employed in this manner (see Case Studies Section 6.4.5). In cases such as TSE, the architecture models defined at the very beginning can evolve into detailed design, ultimately identifying specifics of component type, cost, performance, and so on.

Architectures and ESE: In ESE, the other dimension of architectures—the holistic one—provides a definition of the emergent potential of the collection of purposeful activities of interest to the sponsor. Just as TSE architectures serve as the equivalent of human anatomy, in ESE the architecture focus broadens to consider the synthesis of those TSE details on the humans as a whole and the potential emergent behaviors family and societal influences can have. The goal of

---

* Please note that the term "system" is used broadly—it includes people, processes, hardware, software, and documentation.

architectures in ESE—or more commonly, Enterprise Architectures (EAs)—is to describe capabilities manifest in the enterprise that may or may not be evident simply from the component systems. Thus, the EA exists at multiple scales. And, at each scale there are architectural artifacts that describe the properties (capabilities) at the scale, as well as details of the components and their interactions at the next finer scale.

An example of emergent potential discovered through the use of architectures may be seen in the study conducted some years back on the Copernicus Architecture. In that study, two modes of responses to queries—one very terse and the other quite more verbose—were compared against the availability of communications capability. The study showed, counterintuitively, that the more verbose communications mode was actually more effective as the communications capacity dropped [5].

Another example is the software capabilities of the familiar Windows desktop: they are almost entirely emergent when compared with the hardware capabilities of the integrated circuits that make up the personal computer (PC).

To sum up, architectures are multidimensional descriptions that may be rendered at various scales and for multiple purposes. At the TSE level, architectures decompose the "system" under study into its component elements maintaining and further specifying the relationships between the components. At the ESE level, architectures decompose within and across the system's environment, identifying and synthesizing the opportunities for emergence and—by extension—innovation. In the example cited above, one of the opportunities for innovation the architecture identified in that enterprise was the ability of using the terse queries to forecast the future needs of the Warfare Commanders afloat with sufficient accuracy to enhance the effectiveness of the future information queries over more limited communications capabilities once underway ([5] p. 68).

6.3 Architectures and the Enterprise

In the previous section, we identified two main applications of architectures—at the TSE and ESE levels. As TSE is the most commonly known and will be illustrated with a detailed case study later on this chapter, this section will focus on how architectures are developed for the scale of the Enterprise through the development of a set of nested architectures that address the different scales of interest to enterprise users.

To handle the scale typical of an enterprise, EAs are developed as a nested set of models. The following illustrations offer two such views: one from the Open Group and the other from the Air Force's Operations Support Enterprise Architecture.

Note the notion of the "Architecture Continuum" in the Open Group's graphic above, showing how each architecture scale tends to support products corresponding to that scale. Foundational ESs, such as those developed by the Federal Government to guide the provision of services to the citizenry, support the definition of key products and services. Contrast that with the details implied by the systems architectures in supporting the development of actual solutions (read designs and specifications).

Architectures for Enterprise Systems Engineering ■ 193

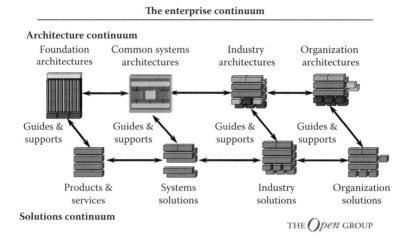

The Air Force's Operations Support Enterprise Architecture provides a better illustration of the multidimensional nature of architectures as it is "scaled up" to address enterprise-wide effects.

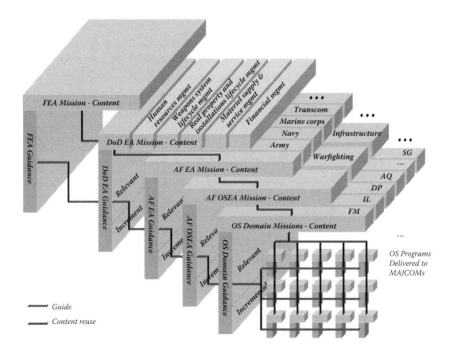

Referring to the figure above, the Federal Enterprise Architecture (FEA) defines the services the government provides its citizens, including those associated with national defense, of which operations support (OS) is part. The FEA "bounds" the enterprise of interest, guiding lower-level, a more detailed architectures in adding further details (dissection if we were using our previous analogy). Note that the next level represents a definition of the Department of Defense Enterprise Architecture (DOD EA) highlighting many of the functional areas representing OS (e.g., Financial Management). Subsequent levels apply these functions to the Air Force at large, the Air Force's operations support domain, and it then breaks down this domain into its different functional members as they are implemented within the Air Force. At the bottom of the cascade are the individual OS programs that realize in hardware, software, personnel, and procedures the architectural models depicted in the previous levels. Of particular interest in this rendition are the concepts of *guidance* and *content* reuse, both of which ensure traceability as you travel up and down the scale. Each level's guidance constrains the one below, focusing the architecture further. The constraints are more that just semantic, as each level's content is reused to the degree necessary for further detailing by the level below. Hence, for example, the Air Force EA cannot cover missions that are outside of the DOD EA.

Nested architectures, scale, and constraints hint at an underlying set of rules or framework that defines the structure and its relationships. Similar to an anatomical map, the framework indicates what the relationships of interest are and—sometimes—how to document them. Most frameworks tend to stop there, similar to the anatomy text that stops at the documentation of which bone, sinew, and so on is connected to which other member. In architectures, there are various frameworks that can be used (e.g., Zachman, DOD Architecture Framework [DODAF], etc.); the critical point is that one should be declared and used consistently if architectures are to be useful for the Enterprise. Below is an illustration of the Air Force's Enterprise Architecture *Framework* and its major elements.

AF Enterprise Architecture
Descriptions

Enterprise
perspective
descriptions

Mission area & cross mission area
perspective
descriptions

Program & node
perspective
descriptions

The perspective focusing on strategic plans, enterprise-wide processes, key information and infrastructure important to the enterprise, and a framework to enable lower-level architectures to be relatable to other architectures together make up the EA.

The perspective focusing on a subset of the enterprise defined by a specific mission, function, business area, or set of capabilities, activities, or shared data. This perspective is primarily operationally focused and is user/operator "centric."

The perspective focusing on an individual system or a group of systems and the interrelationships with other systems. This perspective is primarily system focused and is program manager or node manager "centric."

AF Enterprise Architecture Descriptions
At the top of this framework are those perspectives that cut across the entire Air Force especially in a way that assists in the discovery of emergent relationships and potential. The top level—as well as the subsequent one— tends to constrain and guide the development of the lower ones. Any framework to be truly useful must acknowledge its place in a larger "framework" or context. For the Air Force, their enterprise architecture framework (EAF) defines its context as shown below in the above two graphics. In the first one, the context is simply generalized as the set of inputs and drivers that trigger the content of the architecture descriptions and the ultimate uses and impacts of the architecture. In the second one, all components of the first are exploded into their most significant detail.

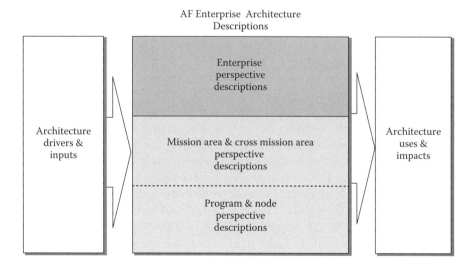

196 ◾ *Enterprise Systems Engineering*

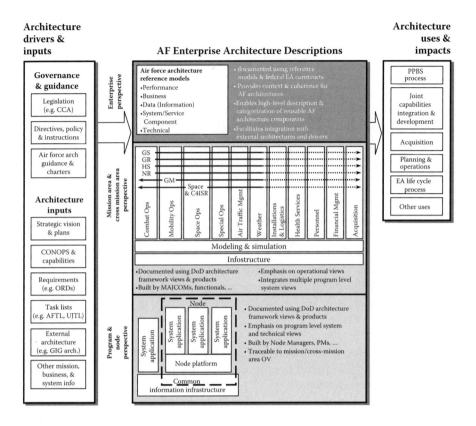

While it is beyond the scope of this chapter to detail each one, note that many of the drivers and outputs are what they should be for what is expected of a public service like national defense (recall the previous discussion on the FEA and its definition of services to the citizenry). On the "driver" side of the flow are statutes and policy guidance from the DOD and other executive agencies; as well as other architectures and architecture-like guidance. On the output or "uses" side are the development budgeting of new Air Force systems and the planning and operations of Air Force forces. Therefore, architectures support the enterprise and—specifically—ESE by enabling the visualization and exposition of the enterprise's contents, in the process highlighting capabilities manifest at the enterprise level that show potential for emergent behavior. This visualization and exposure are not just graphical artifacts, but reflections of an underlying rule set or framework that defines the contents of interest to be described at each level of interest. If done correctly, enterprise architecting can help realize the promise of ESE by providing a complete map and context for the enterprise that enables all of its members to employ it to their greatest advantage. As we bring to a close this highly abbreviated discussion of architectures and their role in ESE, it is important to reiterate what was said in the beginning—the architectures' role in ESE is *promising* but yet to be fully realized.

We hope this discussion has given you enough of what architectures are and their role in ESE, why are they disliked which is also why they are (potentially) such a good tool for ESE.

6.4 Case Studies of Architecture Use

Thus far, we have defined architectures, their potential, and how they are perceived by many of its users. Rather than describe those issues further, we instead focus on showing architectures and their use in an informal case study "sampler." As noted in Section 6.1, the cases in the "sampler" are a picture in time, reflective of the time, effort, and availability of the contributors to this edition, and in no way should be construed as representative of the state of the practice or its future. Further, we encourage investigating the various references cited throughout the volume as they provide additional examples.

6.4.1 Architectures for Procurement Transformation

Summary: As part of the Procurement Transformation effort, the Secretary of the Air Force for Acquisition (SAF/AQC) initiated the Enterprise Architecture for Procurement (EAP) "To-Be" development in FY02 (fiscal year 2002). The driving vision of the effort was to improve mission responsiveness, by leveraging world-class processes and web technology. The vision was to develop a commodity acquisition approach where 98% of sourcing actions were accomplished in a web-centric environment. The initial focus was on developing the operational architecture and governance for Commodity Councils (CCs), which was later integrated into the Purchase and Supply Chain Management (PSCM) operational architecture.

Impact: Architectures became the visualization tool that enabled Air Force leadership to realize the common objectives among otherwise different procurement initiatives, consolidate efforts where sensible, and begin to "lean out" processes to exact additional efficiencies and synergies. Probably, this is one of the strongest examples of using architectures for ESE. In one case, the use of the architecture identified that 14–16 days could be shaved off the average purchase request (PR) processing time a savings that, at the architectural scales involved would net $60–$68 million in real savings across the entire Air Force.

6.4.2 Architectures for Air Force Command and Control— A Practical Illustration of Scale and Architectures

Summary: Three years into its establishment, the Air Force's Air and Space Command and Control, and Intelligence, Surveillance, and Reconnaissance (C2ISR) Center began the development of a set of architecture views or perspectives of the operational nodes, processes, and roles employed in C2ISR of air and space forces across the full spectrum of Air Force-supported operations. The vision was to highlight Air Force's contributions in C2ISR to the nation's defense, contributions that would enable a better appreciation of the Air Force's needs to execute these missions.

Impact: This case highlights how the architecture team wrestled with the issue of scale and avoided the pitfalls of our friends from Indostan and how fickle is corporate memory and commitment for architects. Ironically, the team in this case study became a victim of its own success, not being allowed to complete the total architecture because their work on part of it had such impact that the customer wanted more in-depth analysis.

6.4.3 Architectures for Multiyear Budgeting

Summary: The Air Force's Air and Space Command and Control, and Intelligence, Surveillance and Reconnaissance Center (AFC2ISRC) recently begun asking its budget proponents to align their proposals with the C2 Constellation subEA as a way of providing a first-order alignment of budget proposals and avoid program duplications.

Impact: The architecture provided the overall macro—big picture—view of both those areas addressed by the budget and just as importantly, all those areas addressed in the architecture for which there is no planned investment.

6.4.4 Architectures for the Royal Saudi Air Force (RSAF)— Applying Architectures to Build a SoS

Summary: As a result of the Desert Shield/Storm alliance and further upgrades under the Peace Shield initiative, the Kingdom of Saudi Arabia (KSA) would end up with a collection of Command, Control, Communications, Computer, and Intelligence (C4I) systems that—while accomplishing individually what they were designed to do—were otherwise incapable of working in an integrated manner. In response, the RSAF leadership decided to develop a comprehensive blueprint to modernize their C4I in phases. The RSAF architectures provided a comprehensive blueprint for every one of the RSAF mission areas showing the participants involved, the systems they used, inputs, outputs, information exchanges, and their relationships.

Impact: This is a classic application of architectures to enable the user to visualize both the breadth and depth of the challenges ahead with enough detail to guide investment. It also includes a novel approach to communicating the architecture—particularly important in a different cultural context—that enabled the RSAF users to obtain the maximum benefit from the architecture. Probably one of the strongest examples of applying architectures to an enterprise context (RSAF C4I capabilities), end to end, was appreciated by its users.

6.4.5 Architectures to Develop Combat Capabilities—The JJCIDS

Summary: JCIDS* is the DOD, downward directed, process for the determination of needs, codification of those needs into capabilities, and development of those capabilities into systems, should other means of satisfying those needs not be suitable. JCIDS employs architectures in two areas: one to support the need determination and the other to support the definition of the interoperability ("net-centricity") of the resulting system, should a system solution be required to solve the need. As a practical matter, the use of architectures for determining the need has fallen into disuse (see Section 6.1.3). The more common use of architectures falls under the second category—to define the interoperability or net-centricity of the resulting system when a system solution is required. In this sense, architectures are being used to support and guide system solutions (Section 6.3). For Air Force systems, for example, this would be consistent with and guided by the "Program and Node Perspective" description of the Air Force Enterprise Architecture Framework (Section 6.3). Because of the resource investment in developing a solution, JCIDS policies are mandatory upon those proposing defense programs. One of those mandates is that the architectures developed to describe the net-centricity of proposed systems strictly follow the requirements of the DODAF, the set rule that defines what describes architectures in DOD. The DODAF defines architectures as the composite of operational views (OVs), system views (SVs), and technical views (TVs) of the system under study (see the list of views and short descriptions in Table 6.1).

Impact: The architectures developed to support combat capabilities under JCIDS are generally an application of TSE, although the policies requiring them continue to evolve and require further evaluations beyond the capability dimension to the larger net-centric enterprises. These architecture views become central to the capability's success as they define the interfaces with the network that will be tested when judging the developed solution a success.

6.4.6 Architectures for Combat Capability Development— The Deliberate and Crisis Action Planning and Execution Segment (DCAPES)

Summary: The DCAPES is a set of software applications that serve as the Air Force's single system to present, plan, source, mobilize, deploy, account for, sustain, redeploy, and reconstitute forces and resources for contingency operations

---

* The governing policies for JCIDS are Chairman of the Joint Chiefs of Staff Instruction (CJCSI) 3170.01E, which directs the use of Chairman of the Joint Chiefs of Staff Manual (CJCSM) 3170.01B, and—for architectures—CJCSI 6212.01D. All policies are available through the Internet from the Defense Technical Information Center at http://www.dtic.mil/doctrine/

Table 6.1 Architecture Products

Applicable View	Framework Product	Framework Product Name	General Description
Architecture views	AV-1	Overview and summary information	Scope, purpose, intended users, environment depicted, analytical findings
Architecture views	AV-2	Integrated dictionary	Architecture data repository with definitions of all terms used in all products
Operational views	OV-1	High-level operational concept graphic	High-level graphical/textual description of operational concept
Operational views	OV-2	Operational node connectivity description	Operational nodes, connectivity, and information exchange needlines between nodes
Operational views	OV-3	Operational information exchange matrix	Information exchanged between nodes and the relevant attributes of that exchange
Operational views	OV-4	Organizational relationships chart	Organizational, role, or other relationships among organizations
Operational views	OV-5	Operational activity model	Capabilities, operational activities, relationships among activities, inputs, and outputs; overlays can show cost, performing nodes, or other pertinent information
Operational views	OV-6a	Operational rules model	One of three products used to describe operational activity—identifies business rules that constrain operation

Operational views	OV-6b	Operational state transition description	One of three products used to describe operational activity—identifies business process responses to events
Operational views	OV-6c	Operational event-trace description	One of three products used to describe operational activity—traces actions in a scenario or sequence of events
Operational views	OV-7	Logical data model	Documentation of the system data requirements and structural business process rules of the operational view
System views	SV-1	System interface description	Identification of systems nodes, systems, and system items and their interconnections, within and between nodes
Systems views	SV-2	Systems communications description	Systems nodes, systems, and system items, and their related communications lay-downs
Systems views	SV-3	Systems–systems matrix	Relationships among systems in a given architecture; can be designed to show relationships of interest, for example, system-type interfaces, planned vs. existing interfaces, etc.
Systems views	SV-4	Systems functionality description	Functions performed by systems and the system data flows among system functions
Systems views	SV-5	Operational Activity to systems function traceability matrix	Mapping of systems back to capabilities or of system functions back to operational activities
Systems views	SV-6	Systems data exchange matrix	Provides details of system data elements being exchanged between systems and the attributes of that exchange
Systems views	SV-7	Systems performance parameters matrix	Performance characteristics of Systems View elements for the appropriate time frame(s)

continued

Table 6.1 (continued) Architecture Products

Applicable View	Framework Product	Framework Product Name	General Description
Systems views	SV-8	Systems evolution description	Planned incremental steps toward migrating a suite of systems to a more efficient suite, or toward evolving a current system to a future implementation
Systems views	SV-9	Systems technology forecast	Emerging technologies and software/hardware products that are expected to be available in a given set of time frames and that will affect future development of the architecture
Systems views	SV-10a	Systems rules model	One of three products used to describe system functionality—identifies constraints that are imposed on systems functionality due to some aspects of systems design or implementation
Systems views	SV-10b	Systems state transition description	One of three products used to describe system functionality—identifies responses of a system to events
Systems views	SV-10c	Systems event-trace description	One of three products used to describe system functionality—identifies system-specific refinements of critical sequences of events described in the OV
Systems views	SV-11	Physical schema	Physical implementation of the logical data model entities, for example, message formats, file structures, and physical schemas
Technical views	TV-1	Technical standards profile	Listing of standards that apply to SV elements in a given architecture
Technical views	TV-2	Technical standards forecast	Description of emerging standards and potential impact on current SV elements, within a set of time frames

and military operations other than war. DCAPES is found in both fixed and deployable configuration and it is an integral part of the Global Command and Control System-Air Force (GCCS-AF) using its communications infrastructure. Architectures were developed to build the net-centric or interoperability narrative and for DCAPES to support its approval for development. The experience highlighted in this short case study is that of the architect attempting to assist the DCAPES proponents in rendering the needs for improvements in capabilities (dubbed Increment 2b).

Impact: This is an example of the implementation of the JCIDS policies discussed in the previous case study. It emphasizes the need to work with the architect from very early on so that the architecture can accurately depict the capability interfaces. It also shows how architectures can be an aid in teasing, clarifying, and rendering multidimensional and difficult issues graphically and precisely, an advantage that can be also detrimental if the architecture's tight coupling to the overall set of capabilities being sought is not maintained.

6.4.7 Architectures for E-10 Program Design Activities— A TSE Perspective

Summary: This is a very practical, hard-hitting illustration of how architecture products were used within a large weapons system program for a number of purposes supporting both overall program goals and program management needs. We close with this example because of its very *practical bent and hard-hitting impact*.

Impact: This example shows the power of the architecture in becoming the holistic representation of the state of the system to date, enabling its users to exercise and evaluate 20 different courses of action. It highlights the novel use of object-oriented views (normally considered to be too detailed) to serve the managerial needs of the program leadership. It shows the direct impact of the architecture and its centrality as a "communications tool" for the program leadership.

6.5 Conclusion

Architectures represent a tool, one that enables visualization of the whole, each part, and their interactions. Architectures represent a means to the end of better systems engineering. Within the broad range of TSE and ESE, respectively, architectures TSE decompose dissect the system into its components serving as a means to document each part of the system and its impacts on all others ESE: Elicit the *potential* for emergence of the whole system, describing capabilities and behavior clearly manifest only at the enterprise level. Architectures assist in ESE by capturing the scale of the enterprise and its emergence potential. The use of architectures in both TSE and, especially, ESE is still evolving ... as a community of practice, we are not there yet. Some in the community continue to have faith in the tool and its potential; others have "waxed poetic" about its pitfalls. We have tried to address its detractors in this edition of the volume by debating their concerns and by providing a set of countervailing case studies where this multidimensional tool shows direct and distinct impact. We close where we started—provoking to be instructive. We challenge our readers to use architectures, to join in the ESE discussion, and that in doing so learn to deal with the whole, each part, and their interactions, much unlike our friends from Indostan.

References

1. Sibly, P. G. and Walker, A. C. 1977. Structural accidents and their causes. In Petroski, ed. *Design Paradigms*, p. 169, Cambridge University Press, Cambridge, UK.
2. Gharajedaghi, J. 1999. *Systems Thinking Managing Chaos and Complexity A Platform for Designing Business Architecture*, Butterworth Heinemann, Boston, MA.
3. Ackoff, R. 2005. *Systems Thinking and its Radical Implications for Management*. IMS Lecture/Boston, February 15, 2005.
4. Saunders, T. 2005. Study Chair, System-of-Systems Engineering for Air Force Capability Development, Air Force Scientific Advisory Board, June 30, 2005.
5. Levis, A., Hearold, S. L., and Perdu, D. M. 1994. Effectiveness of two modes of information pull in the Copernicus architecture. In *Science of Command and Control: Part III Coping with Change*, AFCEA International Press, Fairfax, VA.

Chapter 7
Enterprise Analysis and Assessment

John J. Roberts

Contents

- 7.1 Introduction ... 206
- 7.2 The Need ... 208
- 7.3 Essential Characteristics of EA&A ... 212
  - 7.3.1 Multiscale Analysis ... 213
  - 7.3.2 Early and Continuous Operational Involvement ... 215
  - 7.3.3 Lightweight, Portable System Representations ... 216
  - 7.3.4 Developmental Versions Available for Assessment ... 217
  - 7.3.5 Minimal Infrastructure ... 218
  - 7.3.6 Flexible M&S, OITL, and HWIL ... 219
  - 7.3.7 In-Line, Continuous Performance Monitoring, and Selective Forensics ... 220
- 7.4 Stakeholder Roles ... 221
  - 7.4.1 Government Role ... 223
- 7.5 Next Steps ... 224
  - 7.5.1 Practical Considerations ... 226
- 7.6 Summary ... 228
- Acronyms ... 228
- References ... 229

7.1 Introduction

This is not a chapter on research. There are no long lists of programs that have performed analysis and assessment at the enterprise level for reasons related to the development or operational use of command and control (C2) systems or networks. There is no statistically based or anecdotally derived set of "tried and true" best practices to generate a few insightful recommendations. Enterprise analysis and assessment (EA&A), as it will be defined in this chapter, is not only a new undertaking, but one that will be necessary to permit thoughtful, proactive, and robust evolution of the C2 enterprise. For background, the concept of the C2 enterprise is discussed in Chapter 2 and Ref. [1].

Why the assertion that there needs to be a new undertaking—a new twist on the historical types of analyses performed in the past in support of various systems engineering goals? There are four main reasons behind this assertion. First, we are moving into an era in C2 where our infrastructure and operational paradigms are changing in fundamental ways. One can argue on if or how much of the global information grid (GIG) exists today, but it certainly does not exist in the deployed form that most envision it to be. As that happens incrementally, and we move from message-based information exchanges between known entities over dedicated communications links to an internet-like approach, we must be able to assess how well our emerging net-centric C2 applications will perform under a wide variety of new operating conditions that neither lend themselves well to current methods of analysis and assessment, nor to the capabilities of the tools that we routinely use today in support of these efforts.

Second, given an emerging GIG infrastructure relying on Internet Protocol (IP) standards, the introduction of highly interrelated C2 services distributed throughout the C2 architecture (instead of traditional, more independent C2 applications) will have a major impact on mission performance considerations. This service-oriented architecture (SOA) approach to enterprise interoperability, leveraged heavily from the commercial information technology (IT) business sector, also does not exist today in any meaningfully deployed way within the Department of Defense (DoD). Individual programs are just beginning to develop and field initial offerings of web-enabled C2 applications representing small pieces of their overall capabilities. Therefore, there has not yet been a burning need to perform analysis and assessment of net-centric C2 enterprise issues. But this need will arise soon, and there must be some thought dedicated now to enable the emergence of possible methods and techniques that will be required by C2 developers and operators in the near future. As an example, all programs will be required to show how they support the Net-Ready Key Performance Parameter (NR-KPP). The NR-KPP will be used to assess required net-ready attributes for the information exchange as well as the resulting end-to-end operational effectiveness.

Third, the fact that our threat of environment morphs becoming more complex overtime will drive us to be more responsive to world events to the point of becoming proactive in examining potential operating conditions. In order to accomplish such a goal, we need a more flexible infrastructure that will enable users and developers to quickly assess the potential of emerging C2 capabilities to address new, potential, and perceived threats.

Finally, our users expect increased functionality and better performance from C2 systems and capabilities overtime. Simply maintaining the *status quo* with respect to the capabilities and execution performance will mean that we have not exploited the benefits of net-centricity. Providing a wealth of new capability options to users is only acceptable with an accompanying improvement in execution performance, otherwise, the enhanced set of capabilities will utilize even more bandwidth and processing resources than they do today.

Therefore, we need a way to ensure the stability, scalability, and robustness of C2 capabilities as we progress toward net-centric operations. However, if every next-generation system or capability will be dependent upon one or more of the other systems or capabilities in a net-centric, GIG/SOA C2 architecture, does this mean that one must model the whole world or tap into every program's assets in order to analyze any issue for a given program or mission? Using traditional systems engineering analysis methods, one could argue that this might be a logical consequence. Of course, this approach would be completely impractical. So, how can analytical needs be satisfied for enterprise scale C2 issues, and what are those issues?

EA&A will be defined not as the ability to analyze the complete inner workings of an entire C2 enterprise at once; rather, EA&A will be defined in terms of an ability to characterize the behavior of entities or capabilities that are immersed within an enterprise construct. EA&A will emphasize a robust "what if?" approach versus the traditional, highly scenario-dependent attempts at "prediction." It is shown to be that a critical need to leverage modeling and simulation (M&S) capabilities, with a key role for real-time operator-in-the-loop (OITL) and/or hardware-in-the-loop (HWIL) capabilities, though not necessarily the traditional systems engineering use of any of these. This chapter will define EA&A for C2 as the ability to robustly analyze and assess potential outcomes derived within enterprise architecture constructs to permit an understanding of fundamental behaviors. EA&A will relate strongly to enterprise opportunity and risk assessment, aiding in the generation of risk management options for C2 capability developers and C2 operators.

The enterprise analysis techniques proposed in this chapter are intended to permit analysis of operational situations and enable more informed subjective assessment of the mission execution from an operational perspective. The results derived from this approach to analysis and assessment will permit the insights produced to be applied across a wide range of specific architectural, technical, and operational conditions for development, acquisition, and operational purposes. Thus, this chapter will identify the expected utility of the insights to be gained through adoption of EA&A principles.

7.2 The Need

C2 EA&A must ask and answer a different set of questions than traditional systems analysis. Emerging paradigms in the employment of C2 capabilities (e.g., net-centric operations), their associated IT infrastructure (e.g., SOA), and the acquisition of those capabilities (spiral development) are significantly impacting traditional acquisition responsibilities such as risk management and test and evaluation (T&E). The diminishing utility of traditional T&E and certification processes and approaches when applied to quickly evolving net-centric applications motivates a need to steer programs toward supplementing these traditional approaches with alternatives to ensure the achievement of desired behaviors at the C2 enterprise scale. The analytical components of T&E, risk management, and other activities must be addressed within an increasingly complex operational environment.

At the enterprise scale, it will be shown to be much more important to characterize behavior across a set of operating conditions, rather than performance according to a specific scenario (or perhaps a few variants) with all the associated assumption and caveats. The idea is to characterize the performance of the overall capability to execute the required mission under a wide range of operating conditions, no matter how they might be envisioned to occur. The analytical challenge is to consider the full range of possible conditions, even remote possibilities that might occur to ensure robustness. This includes dealing with adversaries that could be acting to defeat the C2 capabilities under assessment. In fact, an environment that not only permits simulated attacks to occur, but actually facilitates this behavior by certain participants will allow users to quickly and concretely understand vulnerabilities of current or proposed capabilities. This example points out the need to have early and continuous operator involvement in the C2 EA&A process. The challenge here is to provide an inexpensive and flexible infrastructure to enable this to occur on a regular basis.

While the need for C2 EA&A is not critical today, it is coming. The current state of GIG and SOA fielding is providing some time to permit EA&A to be developed and better understood. For example, there continues to be debate about the ultimate requirements for the GIG when it is fully implemented. Development issues within the Joint Tactical Radio System family of systems have also stretched out the fielding of net-centric capabilities. Also, the transition of net-centric technology from the commercial workplace to the DoD has encountered some bumps in the road because of real-world issues such as nonassured military communications. Nonassured communications was never a key commercial technology consideration because of the massive telecommunications infrastructure available to financial institutions and other large corporations, who are the major clients driving commercial IT standards.

However, even with the large issues described above, key stakeholders are making progress toward IP-based network operations. Surrogate, near-term, scaled-down concepts, and initial implementations of future wideband IP networks (e.g., Interim Capability for Airborne Networking, Tactical Targeting Network Technology, etc.) are blossoming in the interim while the major building blocks of the GIG are in development. C2 programs (e.g., Theater Battle Management Core Systems, etc.) are incrementally web-enabling certain portions of their capabilities. Finally, industry is moving ahead with concepts such as Enterprise Service Bus prototypes. Within the next few years, there could be enough fielded SOA-based capability riding on a "GIG-lite" infrastructure that C2 EA&A issues could begin to be thrust to the forefront within both the operational and acquisition communities.

So what? What are the advantages of understanding how capabilities within a C2 enterprise behave under normal and unusual operating conditions? How will this knowledge help us? First, there are currently a plethora of strategy, policy, and guidance documents that describe how to design and implement net-centric C2 capabilities. Many of these documents are quickly outdated, or written at a very high level, or even contradictory in places. The sheer volume of this guidance, while well intentioned, becomes difficult to comprehend, apply, and arbitrate at the C2 enterprise level. We *do* need a way to evaluate compliance with key design principles, but we *do not* have our acquisition personnel sufficiently enabled to identify these "needle in the haystack" key implementation strategies. A process and accompanying analysis environment to allow the most successful strategies, policies, and guidance to emerge and be identified will be critical to the evolution of net-centric C2. Those applications that can work effectively under a wide range of operational conditions will have embodied the key principles of net-centricity and probably avoided many of the unnecessary ones. It will become increasingly important to down-select within our continually growing set of strategy, policy, and guidance documents to convey to developers what is critical and what is not. It will then be possible to evaluate compliance with only the most fundamental principles.

Second, if operational personnel can begin to understand under which sets of conditions they will experience difficulties executing their missions, they can proactively develop temporary, emergency work-arounds, and explore other means of accomplishing critical tasks within an operationally safe environment. During subsequent deployments, not only will they be better prepared for the uncertainties of real-world operations, but they might even be able to identify impending problem situations to become more proactive and less reactive.

Third, EA&A might ultimately enable the evolution of an entirely new business model for the acquisition of future DoD C2 capabilities. With the distinct (and perhaps likely) possibility of numerous web services being developed with overlapping functionalities, it will be almost impossible to continually dictate which are the "preferred" or "mandated" services to be used. On one hand, a set of services might perform the exact same function others (e.g., targeting, data fusion, resource tracking, etc.), but some might be extremely inefficient for a particular community to use. On the other hand, another set of services might appear to provide the "gold standard" of capabilities to a wide user base, but end up providing lots of extra complication and headaches for users. Stable, basic services built on trusted, legacy software architectures (e.g., wrapped legacy code) might win out over less stable, embryonic services built on a more robust and extensible software backbone in the short term. However, the balance might very well shift overtime as the legacy-based applications run into increasing development problems (cost, schedule, and performance) as they climb the tough hill to adapt to new operational paradigms as underlying technologies evolve. Eventually, an EA&A environment could lead to a competitive situation where services are posted in an operationally realistic evaluation arena and users migrate to those that work best to satisfy their needs. Those services that are widely used (or are needed by high-profile users) would thrive and receive additional resources for evolution, while those that are seldom used would be left to wither.

All engineering-based product development requires some form of analysis, often multiple forms, and often at multiple points during the lifecycle of individual systems. Systems engineering, risk assessment/management, and T&E support are often the major drivers of systems analysis activities. There will always be a need for some amount of highly detailed systems analysis within individual programs.

Time and cost are critical considerations. M&S-based analyses can shorten analysis timelines because system model development can precede system development. In fact, many analyses performed in support of system level issues to date have already been M&S-based, to one extent or another. However, these efforts have not typically been well coordinated from one phase of program acquisition to another, with an implication of little M&S reuse due to disparate development and domain advocacy for unique toolsets. There is typically even a greater disconnect between those M&S-based analysis activities performed in support of acquisition and those performed in support of operational activities (especially training).

DoD guidance is another important driver for systems analysis. There are a number of areas within the realm of systems analysis that are amenable to capabilities-based analysis and architecture-based analysis. These include basic interoperability assessments, information exchange requirements, function use/utility evaluations, and so on.

However, there are significant inherent limitations within the most widely used commercial architecture tools. A number of system architecture tools that have the ability to run dynamic executable versions of use cases typically employ static input data files or statistically generated message input data to drive information through the architecture to identify information bottlenecks, calculate throughput statistics, show high-/low-use areas, and so on. This type of analysis might be useful as a first cut at systems analysis, but it lacks the ability to take the analysis to the next level of sophistication, including real-time operational scenario execution with operator participation as well as the ability to efficiently inject realistic system functionality (e.g., processing logic for C2 systems) because of the lack of environmental considerations as well as nontrivial runtime implications.

Finally, systems analysis activities have usually been focused on addressing questions that are of critical importance to program milestone decisions. The scope of these activities has traditionally depended upon the analysis tools and capabilities resident within a given program, including the associated contractor (or contractor team). Limited analysis resources within a program can often restrict the application of analysis capabilities to only those topics that can be studied over a period of a few months. Normally, collaboration among program offices to address questions unique to only one program's acquisition milestone needs is rare. It is extremely difficult to match up milestone dates across program with sufficient lead time to coordinate the collaborative execution of major tests and demonstrations.

As a direct result of the realities of tightly focused analyses needs within C2 programs, for example, the following are common practices observed from the conduct of systems analysis efforts across a number of such programs, with benefits and drawbacks (the "cons" associated with each class of systems analysis effort make them inappropriate for C2 EA&A because they impact the inherent flexibility necessary to accomplish the EA&A objectives discussed earlier) indicated for each.

- Emphasis on specific use cases and associated scenarios
  - *Pro*: they help flesh out particular Concept of Operations and Concept of Employment issues
  - *Con*: they are required to undergo a lengthy accreditation process and are difficult to change once approved
- Trust placed in only verified, validated, accredited, and/or certified C2 software applications (including associated system models that are usually not shared outside of the program)
  - *Pro*: supports program configuration management objectives and T&E requirements/guidance
  - *Con*: trust mandated no matter how flawed the processes are known to be
- Mapping of numerous measures of performance (MOPs) to a smaller group of higher-level measures of effectiveness (MOEs)
  - *Pro*: identifies data collection requirements
  - *Con*: consensus process places little analytical rigor behind the choices of mappings
- Model-based system performance predictions to fill gaps in requirements definitions (i.e., eliminate "To Be Determined" system requirements)
  - *Pro*: provides concrete guidance to contractors and defines analysis tool development/integration needs
  - *Con*: produces highly scenario-dependent analysis results with numerous caveats, leading to the inability to extrapolate results or utilize them to gain an understanding of alternate employment concepts

212 ■ *Enterprise Systems Engineering*

- Large HWIL node integrations to support T&E needs in interoperability arena
  - *Pro*: establishes long-term working arrangements with key partners
  - *Con*: expensive to achieve, leading to ever-increasing resistance to change for other uses (once integrated and working properly)
- Large-scale, infrequent joint experiments/demonstrations to look at emerging capabilities
  - *Pro*: program office team learning associated with preevent integrations
  - *Con*: demonstrations (and experiments structured like demonstrations) provide little "leave behind" to leverage and evolve once the event is completed
- Downstream insertion of third-party hardware and software instrumentation "probes" for measurement or diagnostic purposes
  - *Pro*: collects data for subsequent performance analysis
  - *Con*: normally causes perturbations when inserted within a system being monitored, affecting overall results
- Measuring whatever we can
  - *Pro*: provides clues to some categories of system performance issues
  - *Con*: difficult to relate lower-level measurements/metrics to desired outcomes

7.3 Essential Characteristics of EA&A

In order to address questions related to the characterization of expected behaviors of C2 capabilities over a wide range of operational conditions, EA&A must deemphasize the utility of comparing detailed metrics against specific individual requirement values, whether the metrics are derived from measurement, simulation, or estimation. EA&A must instead look for break points where capabilities are either significantly enhanced or totally disabled.

Since EA&A must identify sets of simultaneous conditions responsible for noticeable changes at the mission effectiveness scale, it must emphasize real-time OITL assessment within an environment that is almost identical to an actual operational setting. Interestingly, EA&A must go as far as actually encouraging proactive, asymmetrical threat attacks to occur under nondestructive, nonlife-threatening conditions. It must also foster a culture amenable to publishing and encouraging the external use of system/capability representations (e.g., models).

Key characteristics of EA&A to be discussed are

- Multiscale analysis
- Early and continuous warfighter operational assessments
- Lightweight, portable M&S-based C2 capability representations
- Developmental software versions available for assessment
- Minimal infrastructure

- Flexible M&S, OITL, and HWIL capabilities
- In-line, continuous performance monitoring, and selected forensics

7.3.1 Multiscale Analysis

In many instances, traditional stand-alone analysis approaches will be sufficient for systems analysis and below. However, these traditional systems analysis approaches are generally inadequate for considering multiple "agent" (user, system, subsystem, etc.) interactions and multiple scales of resolution within the enterprise. They often do not effectively consider the significant effects of human interaction with systems. As systems evolve over time and interact with other systems to generate streams of information for human decision makers, we are increasingly observing that phenomena emerge that could not be anticipated through the use of traditional systems analysis techniques. In fact, even the differences in scale/scope of enterprise-level analyses will drive the need for multiple analytical approaches. Emergent phenomena will require a synergistic application of different approaches to address multiscale analysis at the enterprise level.

What is the difference between multiscale and multilevel with respect to a C2 enterprise, and why is the notion of multiscale analysis most appropriate? The idea of multiple levels of analysis relates strongly to systems analysis, in particular, to analysis of noncomplex (i.e., linear) systems. Very briefly, linear systems can be successfully analyzed through more traditional systems engineering techniques, such as functional decomposition. Analysis can be performed at the "piece part" level and integrated at higher levels to properly characterize overall system performance. A C2 enterprise does not behave according to the rules of linear systems theory. A C2 enterprise is an inherently complex system and, therefore, should be analyzed as such. A much more in-depth treatment of this subject can be found in Ref. [2].

The primary focus of EA&A is on the highest scale (i.e., level of resolution) applicable to C2, which is at the mission effectiveness scale. At this scale, the emphasis is on nontraditional analysis issues such as robustness, flexibility, fitness, and so forth. A critical objective is to identify operating ranges for systems and capabilities operating at lower scales that enable acceptable mission execution. In other words, there are normally wide ranges of system/capability performance at lower scales within which no discernable effects can be observed at the highest scale or resolution (mission effectiveness scale). Even significant deviations in communications throughput, subsystem reliability, platform processing speed, node architecture make-up, choice of web service provider, and so on, will often not noticeably affect a user's ability to execute a particular mission. Key questions to ask at the mission effectiveness scale are, for example:

- What behaviors emerge under stressing conditions?
- What are the minimum system and operational architectures required?
- How can errors be overcome?

It has not yet been established how many distinct scales exist within a typical C2 enterprise, but agreement on this concept (e.g., the Open Systems Interconnection 7-layer Reference Model of communications systems) would facilitate greater understanding of C2 enterprise constructs. Table 7.1 shows a candidate set of C2 scales and associated examples of metrics that could be collected at each scale.

It is postulated that most metrics will only map clearly one scale higher or lower than the one at which they can be collected, simulated, or estimated. In fact, this may be why some traditional mappings of MOPs to MOEs have been so difficult to understand intuitively. Mappings that span more than two of the scales in Table 7.1 might be very difficult to trace. Also, it might be impossible to usefully decompose operational effects into a unique set of detailed metrics at lower scales/resolution, since there are likely multiple causes for many observable effects.

A potential implication is that performance metrics at the first scale down from the operational effectiveness scale might be the most important to capture from an EA&A perspective. Another possible implication is that some lower level measurements might not be worth collecting at all, or at least only in very stressing circumstances. However, as indicated in the last column of Table 7.1, some lower scale conditions may transcend many others to have major impacts at even the highest scale. This implies that there is a need to identify catastrophic or cascading failure combinations that can ripple up to affect mission execution. Most likely, there will also be a need to identify combinations of lower scale situations that enable significantly enhanced operating conditions. This could be another critical application of EA&A.

Table 7.1 Multiscale C2 Metrics Relationships and Failure Mode Impacts

Scale	Operational Metrics	Acquisition Metrics	Catastrophic Failures
Operational effectiveness level	Mission success rate	Net-ready key performance parameter	Mission abort
Software applications level	Automated function execution performance	Development cost & schedule	Unanticipated failure mode
Hardware platform/ operating system level	Information processing speed	Version stability; backward compatibility	Hard drive crash
Network management level	Unauthenticated accesses denied; throughput rate	Host system integration complexity	Router overload
Radio frequency/ physical link level	Link closure statistics	Number of radio variants	Link outage
Transistor/ computer chip level	Sub-assembly operating temperature	Power usage; size; weight	Thermal breakdown

7.3.2 Early and Continuous Operational Involvement

Warfighters are often overloaded just trying to cope with the operational demands of each day in the field. When C2 operators have been extracted from deployed locations to participate in forward-looking wargames and exercises, many have experienced the unfortunate circumstance of trying to do today's jobs in a next-generation environment, instead of helping evolve future requirements, concepts of operation, or concepts of employment. It is difficult, without sufficient time and training, to expect operational personnel to be able to properly critique new capabilities.

An enabler for early operational involvement in EA&A is to reduce the need for extensive training, effectively lowering the bar to useful participation in wargames, experiments, and operational evaluations. The use of familiar or uncomplicated user interfaces, especially for new or initial capabilities, will permit the operational personnel to focus on the evaluation of the capability by hiding the complexity of the applications.

Another key to enabling early operational assessment in the acquisition process is to support remote participation in an interactive manner, including capitalizing on significant DoD investments in training systems to support acquisition activities. There are large networks of active duty and recently retired personnel that could be tapped for participation in loosely structured, ongoing activities to assess proposed enhancements to capabilities in the field. If personnel at any of a number of locations could log onto periodic, widely announced evaluation activities at their convenience, the likelihood of a wide range of inputs over a period of time would be substantially increased. The commercial internet gaming industry has been particularly successful in this area, permitting thousands of simultaneous online users to role play in highly realistic, real-time, distributed combat simulations.

With respect to test, early involvement in development activities by any of the means cited above will begin to get operators more familiar with evaluation and assessment within a C2 enterprise. The operators can then help identify and focus on critical test issues to enable more effective achievement of test objectives. This can also serve to help establish confidence in test results as well as permit some difficulty to test situations to be better addressed, especially those requirements related to operational availability, which can only be tested over long durations.

One important component to enable continuous operational assessments within a C2 enterprise is the existence of operational scenarios and use cases. However, at the enterprise scale, it is much less important to get it right with any particular scenario than it is to devise ways to understand and characterize the fitness of the enterprise through constant exposure to diversity. EA&A at the C2 enterprise scale will certainly require injection of pieces of scenarios and use cases for increased realism, but the emphasis must be on more generic, highly flexible representations of wide ranges of employment options, from austere to robust, with typical uses for stakeholder systems incorporated into the mix. Repeated exposure of the C2 enterprise to complexity is critical to effectively characterize the fitness of the enterprise, especially its ability to adapt to stressing and/or unforeseen circumstances. Practically, this can be accomplished via injection of a wide range of operating conditions (as implemented in scenarios and use case vignettes) into online operational situations if those situations are not regularly occurring, such as major attacks by hostile forces. This analytical practice will not only enable characterization of the fitness of the enterprise to day-to-day situations, but to stressing situations as well.

Combined developmental/operational test (DT/OT) events and operational assessments are starting to replace the separate DT and OT event paradigm. EA&A will require the extrapolation of this trend to an even greater degree. When viewed from the right perspective, traditional DT and OT events (often onetime activities) can be morphed to a more periodic (and ultimately nearly continuous) set of operational assessments that occur over a relatively short time.

7.3.3 Lightweight, Portable System Representations

Despite the best efforts of an entire program or project team, it is not possible to identify and unambiguously state all of the requirements of a C2 system in a specification prior to the awarding of contract, due to the inherent complexity of C2 systems. At the enterprise scale, with dependencies among many programs needed to provide important capabilities to the warfighter, it is critical to develop effective and efficient mechanisms for collaboration among key stakeholder programs within the enterprise. Some analytical methodology must be in place to deal with requirements uncertainty.

Experimenting in novel ways by using lightweight and portable representations of C2 systems that can be rapidly accessed by the development and test environments of peer systems is highly desirable. Such a capability would provide opportunities to understand issues and identify opportunities for collaboration as early as possible in program acquisition and fielding schedules without the need to synchronize activities of different programs, which is practically impossible. In order to make this work, the integration times and annual costs must be kept to a minimum. Large, highly detailed models with complex interface requirements and extensive rehosting issues squelch opportunities for collaboration. Early, small footprint models or prototypes of a system can give other programs insight into the evolving functionality of peer systems.

Such lightweight, portable representations are very appropriate for certain classes of programs, including sensor systems and communications systems. For C2 capabilities with large operator populations such as an entire intelligence processing center or an air operations center, the representations would likely be constructed to capture the salient characteristics of specific web services. For examination of future enhancements within a large C2 node, the insertion of lightweight models of specific, proposed capabilities into the operating environment might be a cost effective and efficient way to rapidly evaluate the potential of proposed functionality.

From a methodology perspective, programs could develop and post lightweight representations of emerging capabilities to make them available to other programs without having to understand in advance which programs might want to investigate or take advantage of these new functions. Many of the reviewing programs could quickly discover what any other program is implementing or planning to implement. This method of interaction could lead to new opportunities for collaboration.

7.3.4 Developmental Versions Available for Assessment

It will be necessary to expose users/consumers of information and services (operational personnel and software applications) to both the diversity of other peers and the novelty of emerging capabilities. The aspect of novelty is a fundamentally distinguishing characteristic from distributed T&E environments and system integration laboratory networks. Traditionally, only official, released versions of software applications are eligible to be used in test or integration events. Experimental venues, while making use of early version of capabilities, have neither the persistence of an EA&A environment nor the breadth of scope to encompass an enterprise perspective.

Having access to the developmental versions of C2 applications within an EA&A environment might ultimately enable the evolution of an entirely new business model for the acquisition of future DoD C2 capabilities. With the likely possibility of numerous web services being developed with overlapping functionalities, it will be impractical to continually dictate which are "preferred" or "mandated" services. An EA&A environment could lead to a highly competitive situation where emerging net-centric web services are posted in an operationally realistic evaluation arena and users migrate to those that best satisfy their needs. Those services that are widely used (or are needed by high-profile users) would thrive and receive additional resources for evolution, while those that are seldom used would be left to wither.

Innovation within a C2 enterprise is critical because the requirements imposed on the enterprise are constantly morphing, and the C2 enterprise must be able to leverage all available options to be able to handle new situations as effectively as possible. Requirements can change quite dramatically and quickly, as was demonstrated by the radical impacts of the 9/11 terrorist attacks on U.S. homeland security requirements. Thus, there needs to be a set of solution spaces constantly under evaluation for potential benefit in future applications. Perpetual experimentation is critical to enabling this innovation to occur within an enterprise. Requirements identification must become more flexible to enable better responsiveness and anticipation of key future needs.

As the C2 enterprise evolves, especially as the DoD comes to rely more heavily on distributed networks for communications and service access, vulnerability assessment will become more critical. The Government needs a solid understanding of the inherent vulnerabilities of deployed and developmental DoD network-centric capabilities. Since these capabilities are being incrementally fielded and/or fielded in potentially wildly different configurations in various locations around the world, it will be necessary to constantly assess military networks for many types of vulnerabilities. Once network vulnerabilities are identified, the Government will need to sponsor parallel activities specifically designed to assess their severity and to experiment with potential solutions. It would be highly desirable to have teams of personnel attempting to thwart proposed capabilities and working this cycle prior to deployment of operational networks. This practice could drive innovation up front, rather than relegating it to a more reactive activity.

Rapid software prototyping has provided an opportunity to look at innovative technologies and novel approaches much earlier in acquisition cycles of programs than ever before. While not intended to be robust (in any of a number of aspects), rapid software prototypes have the advantage of being able to convey complex concepts inherent in new technologies to their potential users quickly through experiential exposure, vice through detailed technical explanations. This also allows the users to get an idea of the potential of specific technologies to address operational needs or desires without the heavy investments of funding and time needed to transform research work into operationally fieldable software in order to obtain feedback.

There are two principal types of innovation, revolutionary and evolutionary, as discussed in Ref. [3]. Evolutionary innovation is the normally occurring type and would be expected to occur even if no synthetic environment existed for EA&A. For revolutionary innovation to occur, there needs to be opportunities for very different types of agents to interact. The exposure of existing and emerging capabilities to the C2 enterprise environment could encourage revolutionary innovation, leading to significant increases in capability.

7.3.5 Minimal Infrastructure

In order to support assessment that enables innovative activities to progress while implementation decisions are being made, either a replication of the operating environment or access to the operating environment would be required. For some systems, this can be quite practical, as evidenced by the existence of many program test beds at Government and contractor locations. However, attempts to replicate the detail inherent in significant portions of operating environments, even for a single mission area (e.g., missile defense) inevitably begin to accumulate heavy logistical requirements (hardware, software, facilities, networks, etc.) and often come with high levels of initial investment and large recurring costs.

An important enabler for continuous operational assessment will be the communications infrastructure. Currently, there is a mix of connectivity available within the DoD, with a wide variety of bandwidths, costs, security levels, and contention for use. From a C2 enterprise perspective, the key is to begin pulling together distributed programs with vested interests into loose collaborative frameworks to address critical operational issues with minimal recurring communications infrastructure costs. The approach of creating a huge, generic acquisition infrastructure, which has been tried several times in the past, often becomes costly to maintain and suffers from lack of a long-term advocate or set of advocates, especially when the infrastructure ceases to be new and begins to need significant upgrades. Establishing a small, high-use, core community of interest (COI) network that can effectively leverage existing connectivity within the DoD and/or short-term commercial leased circuits will provide both necessary connectivity among key stakeholders, as well as flexibility for future endeavors with other partner organizations within the enterprise.

Some ways to avoid these issues include loosely linking existing development environments among programs as well as having the ability to switch from online operations to off-line experimentation while taking advantage of selected imaging of local operational information exchanges (versus developing these databases). Some care needs to be taken to ensure that information that could be published back into operational databases is either filtered, tagged, or held locally to avoid operational complications.

7.3.6 *Flexible M&S, OITL, and HWIL*

At the enterprise scale, it will be impossible to predict which critical issues will need to be addressed in which particular order (or in parallel). Therefore, the development of a flexible and extensible analytical framework is the most important consideration. Pursuing a goal of operational breadth first, then technical depth only on an as-needed basis will provide capability earlier and permit evolution of that capability overtime that will be tailored to its true use.

Many large-scale simulations, and especially federations of simulations, have experienced run time performance issues as the fidelity within models increases overtime and/or additional higher fidelity models have been added to the initial federation. High-fidelity models (especially at the system level) will have a significant analytical role. However, for maximum flexibility at the mission effectiveness scale, it will be critical to establish analysis methodologies that can capture results of high-fidelity models in effects-based models.

An example of a large federation that has persisted overtime is the Joint Training Confederation, which has stayed focused on supporting a particular user community whose stakeholders all have maintained interest and support. A key to the success of this long-standing capability is that this federation was not pulled together to support an event and then subsequently dismantled, only to be reconstructed in some variant configuration at a later date. It addresses enterprise training issues even today. There are other examples of long-term OITL successes within the training arena, such as the Air Force Distributed Mission Operations program to train pilots in cockpit simulators, reducing the amount of expensive flying hours required to certify pilots. However, the success of these M&S efforts has depended heavily on the single use paradigm and the inherent stability of the training infrastructure from year to year. EA&A will not have the luxury of dedicated users and single source funding.

The real utility of a loose federation of C2 EA&A capabilities would be to analyze the performance, emergent behavior, and characterize the fitness of enterprise capabilities. In order to get at these issues, it is critical that the M&S, OITL, and HWIL be flexible and capable of looking at a wide variety of issues. Analytical depth should be selectable in a "plug and play" manner used in concert with a common core capability, such that only the aspects of the C2 enterprise relevant to the issue at hand are part of an analysis infrastructure.

C2 HWIL labs run real equipment with live operators in real time and produce information in operational formats. As we move to an IP-based GIG concept, this will effectively lower a key barrier to M&S interoperability because M&S developers will need to pay less attention to the myriad, constantly changing, often misinterpreted, almost never fully implemented joint message standards for various tactical data links. Today, for a C2 HWIL capability to talk to a simulation, operational messages are usually translated into Distributed Interactive Simulation Protocol Data Units. From an EA&A perspective, the M&S assets must be able to interoperate with HWIL using operational information standards (i.e., act like a real system or web service).

7.3.7 In-Line, Continuous Performance Monitoring, and Selective Forensics

As more work is dedicated toward defining and understanding the relationships among the various scales of C2, it will become increasingly apparent that much of what we are able to easily measure and collect today in tests and experiments will not be useful for EA&A purposes. In order to gain insights into issues at the enterprise scale, it will be necessary to identify and capture critical performance measurements automatically, routinely, and on as much of a noninterference basis as possible.

The first step in this process will be to accomplish the collection of this information in such a manner that it has little or no impact on C2 application execution. Today, we routinely insert third-party hardware or software probes into our infrastructure to detect what we perceive as important events. One result of these intrusions is a perturbation of the actual flow of information itself, including generating additional data traffic, creating other ripple effects within the network. Another result is the problem of trying to make the logical connection between measurements and situations that are only indirectly related to these measurements. Also, information is often collected downstream from where it is generated, having been manipulated along the way in some manner. Data that can be automatically archived where it is produced (within each C2 application) and collected either later or off-line, in an operational sense, for analysis is highly preferred.

The second step will be to develop an analysis capability to permit forensic investigation of problems or opportunities. As operations proceed normally, some automated statistics generation over an extended period of time would serve to characterize the normal operating ranges of component systems and capabilities, probably requiring the development of some new tools or augmentation of existing tools. When problems or opportunities are observed at the mission effectiveness scale, forensic analysis would then be required.

The third step will be to identify the sets of critical situations, states, activities, parameters, and so on, which contribute to noticeable impacts on mission performance, either positive or negative. Changes in trends, performance metrics, and operational architecture, as well as coincidental circumstances would be evaluated as potential causes of mission impacts. Key to this activity will be the establishment of "tripwire" values for certain metrics that are suspected of causing changes to occur to mission effectiveness. Overtime, this list will be tailored in one way or another. For example, it might only be necessary to collect certain information when specific thresholds are reached or when other indicators suggest that other types of information are needed to diagnose a problem. The only time that information about performance should compete with operational traffic is when alerts are generated. Otherwise, the architecture should accommodate off-line data collection, analysis, and assessment. This analysis would require a toolset that is not yet fully developed and would likely require additional study to define.

Finally, the results of this analysis and assessment must be made available in a digestible form to the key decision makers. Summary performance parameters must be collected by the applications themselves and made available off-line for analytical purposes. The increasing, distributed development of C2 web services across the enterprise at unpredictable intervals will drive a need to stay current with emerging capabilities. Service orchestration issues within an SOA framework will need constant reassessment as different ways of accomplishing mission tasks are uncovered and understood. The net result will be the development of insights and characterizations of C2 capability robustness and the recognition of critical underlying situations that must be watched closely as they emerge either slowly or suddenly. This information could become critical in understanding whether or not complex systems are adhering to fundamental implementation and standards guidance. Thus, EA&A could serve a vehicle for policy compliance within the C2 enterprise.

7.4 Stakeholder Roles

In recent years, the role of major defense contractor acquisition responsibilities has steadily increased. This has been implemented in the form of total system performance responsibility, various types of integration contract awards, and the scoping of overarching C2 contracts (e.g., Army Future Combat System). In parallel,

Government engineering talent has been lost through retirements and repeated reductions to both military and civilian staffing levels. With this situation as a backdrop, both the Government and the contractors are experiencing difficulties working complex C2 problems in today's acquisition environment. Historically, the Government has had a stronger engineering presence to be able to have a greater influence on the course of C2 acquisition activities. Effective EA&A will require the development of very knowledgeable Government teams able to have unique insights into the technical and programmatic issues of the acquisition often before the contractors do because they have the tools at their disposal to perform system analysis and assessment independently from the contractors. This will directly improve the ability of program management to avert issues.

It will be critical for the Government to maintain experienced cross-domain teams to work enterprise issues that span programs and capabilities. It will also be important for the Government teams to have access to tools and virtual environments that can permit examination of critical issues at this scale. Instead of imposing strict control over models and virtual environments, as is routinely done today, a C2 enterprise analysis Government team must possess a very flexible and fluid C2 enterprise toolset. The exact composition of the Government teams must be able to vary based on need, and the makeup of the teams will need to include operational personnel. A C2 Government analysis and assessment team (e.g., for a particular mission area), as well as the associated synthetic environment, must persist past the award of contracts for individual programs. A persistent Government enterprise assessment team must have specific analytical roles during all phases of current and future C2 contracts.

A possible methodology for addressing C2 EA&A is to hire a contractor to do this thinking, implementation, and execution, much like the lead system integrator model for program execution. While input from industry is essential, this overall approach appears problematic for a number of reasons. For example, despite the consolidation within the defense industry, individual contractors still generally focus on a portion of the C2 enterprise, and there can be difficulty sharing information internally across projects within their contract portfolios. The Government must have the ability to look beyond the sight lines of any particular defense contractor, no matter how large or distributed. However, contractors are increasingly trying to take advantage of the knowledge that they have gained as companies have merged to the point that they have begun to market their integration expertise. System integration experience provides opportunities for depth of understanding of a particular domain or mission area. These companies are in a position to provide depth to the analysis infrastructure for the systems they develop, as needed.

Certainly, a role for the Government is to advocate for key stakeholder participation and, ultimately, even wider participation to achieve maximum effectiveness. Government laboratories and Federally Funded Research and Development Centers are in a unique position to provide important technical support to Government C2 EA&A activities into the future by identifying opportunities for collaboration and helping to bring programs from across the services together. Technical excellence and breadth brought to bear on such a Government team will pay dividends. The Government must also advocate for proper resource allocation.

7.4.1 Government Role

From a C2 system acquisition perspective, the key Government players in EA&A are the project and portfolio managers, chief engineers, and staff engineers. Each has a vested interest in staking out roles in the analysis and assessment arena. What are those roles and why are they important?

First, from the viewpoint of a project manager, as someone who has direct contact with a customer (or set of customers), a primary goal is achieving a high level of overall customer satisfaction with respect to the quality of support and the products that are delivered to the operational users. As C2 systems become more functionally complex to address the emerging sophistication of threats, and as they move toward net-centric operations, it is becoming increasingly difficult to demonstrate the performance of any individual system to a customer or sponsor without the need to show added value in the proper context as well as appropriate interoperation with peer systems. In fact, increasingly, such peer systems will not just be those fielded by a military service (e.g., the Air Force), but by other services and Government agencies as the focus of military operations shifts to the employment of truly joint capabilities. Therefore, Government program managers and their chief and staff engineers will need to address how to establish collaborative approaches to demonstrating the functionality and utility of developed capabilities with other partners in a particular COI, at the very least. Thus, system assessment and evaluation approaches must consider significant enterprise partnerships right from the start of planning opportunities.

Second, from the viewpoint of the Government engineer supporting a particular project or program, the idea of C2 analysis and assessment at the enterprise level can seem overwhelming and even out of scope from an individual tasking perspective. However, it is often at the engineer level that opportunities for collaboration among systems and their developers can be identified and solidified. While there might be an intent for a group of programs or projects (even as few as two) to work together to demonstrate progress toward achieving a new or enhanced capability for a set of users, until the Government engineers from the different programs or projects get together to "peel back the onion" to uncover realistic methods of interaction, significant progress cannot be made. Ultimately, a sufficient amount of this type of collaboration among multiple partners can lead, through analysis, to the uncovering of potentially more effective and efficient means of interoperation at the enterprise level. In turn, this can lead to opportunities to validate proposed approaches via the prototyping of new technical frameworks by technical staff. This crucial step, from pair-wise interoperation to a realization of more general concepts for flexibly providing capabilities, enables desirable enterprise level effects to be produced. Another important contribution at the staff engineer level is the development of appropriate metrics for the customer, service, COI, and so forth, to measure the progress of fielding C2 capabilities. Only at the staff engineer level can such metrics be established, evolved, and quantified. Integration of operational users in the design and requirements definition process is critical to acquisition success.

Another consideration is the skill set required within a Government team of personnel formed to address C2 EA&A. The overall, team skill needed includes the following qualities:

- Big picture perspective
- Analytically oriented
- Technically competent within the COI to be assessed
- Programmatically astute across multiple programs
- Operationally knowledgeable on C2 issues
- Comfortable working in collaborative environments
- M&S background or experience, including distributed simulation
- Aware of applicable technologies

7.5 Next Steps

The vision for C2 EA&A is that users, developers, program managers, chief engineers, engineering staff, contractors, and decision makers within the DoD have a stake in C2 EA&A. The following are some key next steps to enable progress in the EA&A arena. First, it will be critical to enlist the support of a few forward-leaning C2 programs to collaborate on a pilot effort. It will be necessary to identify stakeholder programs with shared capability dependencies. In order to facilitate this, a key early effort will be to develop an implementation guide with practical suggestions for how to proceed given the constraints of existing programs. Such a toolkit will enable staff and management on a particular program to understand how to get started down the EA&A path in a practical manner. It should include suggestions for a variety of concerns, such as keeping infrastructure costs to a minimum.

Second, it will be important to leverage the lessons of DoD M&S-based training programs and commercial computer-based gaming over internet. These activities have focused on breaking down barriers to user participation. They have also been able to make strong connections with user needs and interests.

Third, it will be critical to emphasize EA&A as a new activity, not a new program. The education of key stakeholders to make them aware of the coming need for EA&A as well as their roles in the process will be fundamental to the success of EA&A activities.

Finally, it will be important to invest in critical technologies to enable proper implementation of C2 EA&A. There are a number of areas requiring further investigation to help enable the implementation of C2 EA&A. A candidate list of these areas is as follows:

- Multiscale analysis
  - How do C2 performance metrics collected at a lower scale relate to those collected at the mission effectiveness scale? How can we identify the sets of conditions producing threshold "tripwires" that produce changes in mission performance when crossed? How should we characterize these thresholds?
- Minimal, parallel EA&A infrastructure
  - What type of distributed infrastructure will allow collaborative investigation of C2 enterprise issues without bankrupting programs? Can this be used for incremental fielding of new C2 capabilities (seamless off-line to online transitions)?
- Lightweight, portable system representations
  - How can programs assure themselves that they have appropriate representations of stakeholder/partner system capabilities to properly perform independent or collaborative analyses?
- GIG and/or Airborne Network modeling/representation
  - What type of representation is required of operational communications infrastructure capabilities to permit robust EA&A by individual programs or groups of programs?
- Embedded analysis capabilities
  - How should analysis capabilities be embedded within C2 systems to permit EA&A to occur continuously, remotely, and without interference to mission execution? Can security issues be addressed in such a manner?
- SOA M&S approaches
  - How should an SOA implementation be modeled? Do commercial modeling initiatives (Business Process Execution Language, Business Process Modeling Language, etc.) or MITRE initiatives (Modeling Environment for SOA Analysis, etc.) show promise for EA&A needs?
- Executable architectures
  - How can voluminous architecture information collected by many programs be effectively used for real-time EA&A? Can current attempts to animate architecture tools scale to EA&A?
- Enterprise scalability analysis techniques
  - How can system capabilities be robustly assessed for the ability to scale to enterprise use? How can the viability of Service Level Agreements for DoD applications be assessed?

- Contracting mechanisms to enable continuous operational assessments
  - How can novel approaches to EA&A be incorporated in future C2 development contracts?

The modeling capabilities commercially available today are only beginning to address ways of analyzing enterprises based on an SOA construct and have focused to date on lower-level metrics and more static than dynamic kinds of analysis. Issues such as web service orchestration for C2 applications and end-to-end mission performance impacts of transitioning from dedicated, message-based, tactical data links to GIG communications relying on IP routing for data and C2 information distribution have not yet been worked from an analysis tool perspective. An appropriate M&S infrastructure approach needs definition and prototyping to permit useful analysis to occur at the C2 enterprise level.

As mentioned earlier, the notion of executable architectures that can be useful at the enterprise level for analysis requires effort dedicated toward addressing linkages to M&S tools and analysis frameworks. Some effort has been started by the Air Force in this area, but more work needs to be done.

The notion of lightweight, portable representations for specific C2 systems operating within an enterprise context must be further explored. In particular, such representations would be appropriate for sensor and communications systems and should be relatively inexpensive to develop (in fact, this would be a requirement). Some rapid software prototyping needs to be attempted for a candidate C2 system to validate the assumptions regarding the degree of utility of this approach for increasing collaboration opportunities across the enterprise.

Also, the notion of continuous operational assessments in support of T&E and other program objectives has implications from a contractual perspective. This concept might need to be inserted into future contracts or modifications to existing contracts through creative means until the benefits of such an approach become clear. If the benefits become tangible and well-understood, some standardization of contracting norms or policy decisions might make future implementations of this concept more robust or explicit.

7.5.1 Practical Considerations

In the process of performing EA&A, there will be a number of considerations that will come into play. In some instances, these will be programmatic in nature, policy, or guidance related, or with regards to the limitations of current technology.

Most likely there will be certain test and certification requirements within the C2 enterprise that will be independent of how the overarching C2 EA&A evolves. Those certifications that are required to ensure human safety or to protect against the possibility of litigation, for example, will be necessary and must be included within an enterprise perspective on assessment. These certifications must be periodically checked as the enterprise evolves, however, the frequency and scope of these checks must be determined over the course of time. With experience, the Government will better be able to realize how to ensure that safety, legal, and other such issues are properly addressed and maintained as changes continue to propagate through the C2 enterprise during its evolution. For the near term, however, the approved, existing methodologies for certifications of various types will need to be accommodated even though they might not align philosophically with emerging C2 EA&A approaches. In the near term, the goal would be to look for ways to use results generated in the different evaluation environments to support mutual objectives. In the far term, the routine incorporation of safety, litigation prevention, security, and so forth, processes and procedures into more continuous enterprise evaluations could actually reduce the overall risk of problem occurrences due to the fact that these certification checks would be performed much more often (or even automatically) as changes are introduced.

Configuration management (CM) and model verification, validation, and accreditation (VV&A) are activities that must accompany the execution of any enterprise capability analysis. Standard software CM processes should be completely adequate for tracking system versions as well as the associated software toolsets.

VV&A is another matter entirely. The often unrealistic VV&A requirements derived from the existing DoD policies for M&S are an impediment to M&S activities in support of EA&A, especially for analysis of future systems and enterprise operations (or existing systems in future operating environments), where these systems and capabilities do not yet exist. In such situations, modeled representations cannot be validated. Even current systems are normally fielded in multiple variants that undergo constant evolution, at best making any snapshot modeled representation and accompanying model VV&A only valid for a specific system variant at a specific point in time.

When attempting to identify and characterize fitness and emergent behaviors within a C2 enterprise, it is more important to have some reasonable representation of each crucial component (with appropriate loading considered) than to have a few highly detailed models that represent very specific instances of systems. A goal of "endorsed" EA&A activities can only be achieved if traditional model fidelity arguments are redirected instead toward discussions such as attempting to identify reasonable characterizations of the behavior of component systems and capabilities to satisfy the intended analytical purpose. This is a significant break from the current VV&A processes and guidance for M&S for systems. Although rigorous standards exist at the system/subsystem level within the DoD (with various service implementations and guidance), the practical implementation of model VV&A activities on every program is always tailored to fit schedule, budget, and technical considerations, such that no program ever implements the entire specified process. The law of diminishing returns applies quickly to current model VV&A practices. Realistic policies must be developed to address VV&A issues at the enterprise scale.

Finally, there is little or no guidance available within the DoD regarding C2 EA&A. Most of the existing policy and guidance is aimed at the system level (e.g.,

DoD 5000 series, Joint Capabilities Integration and Development System, etc.). In addition, there is no documented experience in performing EA&A from which to derive lessons learned. The practical impact of this fact is that there will be resistance to embarking upon a new activity for many programs. Funding for some of the precursor efforts will be a challenge. The lack of guidance in this area is part of the rationale for developing the vision for future C2 EA&A contained in this chapter. It is envisioned that early awareness on the part of a few key programs could spur some initial activity in the EA&A domain.

7.6 Summary

The vision for C2 EA&A is that users, developers, program managers, engineering staff, contractors, and decision makers within the DoD have a stake in C2 EA&A. An affordable, flexible, multiscale analysis capability will:

- Encourage continuous experimentation and virtual gaming, leading to rapid innovation
- Examine broad trade spaces of potential operating conditions to ensure robustness
- Create stakeholder understanding of expected C2 behaviors to evolve concepts of operation and positively influence the evolution of C2

Lightweight representations of emerging C2 capabilities can be published to readily accessible servers and made available to authenticated users for assessment when and where these evaluators deem necessary. Online/off-line context switching for operators will enable users to easily switch from their day jobs or training activities over to an environment that looks very close to their operational environment, except that the applications being exercised are next-generation C2 applications. The evaluation environment will use selectively imaged operational databases, augmented with simulated events to permit "apples-to-apples" comparisons of even such offerings as competing web services and service orchestration technologies.

Acronyms

C2	Command and Control
COI	Community of Interest
CM	Configuration Management
DoD	Department of Defense
DT	Developmental Testing
EA&A	Enterprise Analysis and Assessment
ESE	Enterprise Systems Engineering

GIG	Global Information Grid
HWIL	Hardware-in-the-Loop
IP	Internet Protocol
IT	Information Technology
M&S	Modeling and Simulation
MOE	Measure of Effectiveness
MOP	Measure of Performance
NR-KPP	Net-Ready Key Performance Parameter
OITL	Operator-In-The-Loop
OT	Operational Testing
SOA	Service-Oriented Architecture
T&E	Test and Evaluation
VV&A	Verification, Validation, and Accreditation

References

1. Rebovich Jr., G. *Enterprise Systems Engineering Theory and Practice, Volume 2: Systems Thinking for the Enterprise: New and Emerging Perspectives.* Technical Report, The MITRE Corporation, MITRE Product 05B0000043, November 2005.
2. Kuras, M. L. and B. E. White. "Engineering Enterprises using Complex-Systems Engineering." *Proceedings of 15th International Symposium, INCOSE*, Rochester, NY, July 2005.
3. Johansson, F. *The Medici Effect*. Boston: Harvard Business School Press, 2004.

Chapter 8

Enterprise Management

Robert S. Swarz

Contents

- 8.1 Recognizing an Enterprise..233
  - 8.1.1 Introduction 233
  - 8.1.2 What Is the System's Context? 234
  - 8.1.3 How Does the System Fit within the Context? 235
  - 8.1.4 What Are the End-Use Capabilities?..236
  - 8.1.5 How Does a System Enable the Enterprise? 237
  - 8.1.6 Summary 238
- 8.2 Strategic Management..238
  - 8.2.1 The Vision..238
    - 8.2.1.1 The Need for Assessment 239
  - 8.2.2 Goals 239
  - 8.2.3 Planning Approaches 242
  - 8.2.4 Recommended Strategic Management Approach.........................242
- 8.3 Strategic Planning 243
  - 8.3.1 Phase A—(Steps 1 and 2) Define Outcomes...............................243
    - 8.3.1.1 Step 1. Plan to Plan.. 244
    - 8.3.1.2 Step 2. Ideal Future Vision 244
  - 8.3.2 Phase B—(Step 3) Measure Progress toward Results 245
    - 8.3.2.1 Step 3. Goals and Metrics..245
  - 8.3.3 Phase C—(Steps 4 through 8) Strategies for Closing the Gap 245
    - 8.3.3.1 Step 4. Current State Assessment....................................245
    - 8.3.3.2 Step 5. Strategy Development 245
    - 8.3.3.3 Step 6. Business Plans...245

	8.3.3.4	Step 7. Implementation Plans	246
	8.3.3.5	Step 8. Plan to Implement	246
8.3.4	Phase D—(Steps 9 and 10) Implement and Assess the Plan		246
	8.3.4.1	Step 9. Strategic Implementation and Strategic Change	246
	8.3.4.2	Step 10	247
8.3.5	Phase E—Adjust to Changes in the Environment		247

- 8.4 Governance 247
  - 8.4.1 Enterprise Management of Capabilities 247
  - 8.4.2 Budgeting and Enterprise Management 247
  - 8.4.3 Roles and Responsibilities—Skill Sets 248
  - 8.4.4 Enterprise Business Analysts—A Key Role 249
- 8.5 Crosscutting the Enterprise 250
- 8.6 Risk Management 251
  - 8.6.1 Introduction 251
  - 8.6.2 Enterprise Risks and Opportunities 252
  - 8.6.3 Six-Step Enterprise Risk/Opportunity Management Process 253
    - 8.6.3.1 Step 1: Prepare the Risk/Opportunity Management Plan 253
    - 8.6.3.2 Step 2: Identify the Enterprise Risks 254
    - 8.6.3.3 Step 3: Identify the Enterprise Opportunities 254
    - 8.6.3.4 Step 4: Assess the Enterprise Risks and Opportunities 255
    - 8.6.3.5 Step 5: Develop Risk Mitigation and Opportunity Pursuit Plans 256
    - 8.6.3.6 Step 6: Implement the Mitigation and Pursuit Plans 256
  - 8.6.4 Program-Level Risk and Opportunity Process Impacts 256
  - 8.6.5 Tolerance for Change 257
    - 8.6.5.1 Scenario A 258
    - 8.6.5.2 Scenario B 259
- 8.7 Security Management 260
- 8.8 Hidden Costs of Enterprise Information Interoperability 262
  - 8.8.1 Synchronization Issues 264
  - 8.8.2 Association Issues 265
  - 8.8.3 Summary 267
  - 8.8.4 Conclusion 268
- 8.9 Configuration Management 268
  - 8.9.1 Introduction 268
  - 8.9.2 Configuration Items 269
  - 8.9.3 Policies and Processes for ESE CM 269
    - 8.9.3.1 Change Management 269
    - 8.9.3.2 Software CM 271
  - 8.9.4 Configuration Management Evolution 273
    - 8.9.4.1 Existing CM Structure 273

Enterprise Management ■ 233

    8.9.4.2 CM Evolution..273
    8.9.4.3 Necessary Steps for CM Evolution...................................274
 8.10 Infrastructure Management 275
   8.10.1 Enterprise Infrastructure Management Considerations.................276
    8.10.1.1 Technology...276
    8.10.1.2 People 277
    8.10.1.3 Process and Planning...278
   8.10.2 Enterprise Service Monitoring and Management..........................278
    8.10.2.1 Techniques for Monitoring the Enterprise
        Infrastructure 279
    8.10.2.2 Monitoring the Border Architecture 281
    8.10.2.3 Decentralized versus Centralized 281
   8.10.3 Beyond ESM—Information Technology Infrastructure
      Library (ITIL)...281
 8.11 Business Planning 286
   8.11.1 What Is a Business Plan?...286
   8.11.2 Example: A Business Plan for a Military Enterprise 286
 8.12 Technology Transition 287
 8.13 Summary 289
List of Acronyms 289
References 291

8.1 Recognizing an Enterprise

8.1.1 Introduction

This chapter lays out a spectrum of activities included in the strategic management (SM) of an enterprise. It explains how to extend traditional processes (e.g., vision, mission, goals, plans, and configuration management [CM]) to the enterprise level and environment. Finally, it discusses a simplified set of concepts to help provide guidance in recognizing an enterprise and potentially a larger enterprise of which it is a part.

 Systems engineers expect to fulfill an end-user requirement with a solution that potentially includes hardware, software, facilities, and people. The requirements are gathered through interviews, documents, and materials collected from the user community, the acquisition community, and the regulatory community. The technical requirements are defined in a specification that can be verified externally. The acquisition requirements are defined in a Single Acquisition Management Plan (SAMP) or other acquisition planning document, along with a host of supporting documents that demonstrate the acquisition office's capability to develop a system solution. The regulatory requirements are documented as references in the technical and acquisition documents, for example, technical standards and the Federal Acquisition Regulations (FARs).

The background and documentation requirements for supporting system acquisition, verifying the system requirements, and fielding the system are governed by documents such as Department of Defense Instruction (DoDI) 5000.2, *Operation of the Defense Acquisition System* and Chairman Joint Chiefs of Staff Instruction 6212.01C, *Interoperability and Supportability of Information Technology (IT) and National Security Systems*.

The concepts for building systems of systems (SoSs) are similar, but more complex. Several systems engineering (SE) and/or acquisition organizations may develop one or more of the systems that need to be integrated to create the SoS.

As a specific example, within a terminal approach air traffic control system, there are a set of related "systems" that need to interact to provide an integrated sensor-to-air traffic controller solution for small- to mid-sized airport terminal air traffic approach control (i.e., an air traffic control radar needs to integrate with an "automation system" to provide an integrated visual solution to the air traffic controller).

In defining an enterprise, the concept is expanded to span multiple organizations and multiple systems. An enterprise is composed of dynamic, heterogeneous entities some of which can be controlled and others which can only be influenced. It is comprised of people, systems, and processes. Furthermore, an enterprise can be made up of a collection of enterprises. This description assumes that an enterprise is an overarching concept that needs to be defined and managed through use of processes that may not be limited to engineering; for example, sociology or other social sciences may be needed to understand and manage the enterprise's behaviors.

8.1.2 What Is the System's Context?

The system's context is the manner in which it supports a mission or a set of missions and how the system interacts with other systems in support of the same mission. The mission is a set of goals and objectives that, when accomplished, will fulfill an operational need.

Extending the preceding air traffic control system example, there is an operational need to provide close air support for an Army battalion moving into the enemy territory. Elements required to fulfill the mission include:

1. An airfield and terminal
2. Air traffic control capability (e.g., radar, automation system, air traffic controllers, radio, or other communications with aircraft, etc.)
3. An air wing, including pilots, mechanics, mission planners, aircraft appropriate to their mission, fueling capabilities, and the infrastructure to support the close air support mission*

---

* The infrastructure is both operational (e.g., local area network [LAN] or computer workstation) and nonoperational (e.g., shelter, test equipment).

4. Mission planning and control information
5. Mission communications
6. Information/intelligence gathering and dissemination

All these elements are used with a mission plan to provide the support to fulfill the operational need.

In this example, the role of the enterprise is to enable the close air support mission using the set of systems (or elements) that support the mission.* The enterprise, in this particular example, is "Combat Support." The goal is to provide seamless close air support to the Army battalion within the larger context of combat support. The enterprise crosses both system and organizational boundaries. For example,

1. The airfield and terminal may belong to a non-United States (U.S.) government ally.
2. The terminal approach air traffic control capability may be provided by either U.S. or foreign air traffic controllers using either a foreign- or U.S.-owned air traffic control system (radar, communications, automation).
3. The air wing is provided by the U.S. Air Force (USAF).
4. The information/intelligence gathering is provided by the Army and the U.S. intelligence community through both on the ground assets and intelligence satellite information.
5. The information/intelligence dissemination is from the intelligence community to the mission planning and control capability.
6. The mission planning and control information relayed from a command post comes from the Army and the Joint forces supporting the Army battalion.
7. The mission communications capability is provided by the Army, any Joint forces supporting the mission, and the USAF.

Other enterprises (e.g., command and control) could interact with the combat support enterprise to fulfill the mission.

8.1.3 How Does the System Fit within the Context?

How the system fits within the context refers to the interactions and interfaces with the other organizations, systems, and subsystems required for the mission. This could be analogous to a high-level operational view with organizational and functional interface backup information (Figure 8.1).

From this information and inputs from the combat support community, it is possible to derive the information exchange requirements (IERs) to more clearly define the relationships among the elements of the enterprise. Traditionally, the IERs are documented in a concept definition document (CDD) that could support

---

* The system in this case could be any system, SoS, or subsystem that supports the mission.

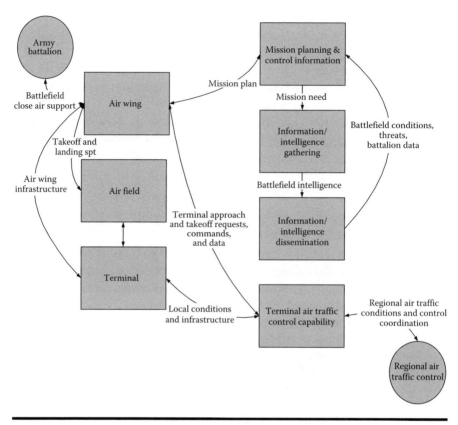

Figure 8.1 High-level interface block diagram for close air support.

a subsystem, system, SoS, or enterprise. A CDD defines the operational needs to be fulfilled by a technical and operational solution—however complex.

The terminal air traffic control capability (or the sum of the systems that implement this capability) relies on a number of inputs from internal and external entities to enable it to ensure safe terminal control conditions for aircraft within its air space; however, it is also clear that the air traffic control capability does not need battlefield information to provide that support. Although the context of the air traffic control capability is local to the airfield, the terminal, and the air wing while in terminal airspace, it still enables the larger combat support enterprise.

8.1.4 What Are the End-Use Capabilities?

The end-use capabilities for a system within an enterprise are those operational needs a system fulfills as it helps enable mission completion. The end-use capabilities for the enterprise are the sum of the operational capabilities that allow successful mission completion.

In the example, the terminal air traffic control capability provides the following:

1. Local air traffic conditions to the members of the air wing
2. Takeoff, landing, and other air traffic commands to the members of the air wing

The terminal air traffic control capability receives the following:

1. Regional air traffic conditions and control coordination from the regional air traffic control authority
2. Local conditions and infrastructure from the terminal
3. Takeoff and landing and other air traffic control requests from the air wing

To fulfill its role within the enterprise, the air traffic control capability must interact with regional air traffic control, the air wing, and the terminal. It must react within three organizations to provide one end-use capability, which is to control members of the air wing within the terminal airspace.

Therefore, the end-use capability for terminal air traffic control within the combat support enterprise is to control members of the air wing within the terminal airspace. One end-use capability of the combat support enterprise is to provide close air support to an Army battalion in the battlefield. Other end-use capabilities for the combat support enterprise are outside of this example.

One challenge in enabling a system or element to support an enterprise is to manage the interfaces among the systems and entities that support the mission. The interfaces can be

1. Physical, for example, power, light, equipment housing, network, communications
2. Data and control, for example, weather, regional conditions
3. Organizational, for example, air wing to air traffic control, USAF to Army, USAF to foreign ally
4. Cultural, for example, U.S. military and civilian personnel interactions with foreign allies

All the interfaces need to be managed successfully, and the infrastructure and entities need to interoperate to enable the end-use capability in support of an enterprise.

8.1.5 How Does a System Enable the Enterprise?

A system enables the enterprise by providing end-use capabilities in support of the operational mission. In the air traffic control example, the end-use capability is to control members of the air wing in the terminal air space.

To provide that capability, the air traffic control "systems" need to process data (e.g., aircraft identity, position, heading, and speed, weather, etc.), disseminate, and display that data in a way that the air traffic controllers can understand and use, and provide a capability for the air traffic controllers to communicate with the pilots to control the aircraft in the terminal air space.

The challenge from the enterprise viewpoint is to allow the elements or systems within the enterprise to interact in such a way that each of their end-use capabilities provide inputs that enable successful mission completion.

8.1.6 Summary

An enterprise results from the interplay of all the elements (systems, subsystems, information, organizations, processes, and people) that fulfill a mission, where a mission is defined to meet a significant operational need. Any number of elements may be used, but mission success depends on managing the interfaces among the elements that are joined to create the enterprise. An enterprise is likely to be dynamic and open to change depending on the geographic, social, and political variables.

In the example used in this section, the combat support enterprise could be quite different if the entire infrastructure for that support was controlled by a single organization. The complexity comes from the variables that require technical and social solutions to enable the interfaces among the elements to interact and provide end-use capabilities that allow mission completion.

8.2 Strategic Management

8.2.1 The Vision

A strategically aligned organization focuses on and measures outcomes, not outputs. Leadership needs to manage change and move an organization forward to achieve the organizational vision; buy-in across the organization is essential (Haines, 2000).

An important part of the SM process is to develop a strategic plan that includes key components—vision, mission, goals, strategy, and actions.

Effective organizations institutionalize the strategic plan and move from a focus on near-term tasks (reactive) to align with the vision. This is the crux of classical strategic planning (e.g., the balanced scorecard approach); however, enterprises are different:

- They are nondeterministic.
- They are capable of complex behavior.
- They exhibit emergent properties.
- Decisions are made according to blueprints of cultural norms.
- They can be counterintuitive (playing the game changes the game).
- They tend to select not the best solution but the one most compatible with the environment.

Therefore a question is, "In this environment does strategic thinking have the same value, and if so, what form does it take?"

8.2.1.1 The Need for Assessment

To work effectively toward an enterprise vision, it is important to have a clear understanding of its current state. One cannot get to a destination without knowing one's current location.

While the statements made above may appear obvious, there are examples of failed endeavors not due to a lack of vision, but due to a lack of self-awareness. The "dot com collapse" that occurred at the turn of the twenty-first century is an example. As the Internet expanded in the 1990s, many saw a new paradigm for doing business that would replace the traditional consumer experience. Numerous start-up companies banked their existence on the notion that consumers would utilize the Internet to purchase goods and services, eschewing traditional "brick and mortar" stores. The theory was, as said in *Field of Dreams*, "If you build it, he will come." Few looked at the current state of affairs at the time. Consumers in general still wanted to go to a store to purchase many items, where they could physically see them, speak with others about them, and walk down the aisle to pick out related and other products while they were there. The Internet for the masses was still in its early stages of growth, with most of its users accessing it through dial-ups, which made transactions relatively slow and unreliable. While Internet usage was expanding rapidly, many of the new users were inexperienced with computer equipment, and uncomfortable with submitting personal information (e.g., credit card numbers) over the Web. There were other factors that contributed to the bursting of the dot com bubble, but the current state of the "Enterprise" was a major one.

8.2.2 Goals

The organization must effectively align resources and activities. To achieve strategic alignment, the organization should manage its activities and processes as projects to make the strategic goals a reality. These goals are listed below.

GOAL 1. Develop and maintain a secure, interoperable net-centric infrastructure.

Description: Plan, develop, implement, operate, and sustain a global information infrastructure to provide secure interoperability and end-to-end connectivity to all our employees and customers. This infrastructure's common architecture and technical standards will ensure that the component of the global information grid (GIG) is a critical element to maintaining interoperability with our partners.

Supporting Objectives:

1. Develop architecture, standards, and protocols.
2. Ensure that the network enterprise infrastructure is interoperable with the GIG and the enterprise architecture.

3. Develop the standards, architecture, and execution plan for server consolidation throughout the enterprise.

GOAL 2. Transform applications and data into web-centric capabilities.

Description: Transition legacy applications into functional area application portfolios to provide an authoritative set of capabilities that will increase effectiveness and efficiency. Create an enterprise portfolio of web-centric solutions to be used across the organization, resulting in operational efficiencies across business activities. Enterprise-wide, cross-functional integration of solutions improves responsiveness, enhances agility, increases the speed of decision making, and enhances mission performance.

Supporting Objectives:

1. Leverage an enterprise portal framework to provide access to enterprise applications, web services, and authoritative data sources.
2. Create a data environment to enable sharing of information across the enterprise.
3. Continue to rationalize, consolidate, and eliminate legacy applications on all networks.
4. Establish and manage functional area portfolios and an enterprise portfolio containing applications.
5. Maximize the use of enterprise agreements in acquiring products and services.

GOAL 3. Provide full-dimensional protection (FDP) that ensures effectiveness.

Description: Protect and defend people, information resources, and critical infrastructures to provide for the continuity of operations. The security and protection of systems, networks, and information depend on the implementation of sound information assurance (IA) concepts and principles across programs and platforms. The implementation of critical infrastructure protection (CIP) measures protects, defends, and secures mission-critical capabilities. The anticipation of threats, elimination of vulnerabilities, and employment of self-defense protection strategies minimize risk to effective network-centric operations.

Supporting Objectives:

1. Develop plans, policies, architectures, and guidance to implement FDP for CIP, IA, and privacy.
2. Ensure that IA is integrated into programs and infused into all communities.
3. Provide integrated IA situational awareness to enable a shared understanding among decision makers for computer network defense (CND).
4. Employ defense-in-depth strategies to defend systems and networks to ensure that access is controlled and that all systems and networks are capable of self-defense.

5. Define data protection requirements for net-centric operations, and develop and deploy robust protection mechanisms across the enterprise.

GOAL 4. Ensure that net-centric capability investments are selected, resourced, and acquired to optimize mission accomplishment.

Description: Select net-centric capability investments that improve capability, mission readiness, and mission performance. These investments are assessed, qualified, and validated as part of the planning, programming, budgeting, and execution (PPBE) process and permit best use of scarce resources. An enterprise implementation plan (EIP) that outlines net-centric capability investment priorities serves as the basis for approval of initiatives and budget allocation. The results of investments are assessed for the contribution to performance goals and objectives.

Supporting Objectives:

1. Implement a net-centric capability capital planning process that validates net-centric capability requirements as part of the budget process and measures the value of net-centric capability investments in meeting mission requirements.
2. Develop and implement a net-centric capability EIP that links vision and strategy to planning, programming, and budgeting guidance and serves as the basis for approving the funding and procurement of net-centric capability initiatives.
3. Develop and monitor performance measures for net-centric capability investments.
4. Develop and implement a standard total cost of ownership model for net-centric capability investments.

GOAL 5. Create processes and integrated systems that enable knowledge dominance and transformation.

Description: Integrate policy, processes, and technology to share knowledge for decision making.

Supporting Objectives:

1. Transform and streamline business processes and supporting information systems.
2. Enable sharing of knowledge among all decision makers.
3. Streamline net-centric governance structures to ensure agile decision making.
4. Develop and refine net-centric policies and processes for fielding integrated and interoperable capabilities.

GOAL 6. Shape the net-centric workforce of the future.

Description: Expand personnel capabilities by strengthening their knowledge, skills, and abilities in managing technology, processes, and information. Information management and IT competencies will enhance personnel capabilities. Professional development improves the performance of the net-centric workforce.

Supporting Objectives:

1. Transform processes used to identify, recruit, manage, and sustain the net-centric workforce.
2. Identify and sustain the required competencies and capabilities in the net-centric workforce to meet current and emerging requirements.
3. Identify, develop, and provide relevant foundational net-centric education and training for all personnels.
4. Provide traditional and nontraditional professional development opportunities for the workforce.
5. Assess and improve workforce health, capabilities, and performance.
6. Apply strategic sourcing management strategies that result in an effective and efficient team.

8.2.3 Planning Approaches

Dr. Russell Ackoff (1984) describes the three basic planning approaches: reactive, preactive, and interactive. An interactive planning approach is recommended. Reactive planning is bottoms-up, tactically oriented planning. What strategy it contains is implicit, a consequence of numerous, independently made tactical decisions. Preactive planning is top-down strategically oriented planning. Objectives are explicitly set, but tactics are left to the discretion of the individual units. Interactive planning is directed at gaining control of the future. It is based on the belief that an organization's future depends at least as much on what it does between now and then as on what is done to it. Therefore, this type of planning consists of *the design of a desirable future and the selection or invention of ways of bringing it about as closely as possible.*

8.2.4 Recommended Strategic Management Approach

High-performing organizations focus on the external, the strategic, and change. Adopt an SM approach that supports all three of these key factors in order to become a high-performing organization. Use an interactive SM approach that supports the current needs of SE and can support future evolution into enterprise systems engineering (ESE). The ideas in *The Systems Thinking Approach to Strategic Planning and Management* (Haines, 2000) were chosen as the basis for the SM process. What follows is quoted or paraphrased from the author, Stephen Haines' work. Haines' intellectual mentor is Russell Ackoff.

"Strategic Management also recognizes that planning and change are the primary responsibilities of senior management with the participation of the whole organization in an interactive planning, execution and assessment process. SM can be defined as the competencies of understanding and achieving three top priority goals:

1. Develop strategic and annual plans and documents.
2. Ensure their successful rollout, implementation, and change.
3. Build and sustain high performance over the long term."

"Going beyond strategic planning into Strategic Management means making a commitment throughout the organization to ongoing strategic thinking and continuous improvement. It means accepting that no one plan can possibly anticipate and resolve every need, that the organization must have in place a Strategic Management process for planning and dealing with all organizational change and growth, now and in the future, and at all levels of management."

"A Strategic Management process is a comprehensive, interactive, and participative system that leads, manages, and changes the total organization in a conscious, well-planned, and integrated fashion based on core strategies. It uses proven research that works to develop and successfully achieve the ideal future vision."

8.3 Strategic Planning

The overall process for enterprise-level strategic planning is shown in Figure 8.2.

Systems thinking is a "new orientation to life." It is about thinking backward from a desired outcome, determining where the organization is now and then finding the core strategies or actions that lead to the desired outcome. This approach employs five critical questions or phases that serve as locator points to clarify this thought process.

- Phase A: Where do we want to be (i.e., our ends, outcomes, purposes, goals, holistic vision)?
- Phase B: How will we know when we get there (i.e., customers' needs and wants connected into a quantifiable feedback system)?
- Phase C: Where are we now (i.e., today's issues and problems)?
- Phase D: How do we get there (i.e., close the gap from C to A in a complete, holistic way)?
- Phase E: What will or might change in our environment in the future? (This is an ongoing question.)

8.3.1 Phase A—(Steps 1 and 2) Define Outcomes

Concentrate on defining outcomes first. Ask the question: "Where do we want to be?" This is the starting point for putting our systems framework into place by focusing on the outcomes our customers desire from us, envisioning the future as if it were today, and then working back to the present.

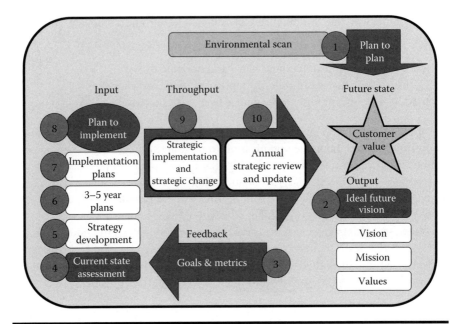

Figure 8.2 Strategic management process—10 Step Model (From Haines, S. G., *The Systems Thinking Approach to Strategic Planning and Management*, CRC Press LLC, Boca Raton, FL, 2000. With permission.)

Outcomes are what the customer wants to accomplish with a product or service. Outputs are what are provided to the customer, frequently defined by formal requirements. How frequently does meeting formal requirements help customers realize their goals? No one says all the time, some say seldom, a few say never! The more focus on desired customer outcomes, not formal requirements, the greater the chance outputs will contribute to customers realizing their goals.

8.3.1.1 Step 1. Plan to Plan

- Set up a core planning team and strategic steering committee to carry out SM.
- Involve key stakeholders in a parallel process to help communicate SM at every step in the process because people will support what they help create.
- Clarify top management's role in leading, developing, and owning the SM process.

8.3.1.2 Step 2. Ideal Future Vision

- Focus on the desired outcomes of your customers to help formulate an ideal vision of the future. Vision is the ideal future that you will strive to realize in the long term by accomplishing the mission daily.

- Define the mission. Our mission is what we are charged to do and what we do daily. It is the starting point from which we will begin our journey of change toward our future vision and the vision is where we will go guided by our values.
- Define core values and align them with the enterprise's values. Values are principles held and qualities cultivated to help realize the mission and vision. Align them with the enterprise's values to serve as guides for behavior at all times.

8.3.2 Phase B—(Step 3) Measure Progress toward Results

Establish a quantifiable feedback system to measure progress toward your desired outcomes. Ask the question: "How will we know when we get there?" It is crucial to develop concrete feedback; we need to define our outcome measures of success. This is how we will be able to gauge whether the implementation of our core strategies is progressing successfully; it is where we will determine the success annually.

8.3.2.1 Step 3. Goals and Metrics

Establish goals to gauge outcome measures of success. Goals are the important things that must be accomplished to complete the mission and realize the vision. Metrics are the means to measure the actual outcome of the goal.

8.3.3 Phase C—(Steps 4 through 8) Strategies for Closing the Gap

Determine where we are now. This is the step in which strategies and actions are designed for closing the gap between our current state and our vision of the future, with specific action priorities necessary to support them. Develop the strategic plan. Allocate actions for business plan development. Develop implementation plans for the following year's work program, including resource allocation.

8.3.3.1 Step 4. Current State Assessment

Analyze strengths, weaknesses, opportunities, and threats (SWOT) and contrast them to the vision.

8.3.3.2 Step 5. Strategy Development

Develop strategies to bridge the gaps identified with analysis between the current state and the vision.

8.3.3.3 Step 6. Business Plans

Work with the business units and functional units to develop 3–5 year plans that achieve the goals by applying the strategies to realize the vision. These business plans should be developed interactively with the strategic plan through a parallel planning process.

8.3.3.4 Step 7. Implementation Plans

Develop annual plans to implement specific action items identified in the strategic and business plans.

- Allocate resources and tasks and set priorities for action during the current year.
- Use the strategic implementation plan as input to the SM throughput process.
- Use the business implementation plans as inputs to operational processes.
- Synchronize the strategic implementation plan and business implementation plans.

8.3.3.5 Step 8. Plan to Implement

The strategic steering committee, with the help of the core planning team:

- Educates and organizes senior management for the strategic change management process
- Develops a comprehensive map to guide the changes directed by the strategic plan
- Develops a rollout and communications plan

8.3.4 Phase D—(Steps 9 and 10) Implement and Assess the Plan

Determine what actions we need to take to reach our vision. Ask the question: "How do we get there from here?" Put plans into motion—tracking, monitoring, reporting, and adjusting as necessary.

8.3.4.1 Step 9. Strategic Implementation and Strategic Change

- Transform organization, guided by the strategic plan, business plans, and implementation plans.
- Monitor and track all plans.
- Report progress and issues to the strategic steering committee.
- The strategic steering committee should review progress and issues at least monthly or more often under exceptional circumstances.
- Make adjustments to plans according to monthly reviews.

8.3.4.2 Step 10

Review and update strategic and business plans by reiterating the SM process annually.

8.3.5 Phase E—Adjust to Changes in the Environment

Continuously scan the environment. Ask the question: "What is changing in the environment that we need to take into account?" This is where the process begins to repeat itself and we concentrate on outcomes to address the environmental changes that affect us and our customers.

8.4 Governance

8.4.1 Enterprise Management of Capabilities

To achieve enterprise integration and interoperability, active governance must be in place, with leadership and ownership across programs and authority to make investment changes. Effective governance at the enterprise level is a necessity.

Systems deal with systems requirements: "what" shall be built. Enterprises deal with enterprise capabilities, that is, measure the value of an investment in a capability to users of the capability. This value can be delivered from one "system," across systems in a portfolio, across portfolios, even across different military services to include Joint and coalition. Governance is required to achieve a capability.

Department of Defense (DoD) and the services have put in place capability management organizations to define the roles, processes, and products of capabilities-based planning, capabilities-based requirements, and capabilities-based acquisition.

8.4.2 Budgeting and Enterprise Management

The budgeting process is complex and experts should be engaged to ensure that required capability needs are integrated into the budgeting cycle. Besides experts to assist with planning, many good references are available to ensure that enterprise-wide planning is used. Enterprise engineers need to be aware of the PPBE cycle to ensure that planning for the enterprise aligns with funding.

The planning, programming, and budgeting system (PPBS) was established in the 1960s when a programming business perspective integrated an existing disjointed planning and budgeting system. Since 1962, resources are systematically allocated and aligned with roles and missions of the DoD. As part of the PPBE, DoD created the future years defense program (FYDP) that provides visionary long-term focus for the defense program planning.

In 1986, a 2 year PPBE cycle was established (also called the biennial PPBE):

1. In an even year program, the DoD produces a program baseline (PB) based on the military services and defense agencies program objectives and budget estimates.
2. In the odd years, the military services and defense agencies reprice their submission and then submit an amended budget estimate to the Defense Department which subsequently develops an amended PB.

In actuality, however, there has been little difference between an even year and an odd year except that the FYDP in an odd year is only 5 versus 6 years in an even year.

In 2001, the Secretary of Defense combined the program objectives and budget estimates into a single submission from the military services and defense agencies. In 2003, biennial budgeting was reinvigorated, creating a 4 year program and budget cycle with two "on years" in which program objectives are submitted and two "off years" in which the services do not create program objectives but can still make critical program adjustments.

8.4.3 Roles and Responsibilities—Skill Sets

Managing the enterprise requires core "technical" competencies to fulfill enterprise roles: relationship management, enterprise integration, capability planning, acquisition support, technology transition, and domain expertise. Supporting the enterprise is a technical role that expands our skill sets beyond best technical solutions, but challenge us to provide a balance of strategic planning, business and investment analysis, horizontal (social and technical) engineering, and technology skills.

An enterprise team needs to consist of a balanced mix of these skills, depending upon the strategic and business plans. Teams are composed of program managers (PMs), domain experts, enterprise engineers, systems engineers, task engineers, and acquisition support specialists.

General definitions of skill categories to support enterprise roles follow:

- Relationship management
  - Collaboration and teamwork across organizations
  - Change management
- Enterprise integration
  - Strategic planning
  - Policy, standards, and guidance
  - Metrics and measures
  - Business planning
- Capability integration
  - Capability planning and analysis
  - Investment planning and portfolio analysis

- Capability roadmapping
- Program Objective Memorandum (POM) focus
- Capability solution studies
- Engineering and architectural analysis
- Acquisition support
  - Capability-based acquisition support
  - Innovative acquisition
  - Technology transition
  - Technology analysis and planning
  - Technology vision and roadmaps
  - Technology assessment
- Domain expertise
  - Warfighter
  - Joint and Coalition
  - Army, AF, Maritime
  - Intelligence

8.4.4 Enterprise Business Analysts—A Key Role

Enterprise engineering organizations need to include "enterprise business analysts (EBAs)." This is a technical role. EBAs have a business orientation and keep their pulse on the business. An EBA understands how to communicate and interact with business partners, probe for key business content, analyze the results, and present the results to the business and technical communities. Representation of results is generally done using a formal methodology (e.g., unified modeling language [UML]). Solutions may be materiel; other doctrine, organization, training, materiel, leadership and education, personnel, and facilities (DOTMLPF) approaches; or a combination. The EBA does much more than write down the requirements, and is responsible for maintaining an enterprise perspective and focuses on each business area as part of the broader whole.

Fred Cummins (2002), in his book, *Enterprise Integration: An Architecture for Enterprise Application and Systems Integration*, calls EBAs "knowledge workers" to address the need to manage enterprise intelligence. It is no longer enough for management to make decisions and provide only chain of command leadership. To manage the complexity of the business and support decision makers, management needs to rely on knowledge workers for enterprise expertise. (This author proposes that the term "knowledge worker" implies a high-end user, and that Cummins identifies EBAs, a technical role with unique skill sets to understand the business and identify business solutions including technology.) The EBA should move us from functional stovepipes to an enterprise perspective.

Cummins says "The enterprise is managed not by functional islands, but by a collective enterprise intelligence that is integrated across business functions. Top managers define the business, provide the vision, and create the incentives, but they do not have the expertise or the time to participate in every decision that affects the operation or future success of the enterprise.... Enterprise intelligence is the integration of people and systems, sharing information, collaborating ... [responding] intelligently."

Cummins is right in that knowledge workers are at all levels throughout the enterprise, identifying and solving problems, sharing insights and combining skills. The EBA is a specialized knowledge worker who is a business and technical specialist working in the enterprise space (separate from but in support of the business of the enterprise), across the enterprise, leveraging and understanding enterprise knowledge and interpreting knowledge into enterprise intelligence. The EBA has a unique and necessary role in the enterprise, interacting between the enterprise and the knowledge workers. We need to understand and staff this role in order to properly manage the enterprise.

8.5 Crosscutting the Enterprise

Duplication of effort is an inefficiency that comes from enterprise complexity. If two organizations within an enterprise need to design a system that performs the same function for each of those organizations, each will typically engineer its own solution without consulting the other organization. In many situations, the two organizations will not even be aware that they have a common need. One of the challenges of enterprise system engineering is identifying such situations and creating the bridges between organizations that will foster common solutions. Crosscutting the enterprise in this way is a part of enterprise management because the most difficult aspects are getting people to talk with each other, reducing the tendency for each organization to protect is own turf, and fostering the sharing of ideas and solutions.

From the point of view of a given organization, the easiest path to a solution is often to constrain the problem by focusing on the organization's own requirements and developing options for comparison, selection, and further refinement. When another organization's solutions are thrown into this process, the human reaction is typically defensive. It can be cumbersome to consider additional solutions, especially if those solutions address slightly different requirements or are at a different stage of the acquisition cycle. This is how stovepipes are born.

Astute enterprise managers recognize the inefficiencies of stovepipes and look for ways to tear down the communication barriers and encourage people to learn from each other and search for common solutions to common problems. If one organization is ahead of another in the acquisition cycle, then there may be opportunities to share lessons learned that benefit the following organization. If the two organizations have been examining different technical solutions, then there may be opportunities for cross-fertilization of their ideas.

A crosscutting solution may not always be the most efficient. Some solutions may be less expensive in the short term and more expensive in the long term, and the reverse may be true for another organization. Business cases are clearly needed. By developing business cases for all solutions, the enterprise manager can determine whether a crosscutting solution is in the long-term interest of the entire enterprise rather than serving more narrow objectives. Sometimes the choice of a crosscutting solution will require compromise on the part of one or more organizations. In these cases, the enterprise manager must thoroughly identify with the positions of all the stakeholders and make them all into advocates for the enterprise even if their own short-term interests are not served.

One dilemma is in deciding *when* to bring together organizations to work on a crosscutting solution. If organizations collaborate at the beginning of development, there is a good chance that stakeholders will be more open to the solutions of others. However, organizations usually do not start developing a particular capability all at the same time and have different development cycles once they do get started. An additional complication is that solutions tend to be broad concepts at first and evolve during the development process to address increasingly specific requirements. In the broad concepts phase, a solution may lack the specificity needed for some types of collaborative problem-solving. The obvious time for crosscutting collaboration is early and often. As soon as a crosscutting potential arises, the enterprise manager should begin bringing stakeholders together—and keep bringing them together until the cross-cutter either fails its business case or has reached implementation.

Crosscutting the enterprise is a never-ending task. External drivers change, organizations and systems within the enterprise change, and people change not only their job positions but also their attitudes toward other stakeholders and their understanding of key issues. The exchange of information among the people spread across the enterprise greases these changes and makes the enterprise run more smoothly.

8.6 Risk Management
8.6.1 Introduction

A critical component of enterprise management is an enterprise risk management program. With proper execution and management of such a program, an enterprise manager should be in a better position to assess trade-offs, capitalize on opportunities, mitigate risks, and make decisions to evolve the enterprise. Within the enterprise risk management program, an enterprise risk assessment process must be implemented. At a minimum, key attributes of the process include: structured, forward-looking, comprehensive, and continuous or iterative. Additionally, the assessment process should allow for the insertion of opportunities as well as risks. Note that an opportunity to evolve the enterprise may pose a risk to a particular program and the converse may also hold true. Thus, feedback of the decisions at the enterprise level need to flow down to programs within the enterprise, and, similarly, programs within the enterprise need to ensure the enterprise level is aware of impacts and potential consequences, as well as advantages.

8.6.2 Enterprise Risks and Opportunities

The unique characteristics of an enterprise lead to risk types and opportunities that may not normally be encountered with systems or SoSs. This is because the compatibility between systems and their mutual interactions creates a dynamic that may be an order of magnitude greater than the sum of the individual system's dynamics. Many enterprise risks and opportunities may occur at the program level also. For example, requirements risks exist at both program and enterprise levels; however, at the enterprise level the probability and impact may be more severe because of complexity and scale. Thus, additional effort may be required to identify resolutions or mitigation strategies to meet the needs of the enterprise.

The complexity of an enterprise may be dependent on the uniqueness of the system(s) supporting it. For example, a system might have a single acquisition program office and primary contractor with a limited number of user communities. SoSs are likely to have different program offices and user communities. Joint programs, where one service dominates the acquisition, and cooperative programs, where services share responsibility and control, are more complex because of the diversity of the user community. International programs add cultural complexities. Thus, because we are dealing at the enterprise level, the number of stakeholders and their objectives vary so greatly that the scale/scope in complexity of the risks are far less intuitive, predictable, and manageable. At a minimum, the risks and opportunities listed below should be considered when assessing the enterprise.

The complexity of an enterprise system may lead to:

- Unexpected costs and inefficiencies
- Unidentified and unexploited opportunities for savings and efficiencies
- Difficulty in assessing analysis of alternatives
- A lack of or uncertainty in agreements
- Difficult decision making due to the potential lack of clear hierarchy and authority
- Lack of a clear or homogeneous stakeholder community
- Having to provide for additional stakeholders (opportunity to expand capabilities)
- Stovepipes or the lack of overarching control
- Conflicting requirements, schedules, and so on, or more interoperability problems or more interoperability opportunities
- An increase in political challenges or an increase in the complexity of all the "ilities"

The relative scale/size of an enterprise system and its systems may lead to:

- Opportunities to support expansion and contraction
- Multiple and possibly conflicting cultural views that must be addressed or the inability to satisfy a broad set of stakeholders
- The opportunity of economies of scale
- Unplanned, and possibly unknown, overlaps or conflicts of implementations
- Creation of high deployment and transition costs
- An increase in sustainment costs or poorly defined or diverse interfaces
- Stovepipe problems
- Interoperability problems

Planning becomes difficult. The unknown and causal relations in an enterprise system may lead to:

- Uncertainty in the cost, schedule, delivered capability, and unexpected consequences of developing and delivering a new system
- Breakage of many traditional SE life cycles
- Continuously evolving requirements or uncertainties of future requirements or uncertainty of schedule
- More room for conflict or unanticipated reconfiguration of systems or the inability to predict
- System dynamics issues or unknown behaviors
- Unimagined capabilities
- More opportunity-like activities
- Capitalize on operational trends
- Changes in leadership

8.6.3 Six-Step Enterprise Risk/Opportunity Management Process

At a high level an enterprise risk/opportunity management process is similar to any other risk management process. However, the emergent behaviors and complexity of an enterprise may surface additional opportunities and risks that may not be considered at the program level.

8.6.3.1 Step 1: Prepare the Risk/Opportunity Management Plan

The first step of an enterprise risk/opportunity management process is to plan the activity. An enterprise manager may begin this process by selecting the team, which should include core representatives from all areas of the enterprise, not just technical experts or the PMs of the various programs included in the enterprise. Representatives should be from across the enterprise, including users, PMs, cost analysts, technical and domain experts, schedule analysts, and contractual representatives. As the enterprise evolves, noncore team members may shift in and out of the risk assessment process. As the complexity of the enterprise grows, it will become important to establish guidelines on who may participate, who may "vote," and so on. This will become more important in a joint or international environment where the perception of fairness is easy to distort. The team sets forth the ground rules, especially with more complex enterprises. Ground rules for even routine activities such as terminology and agendas should be agreed by all the participants. If, for example, a contractor uses a different tool than the government, there should be agreement on how to interface the tools and share the information so that everyone is addressing the most pressing risks and significant opportunities. Further rules should be set on timelines for assessments, the meanings of priority settings and consequences, and rules for handling risks/opportunities that apply to only a portion of the enterprise. The enterprise manager, along with support from the team, identifies the key enterprise objectives or requirements that should assist the team in identifying risks. Enterprise objectives or requirements may conflict or not support individual program objectives or requirements. Everyone should be aware of the possible conflict of interests and be prepared to discuss. In the end, mitigating a risk or taking advantage of an opportunity at the enterprise level may be beneficial to the enterprise but extremely damaging to a particular system(s) or program(s). Enterprise assessments highlight these potential conflicts and provide a forum to discuss the pros and cons before a decision is made.

8.6.3.2 Step 2: Identify the Enterprise Risks

A facilitator leads the team through a structured brainstorming process to identify enterprise risks—those risks that cut across/impact the enterprise. Additionally, enterprise risks may be input to the process from program-level risk assessments that have the potential to impact the enterprise. Collaboratively, the team identifies complete risk statements for each risk.

8.6.3.3 Step 3: Identify the Enterprise Opportunities

The Industrial Research Institute defines opportunity as "A business need or technology gap that a company or individual realizes, by design or by accident, that exists between the current situation and an envisioned future in order to capture competitive advantage, respond to a threat, solve a problem, or ameliorate a difficulty."* Identifying opportunities at the enterprise level will be a complement

---

* Industrial Research Institute, Inc., 2200 Clarendon Boulevard, Suite 1102, Arlington, VA 22201.

to identifying risks at the enterprise level. Opportunities are identified by those at the strategic and/or tactical level of the enterprise and then managed by the enterprise manager in accordance with the enterprise risk/opportunity management process. It is important that the manager understands opportunities when they arise. With robust system engineering, systems will be designed to accommodate growth in opportunistic capabilities that are identified throughout the life cycle. Some opportunities may be identified as the enterprise is planned and developed; however, it must be expected that there will be unanticipated, emergent events that present both opportunities and risks. Opportunity identification must take into consideration a spectrum of economic, cultural, industry, corporate, scientific, and leadership considerations to capture as many opportunities as possible. Naturally, not all opportunities will be acted on, yet the list of all potential opportunities can be reviewed again at a later date when the climate may have changed.

8.6.3.4 Step 4: Assess the Enterprise Risks and Opportunities

The risk/opportunity assessment is dependent on the comparative value of risks and opportunities, but the assignment of value is highly subjective and dependent on the environment. Enterprise opportunities may pose a significant risk to the programs involved. The enterprise manager must be prepared to provide incentives to the individual programs in order to cause the opportunity to outweigh the risk. For example, an enterprise opportunity may be to enable networking throughout the enterprise, but adding network capability may be very costly for individual programs. If the program-level opportunity-to-risk threshold is not met, then the program will not implement the capability. However, if the enterprise can provide all or part of the cost to the program, the opportunity-to-risk threshold will be met and the capability will be implemented. Of course, this poses further cost risk to the enterprise. Effective risk and opportunity management requires an approach to determining which of the identified items deserve more or less attention. Thus, the next step in the assessment process is to rank the risks and opportunities. The method of ranking depends on the creation of a metric that gauges the "importance" of each item. There are a number of parameters that can be considered for inclusion in the ranking metric:

- Dollar value of effect
- Strategic value of effect
- Schedule value of the effect
- Probability-weighted value
- Time value of the effect
- Cost of mitigation

The selection of the correct metrics for ranking purposes will depend on the priorities of the enterprise.

8.6.3.5 Step 5: Develop Risk Mitigation and Opportunity Pursuit Plans

The enterprise assessment team defines mitigation strategies for each risk and periodically reevaluates risk priorities. If the mitigation strategy is successful, then the risk priority decreases over a period of time. A similar process may be employed with enterprise opportunities using the probability of opportunity occurrence as the metric. For an enterprise opportunity that increases risks on the individual program components of the enterprise, incentives provided by the enterprise that reduce risks to the programs will increase the probability that the opportunity will be realized. The programs themselves may also discover some benefits that were not initially apparent. Note that the enterprise-level risk management program must have bidirectional interfaces with the programs' processes to determine the effects of pursuing enterprise opportunities and to ensure the realities of the current funding practices are taken into account.

8.6.3.6 Step 6: Implement the Mitigation and Pursuit Plans

At an enterprise level, cultural changes must be initiated to create an environment that rewards those seeking to capitalize on opportunities and mitigate risks that most benefit the enterprise as a whole. Business performance must be adjusted for risk and opportunity. The gap between development of plans and implementation must be bridged. Champions must be identified for each risk and opportunity being mitigated or pursued. A balanced approach must be taken so as not to oversimplify or plan too deeply. A baseline must be established so that progress or lack thereof, may be readily assessed. Desired results should be stated to keep focus and allow for the handling and reassessment of emergent or unintended outcomes. Periodic tracking of both risks and opportunities at the enterprise level should be implemented.

8.6.4 Program-Level Risk and Opportunity Process Impacts

In order for the enterprise-level risk/opportunity process to be successful, modifications are needed to the existing program-level risk management processes to support it.

With the DoD shift to a "net-centric" paradigm and the concomitant focus on publishing and consuming data, the PM of a "classical" system must deal with additional risk.

The PM needs to:

- Shift from the mindset that he/she is
  - Fully knowledgeable and in control of the environment
  - In control of all the resources needed by the program

- Focus on consuming data and services provided by other agents not under his/her control
  - Trust that these agents are reliable

All of the above involve assuming risk. Classical PM behavior is to be risk averse, to develop strategies to avoid or mitigate risk. This ingrained behavior will not produce the desired social change needed to encourage adoption of the new net-centric vision. Mandating behavior has not been particularly successful in the past nor is it likely to be in the future. A better strategy is to reward the behavior desired, in addition to implementing and enacting consequences to undesirable behavior, that is, the carrot and the stick.

The incentives must reward the PM for assuming risk, for depending on other agents to produce data and services on which he/she will depend. The consequences for not doing so should be equally clear.

However, no amount of rewarding will be effective if the risk is perceived as too great. To enable change, therefore, the challenge for the DoD will be to establish a set of reliable, consistent, available, and ubiquitous capabilities that will form a solid, trustworthy foundation on which PMs and system builders can rely.

Things that we need to think about:

- Stop thinking of building closed systems or SoSs.
- Complete knowledge/control over the environment is impossible.
- Flexibility and adaptability is the key.
- We are a small part of a heterogeneous, constantly changing, and adapting environment.

What we can do:

- Focus on doing a small constituent part very well.
- Expose interfaces and outputs in standard format.
- Maintain interfaces and standard products invariant (or backward compatible).
- Make no assumptions on how others will use the capabilities you provide.
- Allow others (and yourself) to aggregate in planned and unplanned ways the constituent capabilities.

8.6.5 Tolerance for Change

With the pursuit of flexible systems comes the need for:

- System release processes that tolerate change
- Decision aids that focus process choices on relevant change characteristics

Technical advances in areas such as commercial application software, architectures such as service-oriented architecture (SOA), patterns, interoperability infrastructure, and business process management have all contributed to the IT industry's imperative to become flexible, agile contributors to mission performance. These advances are bearing fruit in the form of software and system infrastructure that is quick to: fit business processes, configure, install, support operations, and provide feedback. However, in the realm of government, safeguards against change prevent getting such products deployed on an intended wide scale, equally quickly. The stall imposed by lengthy release tests was appropriate in times when quality, for all practical purposes, had to be tested in. Now the time devoted may offer negligible additional quality. This is not to say that formalized release procedures are never of use, but that management needs justifiable alternatives.

Armed with flexible systems and slowed by an aversion to the risks associated with change, how do PMs make sound cases for how much or how little procedural formality should precede a worldwide release? This section considers the factors of change frequency, likelihood of ripple effect, and release processes, to be juxtaposed with program characteristics for business process, system architecture and technology, and degree of formality in release processes. The result supports a management position aligning the project characteristics with the choice made between lengthy, formal release procedures and simple, immediate releases. While the need for judgment remains, the ability to envision the situation and commensurate actions for trade-offs and program risk assessment has greatly improved. The following day-in-the-life scenarios have been drawn from recent events.

8.6.5.1 Scenario A

1. The PM and independent test director debate the relative merits of the test plans for a change to the source and receiver for a system-to-system interface.
2. The underlying technology is well architected such that the new source and receiver can be implemented via a configuration change to the source and receiver internet protocol (IP) addresses directly in the production environment.
3. The reason the architecture was set up this way is because transitions between data sources of a legacy nature and a modern enterprise resource planning (ERP)/SOA nature were expected. There are many of these transitions coming and their arrival could peak with several changes expecting to be processed in 1 week.
4. The test director makes the case for the standard release process for each change, subjecting each change to serial tasks that take approximately 2 months.
5. The PM is at a loss to justify why this change does not result in the risk that the test director perceives.
6. How to get the two opinions closer together? Use the "Tolerance for Change Decision Map" to evaluate characteristics of the change with the strengths of the architecture and technical solution. If there is "good" risk alignment, in

other words, the change is in the realm of those anticipated by the architects who responded by containing the ripple effect of such changes, then risk is low and a corresponding low-formality release is indicated.

8.6.5.2 Scenario B

1. PM is implementing a commercial off-the-shelf (COTS)-based solution that has been modified only within the confines of the original equipment manufacturers (OEMs) configuration methods. The integration of the COTS product in the organizations' technology framework has been performed in a parallel task to speed deployment and reduce the risk that an integration problem will surface late in the process. The integration project went well and tested well, resulting in few issues (all resolvable well ahead of plans for worldwide fielding). The out-of-the-box COTS product is now available as a "framework service" in accordance with the architect's intentions for how such capability should be provided across the enterprise.
2. The customer for this capability has made it clear that schedule demands will not be eased. The strategic decision to pursue a COTS solution, shown in industry to have been implemented in months rather than years, was no accident. Industry had set the standard for government performance. The strength of the OEM software is well documented with over 25,000 installations worldwide, representing 300,000 users.
3. The PM is planning the release of the "configuration," that is, the update of the out-of-the-box capability with the specific workflows and data requirements of the first user group to employ the new framework service. The PM contemplates how much time to allot and "how much" process in the form of tests and reviews to accomplish as part of taking the configuration operational. Does the original schedule estimate still hold up under the current circumstances?
4. The PM plots the characteristics on the "Tolerance for Change Decision Map" and notes that data migration is involved, that user interface labels have changed, that data-value-lists have been constrained with additional filters, and that other changes are in concert with the OEM's expectations for change. Further he notes that only two of the ten enterprise application archives of the physical software installation package are impacted. On the other hand, the data migration software is custom, although generated with the aid of an extract, transform, and load (ETL) tool, and some uncertainty remains as to the worldwide quality of data to be migrated. The "mapping" drill results in indications that perhaps two paths of testing and release are in order, that could be performed in parallel, and that since the queries underlying the data-value-lists teeter at performance thresholds, priority should be given to a performance checkout aimed at data-value-lists versus a full-blown performance evaluation (previously performed by the OEM).

5. The PM presents a well justified plan for prioritized, parallel paths of release that are commensurate with risk and schedule pressures. One addresses the data migration effort; the other addresses the release of the configuration to the framework service.
6. Since the independent test director has recently adopted ways and means of performing certain tests and reviews in parallel, he or she agrees with the PM's recommendation. They agree to a scaled up "dry-run" of the data migration effort and begin the targeted performance assessment on schedule. The release goes out on time.

8.7 Security Management

IA is critical to the successful implementation of any system processing national security information for the DoD. A system is not, and cannot ever be, truly networthy without an evaluation of its vulnerabilities, threats, and residual risk.

IA strategies for net-centric enterprise services (NCES) programs, in accordance with the Clinger-Cohen Act (CCA) of 1996,* states that all acquisition programs acquiring mission critical/mission essential IT must have an IA strategy that is consistent with DoD policies, standards, and architectures, to include relevant standards.

The key is to provide the right level of IA to protect the enterprise given the varying levels of trust of users, varying levels of trust of IT components, and varying levels of sensitivity of information and services. In addition to the U.S. Joint Services, the enterprise community may include coalition users with varying degrees of clearance and the need to know. For NCES to be successful, interoperability and sharing (e.g., achieve optimal information sharing and access at acceptable risk) must occur across the entire enterprise community.

The term coalition refers to a cooperative arrangement between two or more nations that share information and operate collectively, with one of the nations assuming a leadership role. The term bilateral refers to an exclusive relationship between the U.S. and a partner nation. It is conceivable that the same term could be applied to the same partner nation in different contexts. For instance, most routine operational and intelligence information could be broadly shared throughout the coalition, while more sensitive information would be shared only on a bilateral basis. An Air and Space Expeditionary Task Force (AETF) may be supplemented by one or a number of coalition nations. The creation of machine-to-machine interoperability with all levels of command and coalition partners will help ensure predictive awareness and facilitate effective planning and execution of air and space operations.

---

* See http://communibuild.com/NII/guidebook.htm#_Toc63068350

To operate in a multilevel security domain (enclave) environment, a shared cross-domain resource (SCDR), that is, a shared guard, is used to provide dynamically defined data transfers between security domains.

The SCDR can be designed to use the services planned for the Defense Information Services Agency (DISA) NCES, the Air Force (AF) common integrated infrastructure (CII), the global information grid enterprise services (GES), and GIG IA architecture. These architectures and services are at different states of maturity. This situation offers risks and opportunities for future SCDR projects.

A security guard is a device or collection of devices that mediate controlled transfers of information across security boundaries, separating one or more security domains. A security domain consists of people, data, systems, and networks all of which are subject to a shared security policy or set of rules governing, among other things, who and what has access to data and services. It is "trusted" to allow sharing of data across security boundaries and enforces a defined security policy. A guard is also known as a cross-domain resource. Today, guards typically support one source/destination organization pair, one data object type (with some limited capability exceptions), and one policy.

Agility and Flexibility—Network centric warfare (NCW) requires flexible communications and data-sharing mechanisms to achieve information superiority. Cross-domain solutions must transition from their current rigid implementations to architectures capable of supporting the dynamic information sharing requirements of NCW.

Speed and Performance—NCW requires the fastest possible response times of mission critical systems. Cross-domain solutions often lag behind current technologies because of the difficulties in upgrading the technology. Cross-domain solutions must transition to architectures that support an affordable infusion of the latest technologies and solutions, a streamlined accreditation process, as well as new concepts of operation and other performance-enhancing modifications.

Robustness and Availability—NCW requires high availability and a built-in tolerance for problems. Cross-domain solutions must transition to architectures that offer affordable robustness with high availability and fault tolerance.

Enclave—Collection of computing environments connected by one or more internal networks under the control of a single authority and security policy, including personnel and physical security. Enclaves always assume the highest mission assurance category and security classification of the automated information system (AIS) applications or outsourced IT-based processes they support and derive their security needs from those systems. They provide standard IA capabilities, such as boundary defense, incident detection and response, and key management, and also deliver common applications, such as office automation and electronic mail. Enclaves are analogous to general support systems as defined in the Office of Management and Budget (OMB) Circular A-130, Management of Federal Information Resources. Enclaves may be specific to an organization or a mission, and the computing environments may be organized by physical proximity or by function independent of location. Examples of enclaves include local area networks (LANs) and the applications they host, backbone networks, and data processing centers.

Interoperability and integration of IA solutions within or supporting the DoD will be achieved through adherence to an architecture that will enable the evolution to NCW consistent with the overall GIG architecture and implementing a defense-in-depth approach.

Layers of technical and nontechnical solutions will be employed to:

1. Provide appropriate levels of confidentiality, integrity, availability, authentication, and nonrepudiation to information and resources within the GIG.
2. Defend the enclave perimeters.
3. Protect all information systems, enclaves, and computing environments (including applications and databases) from external and internal threats.
4. Use supporting infrastructures such as common access card (CAC), public key infrastructure (PKI), biometrics, modernized cryptographic capability, and key management infrastructure (KMI) to enforce IA requirements.
5. Implement a protected IA architecture for incident identification and response capabilities.

8.8 Hidden Costs of Enterprise Information Interoperability

Enterprise information interoperability (EII) implies, among other things, that a large amount of the enterprise's systems (including its processes, applications, and services) are able to seamlessly exchange information. A bare necessity for this is to have a cooperative, enterprise-wide information perspective that unambiguously defines the information that is shared among the enterprise's internal and external parties. A conceptual model can be used to represent this global understanding of information from across the enterprise. Once a conceptual model for an enterprise is available, it can be used to facilitate many important activities that the enterprise might want to conduct, such as:

- To document the enterprise-wide understanding of the meaning of data from across the enterprise
- To guide information integration from its disparate (stovepipe) component source systems, for example, as a precursor for an enterprise data warehouse (EDW) effort or for an ERP effort
- To provide a common picture of operations throughout the enterprise

The creation of an enterprise-wide conceptual model reflects a composite view of the enterprise data (or subviews by selected topics for ease of comprehension or ease of analysis) even though its data might originate from multiple, autonomous component, transactional systems. Once developed, the conceptual model can easily be used to identify two hidden costs within a data integration effort. The hidden costs are those associated with either (a) identifying and preventing several subtle forms of diminished data quality or (b) not addressing them adequately and accepting the consequential costs of inferior decision making based on the diminished data quality. These costs (of prevention or of consequence) appear whether the integration effort is aimed at a centralized database or a federated database.

A variety of enterprises may pursue EII efforts for one reason or another. In government, for example, the Department of Homeland Security (DHS) must patch together data from a variety of sources to get a composite list of known or potential terrorists. This task is similar to the merger of databases among banks. However, the DHS integration effort can be much more difficult. This is because banks are generally in the same kinds of businesses and, therefore, view the objects in their domain in a comparable manner, but DHS must merge the diverse views of border protection and enforcement, for example, with immigration and transportation. The more diverse the views being integrated, the more valuable a conceptual model can become for locating the areas that produce the subtle erosion of data quality hidden in the integration effort.

A conceptual model for an enterprise identifies and defines, in unambiguous terms, the following:

- The objects of interest to the enterprise
- The relationships among those objects
- The attributes associated with each object (and, optionally, each relationship)

The current source systems within the enterprise may have historically viewed the objects within the enterprise in different ways. The systems could be maintaining different data for the same incidence of objects within their own domains; but, they might not maintain incidence of the same objects across their domains. For example within the DoD, some systems keep data about the active and reserve forces, others keep information about government employees and contractors, while others keep tabs on all of these. Therefore, when considering the combination of "person" data from all of these sources, they are likely to be keeping different information about them for their own specific purposes. For example, the personnel systems maintain aspects about a person related to terms of service, benefits, and payroll; the training systems can be maintaining aspects related to education qualifications and certifications; and the medical systems can be addressing aspects related to healthcare. The conceptual model can be used to identify the disparities across the source systems.

The conceptual model for an enterprise is used to map attributes of an object in the enterprise model to one or more source systems that maintain the same or comparable data for those attributes. When the data from these sources are viewed in a centralized or federated database via the conceptual model, two issues may arise: one related to the *synchronization* issues within the model and another related to *association* issues within the model.

8.8.1 Synchronization Issues

The synchronization issues, as seen in the following illustration, arise from the disparate context of the source systems. They may be maintaining their instance data differently because of their own contextual views. The following is a partial list of potential context view differences among autonomous components (with examples from the USAF):

- Place differences (e.g., concepts are at the base level, rather than at the major military command level or vice versa)
- Time period differences (e.g., concepts are related to plans, rather than actuals or vice versa)
- Justification differences (e.g., concepts are estimates, rather than validated or vice versa)
- Usage differences (e.g., concept are discrepancies related to operational tempo, such as, battalion versus flight)
- Granularity differences (e.g., concept dissonance of trajectories)

The synchronization issue is illustrated in Figure 8.3. The issue ultimately gets reflected in the database when combining the data from multiple sources for a

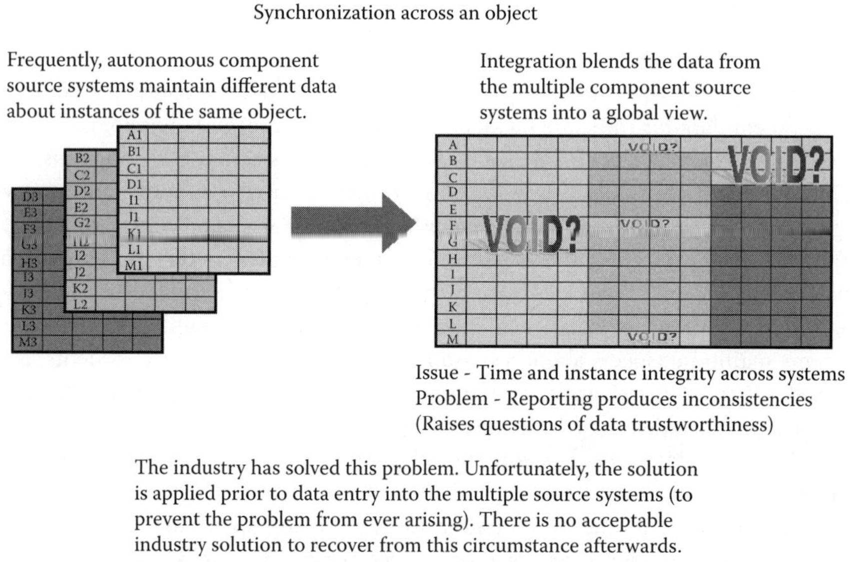

Synchronization across an object

Frequently, autonomous component source systems maintain different data about instances of the same object.

Integration blends the data from the multiple component source systems into a global view.

Issue - Time and instance integrity across systems
Problem - Reporting produces inconsistencies
(Raises questions of data trustworthiness)

The industry has solved this problem. Unfortunately, the solution is applied prior to data entry into the multiple source systems (to prevent the problem from ever arising). There is no acceptable industry solution to recover from this circumstance afterwards.

Figure 8.3 The synchronization issue.

single object. The synchronization differences between the sources/systems might show up in a database report, for example, that shows an airplane being refueled in flight while simultaneously awaiting repairs of its engine at a base. If this is seen in a report, the trustworthiness of the data will be subject to question and decision makers may doubt the value of the integration investment and/or the competence of the designers and implementers. This can be considered a nontangible cost.

Differences can be more subtle and not be easily apparent in a report. If data elements do not distinguish in an obvious way that they are related to the planned activity on an item versus the actual status of an item, then decisions may be made based on false assumptions as to what the data are reflecting. With enough of these mistakes, decision maker disuse of the enterprise database will be followed by searches for liability. The pedigree for the mistake can eventually fall on the constitution of the database design, its implementation, and/or its operation, as rightly it should. This loss of trust by a decision maker can also be considered a nontangible cost.

The synchronization issues can be detected by using a conceptual model for the enterprise, mapping the attribute contributions from the source systems, and locating each whose attributes map to source systems that have any of the following: different contexts, different update frequencies, or different populations. There are strategies for avoiding the consequences that will arise from these types of synchronization issues.

8.8.2 Association Issues

The association issue arises when objects within the enterprise's conceptual model have new logical associations that the enterprise is interested in exploiting, but which have never been explicitly implemented across the current autonomous source systems. There may have historically been an implicit attempt to maintain the desired logical associations by having users of the two systems perform comparable updates to each system. This can eventually be futile, because keying inconsistencies can occur as the user enters data into the separate systems or because the systems have different contextual views among the systems. This situation is comparable to those as discussed above regarding synchronization issues. The nature of associations among objects can be seen in Figure 8.4.

As shown in Figure 8.4, EII usually requires that objects in separate databases (i.e., Objects X and Y above) be logically related according to certain enterprise business rules. In Figure 8.5, the voids indicated in gray in the cross-reference table of the figure must be checked against the business rules for the enterprise. These "new" business rules will usually have never been validated by the individual source systems before the federation effort, because it would have required the source systems to communicate with one another on a near-real-time basis. If the voids in gray are valid, such as a customer can exist without an account, then the void is

266 ■ *Enterprise Systems Engineering*

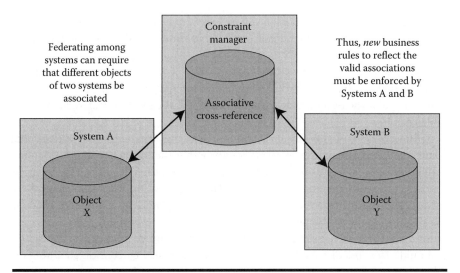

Figure 8.4 Associations among objects.

valid. If a void is not valid, then the transaction attempting to perform such an update must be prevented. They cannot feasibly or reliably be reconciled at a later time. If they are allowed to be inserted into the database, they will cause referential integrity problems and applications using the business rules to report information from the database may experience unpredictable behavior and erratic results. Decision makers using results from this environment will find the situation intolerable. This can be considered a nontangible cost. However, the prevention can be a costly endeavor, as shown in Figure 8.5.

The association issue shows up in a federated database when invalid associations among objects are erroneously permitted. Validation against business rules must be performed *prior to a source system allowing* such a transaction to update its own database. When disparate source systems contribute to an enterprise database, they participate in a database federation. An aspect of this participation can be that they must explicitly conform to certain business rules regarding logical associations among the various source systems that they had only previously implicitly enforced on a best effort basis as discussed above. Explicitly enforcing the association business rules will be an added cost to the source systems.

The source systems must, upon receipt of a transaction and prior to committing the transaction to their own database, determine which association rules apply to that transaction. If a rule applies, the source system must check the other applicable source systems for rule compliance regarding that transaction. Access to the other source systems is through the cross-reference. After validation across the other applicable source systems that are involved in that transaction, then the source system is allowed to commit the transaction to its own database.

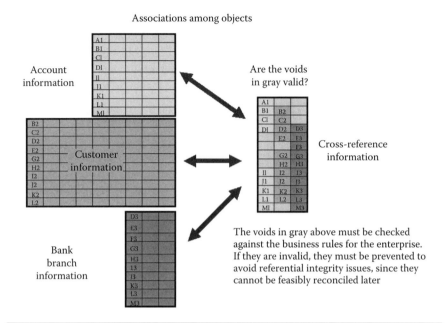

Figure 8.5 The association issue.

If the transaction is invalid, the user who was attempting the transaction is notified of the error and the transaction is rejected. The user error is more often than not a keying error that would have otherwise gone undetected. To assure successful resubmission of the transaction with a less simple solution, the user with the rejected transaction would potentially follow certain additional business practices to correct the transaction prior to resubmission.

New business rules must be formalized and implemented among the source systems regarding which of the source systems is allowed to create, update, and delete entries in the cross-reference table and under what conditions.

There will be cost to all source systems to implement the enterprise business rules where they are to participate in explicit associations with one another. The conceptual model can be used to identify where these associations occur and what are the applicable business rules.

8.8.3 Summary

There is a general lack of appreciation for the cost to federate source systems. Two issues in particular, a synchronization issue and an association issue, are not that apparent when considering the costs of federating source systems. A conceptual model can aid in identifying the issues that drive those costs. In particular, without a conceptual model, there is practically no way to quantify the costs of source system federation. The conceptual model makes these two issues visible and quantifiable for costing purposes. Furthermore, with a conceptual model, it is possible to forewarn of the costs, or dangers, of not forthrightly addressing these two issues with adequate investment funding. These costs are referred to herein as nontangible costs related to diminished data quality, decreased database value, and reduced decision-maker trust.

8.8.4 Conclusion

When considering an EII effort, use a conceptual model to align semantic differences among components in the federation and to anticipate the degree of effort to accommodate the emergent properties.

8.9 Configuration Management

8.9.1 Introduction

In the context of SE, CM is defined as "a discipline applying technical and administrative direction and surveillance to: (1) identify and document the functional physical characteristics of a configuration item, (2) control changes to those characteristics, and (3) record and report change processing and implementation status" (Kossiakoff and Sweet, 2003). The activity of CM is often performed under the guidance of a CM plan. While the specifics may vary, most CM plans answer the questions of:

- What are the CIs that will be managed?
- What is the formal scheme for identifying CIs?
- What are the CM procedures? Be sure to include change control, version (baseline) control, and auditing of the CM process
- Who is responsible for various CM procedures and policies?
- What records (the CM database) must be produced and maintained, in what form, and using what tools?
- How will CM information be made available to those who need it?

Over the past several decades, these questions have been well addressed in the realm of SE. Moving into the realm of SoS added layers of interaction and interfaces have complicated the answers, but the methods remain largely unchanged. In the realm of ESE the answers to almost all of these questions will change significantly due to the complexity and emergence character of enterprise systems. There will also be new questions that must be addressed for effective CM in the enterprise realm. The purpose of this section is to propose some answers.

8.9.2 Configuration Items

This section will identify which items should be considered for control under an ESE CM plan. It will also consider related issues such as naming conventions for CIs, the structure of CM databases, the communication of CM information, and the tools that can be applied effectively in the realm of ESE. This section will consider such questions as the level to which people, organizations, processes, and their interactions with one another and systems need to be considered CIs in the realm of ESE.

8.9.3 Policies and Processes for ESE CM

The institution and implementation of policies and procedures in the realm of ESE is complicated by the scale, the large number of interacting systems and organizations, and the complexity and emergence character of the ESE realm. This section will provide insights into how CM policies, processes, and their requisite organizational structures can be effectively implemented in the ESE realm.

8.9.3.1 Change Management

CM in the realm of SE includes change control, a process that is passive, reacting to the delivery of new CIs and change requests. This section will explore the meaning of change control in the realm of ESE. It will consider the need for a more active and "forward leaning" form of change management—a superset to change control.

Transformation from a "system" to an "enterprise" perspective involves change—of people, processes, and technology. In the past, we focused on CM—identifying systems CIs, and collecting them into a change control process. Enterprises are more complicated. If we envision an enterprise to be a network with a collection of hardware devices and software services, then we have a new challenge of managing change. We have to understand how people use and access the network as well as manage change of unknown combinations and collections of services.

Examples of enterprise hardware services that require change management might be server farms, CPU platforms, and hardware devices. When we separate the hardware layers of a server, we may acquire devices as requirements emerge (not acquiring stand-alone servers). Similarly, depending upon the capability desired by the user (that might be determined at run-time) different combinations of data and application services might be accessed. This requires robust configuration/change management across services to produce predictable results.

Cummins (2002) tells us that we need to maintain a configuration model that includes the following items, "along with their relationships and version dependencies: network nodes and links, firewalls and gateways, computing hardware, systems software, application services, applications, documentation, users. The model must support transformation to new configurations." The complexity of a network may require even more CIs to manage.

Cummins also proposes that the change management organization have defined processes as follows:

- Capacity planning—plans for growth or for new applications and services
- Hardware and network configuration—managing change to physical configurations based on capacity plans and requirements, includes changes to topology
- Systems and services installation and planned upgrades—managing compatibility
- Application deployment
- Application changes
- Process improvement and workflow management

These processes are dependent on one another. The role of the change management organization is to identify and manage the changes. In addition, the change management organization must continue to manage the change in an evolving enterprise, and identify new processes that need to be included in change management. CM of the enterprise elements as well as managing people and process change are keys to the success of the enterprise.

8.9.3.1.1 Change Management and People

The previous section discussed change management in the sense of an active, forward leaning process. This is a departure from the classic change control of SE. Another type of change that ESE may require is *cultural* change management—the process by which changes can be made in the attitudes and actions of people.

The people aspect of change management may often be the most challenging. People are often resistant to any type of change, when they do not have some level of control. According to Bernard (2004), "Change Management is the process of setting expectations and involving stakeholders in how a process or activity will be changed, so that the stakeholders have some control over change and therefore may be accepting of the change."

Whenever an organization imposes new things on people there will be difficulties. Participation, involvement and open, early, full communication are important factors. Change can be viewed as a personal attack when people are confronted with the need or opportunity to change, especially when it is "enforced," by the organization. Increasing the stakeholders' level of control helps to successfully manage change and can be accomplished in many ways:

- Regularly and effectively communicate enterprise engineering activities to stakeholders
- Plan enterprise development in appropriate achievable measurable stages

- Involve stakeholders in the enterprise management—enable and facilitate involvement from people, as early and openly and as fully as is possible
- Allow for stakeholder input to planning and decision making
- Manage stakeholder expectations

In *Slack*, Demarco (2002) describes a management approach that is different during change. No longer is there a clear manager/subordinate relationship that many of us take for granted. It may become hard to tell the difference between someone who is trying but failing to adjust to change and someone who is resisting it. Likewise, we are used to being rewarded after successful performance (e.g., by a paycheck). During change, reward must come before performance. This is done by relying on staff who have gained your trust through a track record of time and enticing them to embrace change. Most importantly, "The key tools of management in the knowledge organization are the tools of change management. Instead of authority and consequence, the best knowledge-work managers are known for their power of persuasion, negotiation, ... and their large reserves of accumulated trust."

8.9.3.2 Software CM

When an organization is developing software, an approach to CM is vital. Currently, open source projects use two popular CM systems, concurrent versioning system (CVS) and subversion (SVN).

CVS utilizes a client–server architecture: a server stores the current version(s) of the project and its history, and clients connect to the server in order to checkout a complete copy of the project, work on this copy, and then later check-in their changes. Typically, client and server connect over a LAN or over the Internet, but client and server may both run on the same machine if CVS has the task of keeping track of the version history of a project with only local developers. The server software normally runs on Unix (although a Windows NT CVS server also exists), while CVS clients may run on any major operating system platform.

Several clients may edit copies of the project concurrently. When they later check-in their changes, the server attempts to merge them. If this fails, for instance because two clients attempted to change the same line in a certain file, then the server denies the second check-in operation and informs the client about the conflict, which the user will need to resolve by hand. If the check-in operation succeeds, then the version numbers of all files involved automatically increment, and the CVS server writes a user-supplied description line, the date and the author's name to its log files.

Clients can also compare different versions of files, request a complete history of changes, or check a historical snapshot of the project as of a given date or as of a revision number. Many open-source projects allow "anonymous read access," a feature that was pioneered by OpenBSD. This means that clients may check and compare versions with either a blank or simple published password (e.g., "anoncvs"); only the check-in of changes requires a personal account and password in these scenarios.

Clients can also use the "update" command in order to bring their local copies up-to-date with the newest version on the server. This eliminates the need for repeated downloading of the whole project.

CVS can also maintain different "branches" of a project. For instance, a released version of the software project may form one branch, used for bug fixes, while a version under current development, with major changes and new features, forms a separate branch.

CVS terminology dubs a single project (set of related files) managed by CVS as a module. A server can manage several modules; it stores all the modules it manages in its repository. The copy of a module that a client can check out serves as a working copy and reflects recent changes other clients commit to the module. Update is the process of acquiring latest changes from repository to the working copy. CVS developed from an earlier versioning system called revision control system (RCS), still in use, which manages individual files but not whole projects. Dick Grune has provided some brief historical notes about CVS on his Web site.*

A group of volunteers maintain the CVS code. Notably, the development of the Microsoft Windows version of CVS has split off into a separate project named CVSNT and has been more active in extending the feature set of the system, even porting the changes back to the UNIX platform under the name CVSNT.

Historically, the relationship between CVS and the GNU project could appear somewhat ambiguous: the GNU Web site distributed the program, labeling it "GNU package" on one page and "other (general public license) GPL-licensed project" on another. This was recently clarified when CVS development moved,[†] with CVS officially relegated to the non-gnu category. On the FTP site, the program has traditionally and still resides in the /non-gnu/ directory.

Limitations:

- Files in a CVS repository cannot be renamed; they must be removed and re-added.
- The CVS protocol does not provide a way for directories to be moved or renamed. Each file in each subdirectory must be explicitly deleted and re-added.

A number of key developers who have worked on CVS are now responsible for SVN, released in early 2004, which aims to replace CVS by addressing some of its limitations. SVN is a version control system designed specifically to be a modern replacement for CVS, which is considered to have many deficiencies.

---

* See http://www.cs.vu.nl/~dick/CVS.html
† From http://www.cvshome.org to savannah.nongnu.org

Commercial options include:

- Microsoft Source Safe
- Rational (IBM)

8.9.4 Configuration Management Evolution

CM as an engineering discipline will change as traditional system engineering evolves toward enterprise engineering. This section compares the existing CM structure to a proposed enterprise structure for the purpose of planning the necessary steps to make this evolution happen.

8.9.4.1 Existing CM Structure

CM is a system engineering process which provides a structured approach for developing the products, which the AF acquires to maintain a capable fighting force. CM is applied to system acquisitions in order to control the technical baseline which drives the acquisition process.

This basis for the existing CM process had been established when systems were built to address specific threats. These systems were built against well-defined requirements developed as threat needs of the day were understood. They performed unique functions which were not designed to be interoperable with other systems performing other unique functions. Usually these systems were of a unique design requiring them to be built from the ground up. Their development took years, resulting in completed products that could not meet new requirements which evolved from growing threats.

It is important to understand this existing process in order to know the point-of-departure baseline for migrating toward the enterprise.

8.9.4.2 CM Evolution

It became evident that products needed to be developed not only to meet evolving requirements but also to take advantage of rapidly developing technology as it became available. COTS products addressed the improved technology portion of the equation. The AF now had a way to manage the implementation of their evolving requirements using evolving technology. This had to be done using a carefully controlled CM process. This was the start of the evolution of the CM process through evolutionary acquisition.

Two important CM themes are: (1) a well-defined CM process is critical in any stage of acquisition, and (2) the structure of the CM process will adjust (evolve) to meet the needs of the evolving enterprise.

The basic CM characteristic of "tight management of the technical baseline" will never change. The method as to how this will happen will be a function of how the enterprise evolves. The following suggests how to approach this uncertain path by discussing basic elements which make up the core of CM.

8.9.4.3 Necessary Steps for CM Evolution

The following represents some of the planning steps needed to understand CM evolution:

1. Characterize elements of the "as-is" process. This can be done by using a "today" acquisition. This could involve determining steps which have occurred between two points in the past, such as changes between unique product acquisition and evolutionary acquisition (involving COTS) discussed earlier. From this, develop a list of expected items to take us from the COTS evolution to a possible next stage. This will not be a final list; it will only help start the process. This list will evolve over a period of time.
2. From (1), determine a "best estimate" order of steps of how this might occur. This will be a "best guess." For example, a specific program (acquisition) may have a specific CM need as to the order of unique functions/capabilities to be implemented. This order would be important to a program already in process versus definition of a CM approach for a new system in the initial stages of development. Different phases of a program may benefit more effectively using an approach best suited for the program needs. An example could be the air traffic management (ATM) program in which aircraft capabilities are implemented on various aircraft platforms at different times. The CM approach used for ATM may be entirely different than for an acquisition program with another set of goals (possibly a new development). It is critical to have an accurate CM approach since it will be used as a tool for certification of aircraft allowing access to specific flight routes which are high in traffic density (involving safety of flight concerns).
3. Determine how the Department of Defense Index of Specifications and Standards (DODISS) artifacts may be implemented in programs with different approaches to their goal solutions. Try to predict their use in the enterprise from their use today using the "delta determination" approach described in (1) above. It is expected that the DODISS artifacts would have a baseline structure across programs, but may be more efficiently implemented differently across different programs.

As an example, ATM may not have to employ the full list of DODIIS artifacts each time a new aircraft platform is upgraded with a new capability if staying within the same airframe family. However, going from one airframe family to another may require the use of a larger subset of the artifacts.

8.10 Infrastructure Management

ESE should be viewed as an expansion in scope of the traditional SE that requires dealing with unpredictable behaviors due to the large number of interacting variables. Interactions of these variables are increasingly difficult to quantify in a meaningful way and demand a change to the traditional system engineering infrastructure. Traditional SE seeks to optimize a design for a specific purpose or a few closely related purposes. It also focuses on the physical laws that govern design and the resulting predictable behavior of the system or SoS.

A continually changing and unpredictable competitive environment has caused a paradigm shift in the way business operates. This transformation has accelerated in response to the evolution and globalization of information and communication. Today's enterprise paradigm is dynamic and multidimensional. An enterprise is characterized by people, processes, and technology interacting with each other and their environment to achieve goals. Infrastructure management is critical to the evolution of the enterprise. Figure 8.6 illustrates the complexities, considerations, and impacts associated with today's multidimensional enterprise.

The management of IT infrastructure has a significant impact on the success or failure of business processes, directly affects the quality of service of business applications, and contributes to the satisfaction of internal and external users.

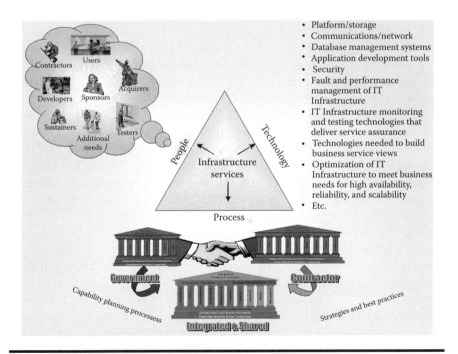

Figure 8.6 The multidimensional enterprise.

Infrastructure management should be considered as a holistic approach that encompasses architectural issues, people, processes, and best practices associated with this discipline, along with the associated network and system management tools.

8.10.1 Enterprise Infrastructure Management Considerations

Enterprise engineering involves all the activities that enterprise organizations perform to improve productivity, gain and maintain competitive advantage, optimize resources, deliver quality products and services, and meet customer expectations and demand. These can include traditional activities such as reorganization, concentration on core products and competencies, niche marketing, acquisition, merger, and new technologies. As the scope and the size of the enterprise grow, there is an increase in the complexity of managing the infrastructure within the enterprise. This drives the need to take a broader view and a more holistic approach to manage the enterprise infrastructure. There are issues that need to be considered in infrastructure design and implementation in a heterogeneous environment. There are also management considerations that must be taken into account in a dynamic enterprise paradigm. Thus, one must account for the fundamental characteristics of the enterprise, that is, technology, people, and process.

8.10.1.1 Technology

Rapid changes in software, hardware, communications, network, information services, and data manipulation have made technology transition in the enterprise infrastructure a necessity rather than an option. When making decisions toward large infrastructure investments, one needs to consider—how much risk is involved in adopting a new infrastructure technology? Does a high technology readiness level (TRL) correspond to a low risk in the investment for that technology? Where is the risk mitigation plan? What is the migration plan and schedule? However, assessing technology readiness or maturity does not accurately capture the risk involved in adopting a new technology. A holistic approach to technology risk includes obsolescing and negative factors that together with the TRL provide a better representative of technical risk for technical transition in enterprise infrastructures.

When managing technical refresh cycles the complexities of individual hardware or software components need to be worked out. This requires discipline and centralized control at an enterprise level. It will make the technical refresh tasks more difficult by quickly changing technology and plug-and-play products without considerations of the technical roadmap, users' needs, and budgeting. One cannot take the traditional engineering approach, "one-size-fits-all configurations and locking users into rigid refresh cycles," in an enterprise environment.

Two of the biggest challenges are detecting emergent behaviors and designing the enterprise infrastructure with enough flexibility so that it can be adapted to the changing environment and the emerging technology at the least cost and with cycle times that are better than any competitors. Characteristics unique to the enterprise are rapid exchange of information between all elements of the enterprise, the flexibility to adapt to rapid changes in the environment/technology, and the ability to exhibit unexpected behaviors while meeting the goals of the enterprise. These characteristics require that the elements of the enterprise be connected and share information. The system engineer must focus on development of ways for the elements of the enterprise to communicate. In designing information and communication systems, this creates a new dynamic for the enterprise systems engineer to adopt a mindset of "need to share information" rather than "need to know." The new point of departure in information system design is that all information in the enterprise is available to every other element in the enterprise so that as the environment or goals of the enterprise evolve, the enterprise can respond appropriately. Filtering should be used selectively (e.g., an individual's salary information). Another part of the enterprise system engineering infrastructure is to develop query tools for requesting information an element needs from any other element in the enterprise.

8.10.1.2 People

The enterprise environment presents challenges in dealing with systems which not only have critical technological components but also have significant enterprise-level interactions and socio-technical interfaces that determine the design, development, implementation, and operation of the system. The traditional engineering approaches for the creation of these systems and their infrastructures are based on a single system orientation without an enterprise perspective. One needs to develop a holistic view that bridges traditional engineering approaches with increasing insights from a cohesive engineering management view. This includes intersections of human activities and technical systems.

Management of the infrastructure is concerned with the design of the whole enterprise; the overall products and end users; the regulations and policies governing enterprise use and disposition; the relationship with enterprise sponsors, contractors, and vendors; the enterprise licensing and maintenance agreements for the enterprise infrastructures; and the training programs and documentation to ensure user familiarity with enterprise applications and infrastructure configurations.

In order to survive in the twenty-first century, an enterprise must make change management an integral, enterprise-wide process. The nation's newest management "gurus" urge enterprise executives and managers to think in radical terms, often recommending dramatic overhaul of entire operations at a single stroke. "Most of [the new gurus] agree that [an enterprise] should organize itself on the basis of process, such as fulfilling an order, instead of functions, such as marketing or manufacturing," writes John Byrne in Business Week. "That takes the enterprise's focus off its own internal structure and puts it on meeting customers' needs, where it belongs. [They] generally agree that time can be squeezed out of every job; that self-managed teams throw more challenge and meaning into employment; and that enterprises sorely need to create networks of relationships with customers, suppliers, and competitors."

8.10.1.3 Process and Planning

Enterprise engineering also includes new techniques and methods such as business process reengineering, continuous process improvement, total quality management, enterprise architecture, and enterprise integration. One thing that distinguishes today's successful enterprise, business, and government, is the ability to adapt to changes in environment, in markets, and in customer expectations. The ESE infrastructure must be able to document an enterprise's performance measures and link them to appropriate elements of strategic plans. This linkage allows quick reaction to changes in environment, policy, or customer requirements. The information requirements that support an enterprise's performance measures must be modeled and translated into a database, data warehouse, decision support system (DSS), and executive information system (EIS) designs. These designs can then be developed and implemented to provide for automated capture, reporting, and analysis of performance measures. Enterprise engineering must develop and use tools to support strategic planning and performance measures and provide the means to link strategic planning and performance measures as near real-time, integral and dynamic parts of the enterprise so that performance can be continuously updated and evaluated to see trends in relation to the strategic plan. Tools must be developed to evaluate the environment, changing policies, and evolving customer requirements.

Enterprises are often overwhelmed by the burden of implementing new infrastructure. Every new technology seems to introduce an exponentially greater number of management challenges, cost/budgeting considerations, integration issues, organization policies, quality assurance, training, security threats, and resources that result from each. It is impossible for enterprises to keep implementing new infrastructure as it becomes available. Enterprises need to develop a technology infrastructure plan that can be implemented in manageable phases suitable for the enterprise architecture. The plan should be scalable and flexible as well as include business strategies for delivering enterprise capabilities in incremental phases.

8.10.2 Enterprise Service Monitoring and Management

The monitoring and management of the health of the enterprise infrastructure is complex and important in order to provide service to the end user. As the enterprise grows, so does the complexity and the challenges.

Enterprise service monitoring (ESM) is needed because support for business services delivered through the infrastructure must be integrated, since end users should not be expected to diagnose and solve the intricacies of network versus server versus application versus database as a root cause for problems they experience. ESM is also needed because only through an integrated system can problems like the following be solved: losing a service and not knowing whether the cause is a network device, a server, an application or a database, and whether it impacted the end user. ESM enables the collection of enterprise-wide performance information to produce trend statistics. If they show performance degradation, remedial actions can be applied. ESM should be at the same level as the service groups, but separate from them to assure that objective measures are being taken and reported for the infrastructure, IT services, and the service groups themselves. It is only through the discipline of administration of processes and measurement of the impact of processes that a process improvement program can be effective and deliver on overall improvements.

8.10.2.1 Techniques for Monitoring the Enterprise Infrastructure

When designing an enterprise monitoring and management system, the requirements of the users and customers should drive the design. Potential customers for such a system are the Service Desk that fields complaints of the services offered through the infrastructure, operational service groups whose responsibility it is to maintain the infrastructure to the highest levels of availability and performance and the end users to whom services are delivered through the infrastructure. That which is monitored and managed are the items required to deliver the services offered through the infrastructure and these items can be workstations, network devices, third-party network services such as the internet, digital subscriber line (DSL), T-carrier 1, digital transmission line, 1.544 Mbps, 24 voice channels (T1), and ATM services, servers, operating systems, applications, databases, and uninterruptible power supply (UPS) devices. Monitoring and measuring the performance of the infrastructure should also consider including monitoring and measuring the performance of the operational support entities that maintain the health of the infrastructure and provide support to the end users and customers.

To provide the highest availability and best performance of the infrastructure, as much automation as is feasible is required. Thus automatically and instantaneously detecting outages and performance degradations in the infrastructure and the services provided through the infrastructure is critical. There are many tools available to do this; however, a serious focus should be on accurate detection (therefore avoiding false alarms) and root cause determination. This can be achieved through multiple approaches to monitoring and correlation of symptoms. Monitoring the entire infrastructure and its services in an integrated way will yield maximum results in accuracy and root cause determination. For example, a monitoring tool such as *SMARTS InCharge Availability Manager** can monitor various classes of items (network devices, servers, applications) and correlate failure symptoms to derive one

---

* See http://www.smarts.com

or a few potential causes, with probability weights, to avoid false alarms, such as a failed server when it is really a network device that is blocking access to the server or a failed application, when it is really the server that has failed.

The multiple approaches to monitoring include resident monitoring agents on devices and remote monitoring of devices. They each have their advantages and disadvantages. There are vendors with products for either approach to servers. Monitoring involves polling devices or scheduled transactions to gather data and the collection of agent, device, or system-generated information. While each vendor tries to provide the same capability for everything, it is not an impediment to efficiency, effectiveness, or support to do a best-of-breed approach as most vendor products are interoperable in a loosely coupled system using log files, simple network management protocol (SNMP) traps, messaging and vendor proprietary methods, and application programming interfaces (APIs) for integration. With multiple approaches to monitoring, one can combine intimate monitoring (such as a resident agent) with end-user monitoring, such as a remote-scheduled transaction. An end-user emulator provides both an opportunity to see root cause as well as end-user impact. As an example, a network monitor could detect a failed network device and an end-user emulator monitor which would determine if the device was on a redundant path, in which case the end user would not be impacted.

Automated triage is achievable if all events produced by all monitors are collected into a central system which identifies their source, reformats them into a unified message format, and correlates the information. Flexible event handling is critical to achieve maximum functionality required for a robust triage capability. Visual access to active events and notification of events to operational service groups and other interested parties is necessary to minimize the time it takes to repair. The practice of acknowledging events to the central system allows positive accountability that a notification was received and/or escalation to a different person or notification via a different medium to be automated. Off-loading notification from the central system to a specialized system allows for interface to various media for notification such as email, SMS, and telocator alphanumeric (input) protocol (TAP). Such systems have the ability to put service group personnel into a rotation with primary and secondary contact information, allow for vacation substitutions, and automated escalation policies.

Capturing events to a database for historical purposes at least for 2 years allows for research in trends and behavior patterns and comparison of month to month and month over month performance.

A system that comprehensively monitors the enterprise will have capabilities to do proactive monitoring, that is, attempting to anticipate problems before they have too great an impact. While this is typically a capability that is exploited long after monitoring is in place, it is nonetheless a very valuable capability that represents a sophisticated use of the monitoring tools and the event information.

In considering an enterprise monitoring and management capability, serious thought should be given to positioning the personnel who design and support such a system. They should belong to a centrally managed group that has a reporting level similar to the operational service groups whom they support and measure. This is important to maintain the integrity of the information contained in the reports that accompany these monitoring functions.

8.10.2.2 Monitoring the Border Architecture

The security concerns that are associated with border systems, de-militarized zone (DMZ) systems, systems connected via virtual private network (VPN) servers, and firewall connected systems constrain the monitoring function. Typically the border servers will have locked down operating systems and applications and a resident monitoring agent will not be allowed. Therefore, remote monitoring will be the only acceptable method for monitoring. DMZ systems will have to be monitored in such a way as the event traffic will be allowed through the firewall without authentication which could be discovered by an outside hacker. Systems monitoring or end-user emulation is somewhat fragile over a VPN because it is usually impossible to set up an automated log on to a VPN service. This would have to be manual and is a consideration when attempting to determine the feasibility of such monitoring. Monitoring firewalls is possible through secure SNMP (V3).

8.10.2.3 Decentralized versus Centralized

Managing the systems and applications that comprise a comprehensive enterprise monitoring system is best done centrally. The monitoring servers should be close to the production servers and core servers, and applications should be hardwired for connectivity to each other so they can continue to provide monitoring information even when the network is down. This means notification should also allow for use of the public switched telephone network (PSTN) as a primary or an escalated secondary medium. End-user emulation robots should be on the subnets of the users they represent. Typically they should be without monitors, keyboards, or mouse devices and physically locked in a network closet to provide security and minimum management. Although these will be managed centrally via remote management software, it will still be necessary to have a point of contact (POC) at the location to fix problems that are not remotely fixable. Providing UPS and network rebootable power strips will minimize the amount of remote assistance that will be required to manage the robots.

8.10.3 Beyond ESM—Information Technology Infrastructure Library (ITIL)

The ESM system covers detection of outages and performance degradations in managed applications, databases, and infrastructure (network and servers); provides tools for proactive problem management of incidents and potential incidents;

and provides service-level management for applications delivered, as well as a trending capability and predictive modeling capability for applications on servers.

The next phase of ESM will be in support of the industry trend of business service management and enhanced customer service through service-level management. Coincidental to this trend is another: the adoption of information technology infrastructure library (ITIL) best practices in managing business services.

ITIL consists of 10 processes describing best practices for service delivery and service support. Among them is service-level management which is a process designed to align IT services with business goals and make IT services delivery and support committed to quality customer service.

Such high-level strategies are relatively easy to communicate but significantly difficult to accomplish, inasmuch as they represent cultural changes, potential organizational changes, and the introduction of enterprise-wide processes and disciplines where frequently there are parochial practices designed and developed in a stovepiped environment.

The establishment of business services creates logical services which are derived from and have a direct relationship to the IT components that comprise their provisioning. Mapping these relationships is a major task, as is creating service monitors and reports similar to the application and infrastructure monitoring achieved in the original implementation of the ESM system.

Because ITIL is a comprehensive integrated framework for IT processes and best practices, it is impossible to see a process in isolation and still comply with the ITIL concept. Therefore, all five service design and delivery processes are integrated with each other, all five service support processes are integrated with each other, and all 10 processes are integrated with each other in terms of what information is shared and what relationships exist between them.

Service-level management cannot be considered without considering its relationship to all other processes. Indeed, the existing ESM system has relationships to systems outside its purview, most specifically the Remedy action request system (ARS), which houses the Help Desk application. It is this system which captures incident and outage information. The ESM alerts are used to create outage tickets. The Remedy system is used to populate ticket numbers for outages to the recorded alerts in the ESM system, providing a common data field for ESM produced data and Remedy produced data. The output of Remedy tickets is used in conjunction with ESM alerts to produce research and diagnostic reports.

Service-level management concerns itself with the delivery and support of IT Services to the customer and the end users. It contains the concept of a service catalog which defines the business services available to end users, as well as the availability of the services, their performance commitments, and support criteria. At the heart of service-level management is the monitoring of services to measure the compliance of the services delivered to their commitments of performance as specified in the service catalog. It is this requirement that ties ESM into the ITIL processes.

To monitor the services requires an end-user emulation monitor, as the commitments to the services are in end-user terms. To monitor the end-user service, the end-user emulation monitor must be able to monitor logical services, which means that it must be able to monitor the IT components that comprise the logical service and relate the IT component monitoring to the business service along with the impact of any detected issues. The relationships are contained in a database known as the configuration management database (CMDB) in ITIL. Since the relationship of the logical services to the IT components that provision them is manually created, the relationship is nondiscoverable and a template describing it must be created. This requires the ability to edit the CMDB for both content and schema.

Due to its broad purpose and utility, the CMDB contains a substantial amount of data, and automatically populating it as well as updating is crucial to its feasibility and effectiveness. Fortunately, vendors targeting their products for the ITIL automation market have anticipated this requirement and offer discovery products that do so to varying degrees of completeness.

Discovery products will locate network devices, servers, applications, and databases. Some will also capture configuration information. They will determine the topological relationship between the various objects and even determine the relationships between applications and databases.

Discovery products are used to populate the CMDB and to detect differences between a populated CMDB and what is actually in the infrastructure at the time a discovery scan is executed. Items in the CMDB are generically called CIs. A CI may be a network device, a server, an application, a database, a person, a software license, or even a renewable contract. CIs are whatever is required to manage the infrastructure which is desired to track the status of, or is intricately linked to the support of the infrastructure, such as the on-call staff.

The five ITIL processes comprising the service support processes are designed to keep a CMDB updated. These five processes are: incident management, problem management, change management, release management, and CM.

The incident management process describes the connection a service request made of the Service Desk (expanded Help Desk) has with the CMDB. This would be a content update manually made by a Service Desk staffer using the incident management automation support software such as Remedy Help Desk.

The problem management process describes the diagnosis and solution to a problem (typically an outage) and the connection the solution has to the CMDB. This would be content update made with automation support software for problem management such as Remedy Help Desk and would represent a change to the infrastructure which was preauthorized, such as swapping a network card out when it is the root cause of a network problem.

The change management process describes the handling of a request for change which will require approval and when implemented will solve a problem with a CI in the CMDB. This would be content update made with automation support software for change management such as remedy change management and would represent a change to the infrastructure requiring approval such as a software update.

The Release Management process describes the handling of a release of one or more changes to the infrastructure. This would be content update made with automation support software for Release Management such as BladeLogic Client Automation (BMC) Marimba software.

The CM process describes the management of the CMDB itself including record types, schema changes, content changes, and federated CMDB management.

Compliance with these processes assures that the CMDB will be kept up to date and assures that changes to the infrastructure will be captured so that a change-induced anomaly, when detected, can quickly be tracked and fixed.

The introduction of the CMDB allows not only the monitoring of the business services to be accomplished but also for impact to be displayed in real time, whether it be impact on end users or financial impact of an outage, based on a formula determined to represent financial impact.

There are four additional ITIL processes that have not been covered in the above, inasmuch as they are more concerned with operations than monitoring. These are: financial management, availability management, capacity management, and service continuity management.

ESE should be viewed as an expansion in the scope of traditional SE that exhibits unpredicted (emergent) behaviors due to the large number of interacting variables. Interactions of these variables are increasingly difficult to quantify in a meaningful way and demand a change to the traditional system engineering infrastructure. Traditional SE seeks to optimize a design for a specific purpose or a few closely related purposes. It also focuses on the physical laws that govern design and the resulting predictable behavior of the system or SoS. Performance is achieved by combining the correct materials in the correct way to achieve a predictable and repeatable result.

Enterprise system engineering is fundamentally different; therefore, it requires a different infrastructure. "Enterprise Engineering," as defined by Visible, involves all the activities that organizations ("enterprises") perform to improve productivity, gain and maintain competitive advantage, optimize resources, deliver quality products and services, and meet customer expectations and demand. These can include traditional activities such as reorganization, concentration on core products and competencies, niche marketing, acquisition, merger, and new technologies. Enterprise engineering also includes new techniques and methods such as business process reengineering, continuous process improvement, total quality management, enterprise architecture, and enterprise integration. One thing that distinguishes today's successful enterprise, business, and government, is the ability to adapt to changes in environment, in markets, and in customer expectations. Another key difference includes the necessity to deal with greatly expanded sets of variables—among them changing and conflicting stakeholder interests, a constantly changing environment, the need to rapidly adapt, and faster and more effective movement of data between elements of the enterprise.

The ESE infrastructure must be able to document an enterprise's performance measures and link them to appropriate elements of strategic plans. This linkage allows quick reaction to changes in environment, policy, or customer requirements. The information requirements that support an enterprise's performance measures must be modeled and translated into database, data warehouse, DSS, and EIS designs. These designs can then be developed and implemented to provide for automated capture, reporting and analysis of performance measures. This argues that SE must develop and use tools to support strategic planning and performance measures. SE must provide the means to link strategic planning and performance measures as near real-time, integral and dynamic parts of the enterprise so that performance can be continuously updated and evaluated to see trends in relation to the strategic plan. Tools must be developed to evaluate the environment, changing policies, and changing customer requirements. The information that is collected has to be put into a high fidelity model of the enterprise, developed by system engineering that can be evaluated by decision makers in near-real time. System engineering must develop and integrate decision-support tools for decision makers.

This may mean that enterprises need to totally "reengineer" how work gets done—new goals, new methods, new processes, new measures, and new technologies. Real challenges abound for SE in the Enterprise. Two of the most challenging are detecting emergent behaviors and designing the enterprise infrastructure with enough flexibility so that it can be adapted to the changing environment at the least cost and with cycle times that are better than any competitors. Characteristics unique to the enterprise are rapid exchange of information between all elements of the enterprise, the flexibility to adapt to rapid changes in the environment, and the ability to exhibit unexpected behaviors while meeting the goals of the enterprise. These characteristics require that the elements of the enterprise be connected and share information in the most effective way possible. The system engineer must focus on the development of the most appropriate effective way for the elements of the enterprise to communicate. This forces the system engineer to design the most effective system for sharing information without filtering it. All information in the enterprise must be available to every other element in the enterprise so that, as the environment or goals of the enterprise evolve, the enterprise can respond appropriately. Filtering presupposes the information needed for decision making and should be avoided. So, another part of the enterprise system engineering infrastructure is to develop query tools for requesting information an element needs from any other element in the enterprise. The enterprise system engineer has to understand the enterprise and design it to be very agile. The enterprise must be designed in such a way that it can adapt to rapid changes in environment, stakeholder changes, government policy changes, and changes by competitors. In short, the enterprise needs to be designed to be a self-organizing and self-regulating machine. Information sharing is critical, but starting with an efficient general purpose enterprise design that can be changed quickly, and with the least possible investment in time or money while being more efficient that its competitors should be the goal of ESE.

8.11 Business Planning

8.11.1 What Is a Business Plan?

A business plan precisely defines a business, identifies goals, and serves as the organizational plan of action. Fundamental components include strategic and tactical plans for key initiatives such as investment, development, operations, and communications within an organization with a clear focus on collective success. It helps determine how to allocate resources effectively and efficiently, handle unforeseen complications, and make good overall decisions. A finely crafted business plan is one of the most powerful tools of any business pursuit—both for acquiring investment, and for formalizing strategic and tactical plans and concepts. It will bring to life a vision in the form of actionable road maps and a compelling story of how the new good or service will create value. It is useful for attracting organizational support and it helps align the interests of partners, customers, and employees. It forms the cornerstone of all launch, growth, and expansion activities. The real value of creating a business plan is not in having the finished product in hand; rather, the value lies in the process of researching and thinking about your business in a systematic way. The act of planning helps you to think things through thoroughly, study and research if you are not sure of the facts, and look at your ideas critically. It takes time, but avoids costly, perhaps disastrous, mistakes at a later stage.

8.11.2 Example: A Business Plan for a Military Enterprise

A business plan for a Military Organization Enterprise Management for the net-centric warfighter takes a broader look at warfighter capability needs to drive delivery of materiel solutions. Enterprise management requires crosscutting strategic and business planning that focuses on business integration, net-centricity, SOAs, data interoperability, and other integration functions, independent of singular goals of system program offices (SPOs). In our experience, defining business functions and activities in a business plan are key to working on transformation activities and for tightening up the way-ahead, which has some identity issues or stovepiped approaches that are not aligned with one another and may not meet the needs of the enterprise. In the military, acquisition program offices have always had a defined identity with clear goals and metrics for success for PMs. However, with organizations which are standing up to address evolving net-centric and cross-program integration, as well as enterprise engineering, strategy and business planning are even more important. Organizations that deal with cross-program goals are transformational in nature and a clear definition of business functions, deliverables and customers, with appropriate agreements with stakeholders and customers, will set the key elements in place for success. In enterprise management, business functions are defined based on a strategic plan. Not all the functions in the business plan will be executable in a fiscal year. Strategic direction, assessment of available resources, and business scope are used to prioritize and drive a prioritized implementation plan. While industry business plans are profit oriented, plans in the military are warfighter-capability and budget-oriented. Business functions are separate from, but should drive, an organizational structure.

8.12 Technology Transition

The enterprise system engineer has a crucial role in the governance of a constantly evolving enterprise. The enterprise system engineer must expand the traditional role of system engineer to include an awareness of the overall enterprise, including people, technology, and processes. Enterprise system engineers must focus on enterprise solutions, meaning that they will make assessments based on their impacts to the enterprise architecture, not individual systems or specific missions. They must understand the purpose and missions of the enterprise, how to fulfill them, and any constraints imposed by environment or policy directives.

Enterprise systems engineers are not necessarily a part of the process that determines needs. These are often obtained from directives; however, they do have a responsibility to determine the practicability of implementing needs, including introducing new technologies, and determining strategies for implementations. The role of introducing new technologies is perhaps the least understood aspect of the enterprise system engineer's work.

Enterprise system engineering dictates that needs and systems be evaluated based on their impact on the enterprise. Understanding the enterprise is a prerequisite to determining how it can be improved or how new missions can be added. It includes understanding the interactions of people and machines in an environment not necessarily controllable, followed by the assessment of limitations, and value-added of addressing needs, and finally, the determination of a practicable solution. This is the up-front work for the enterprise system engineer, but it is only the beginning. What follows the selection of a solution is the attempt at implementation. A good solution may be sabotaged by a bad implementation. The final product must be usable, satisfy a mission need, and be compatible with the enterprise. After providing a validated architecture, it is equally important that the ESE be involved in the implementation process, and to do this will require a monitoring (test) plan that makes the development visible.

To validate a design solution, the enterprise systems engineer will need to perform several analyses. Where a new technology is involved, the decision to proceed is a trade-off between the importance of the need and the cost of implementing it. Importance is often influenced by cost, and cost is an important part of the assessment of alternative (feasible) solutions. These analyses do not happen overnight, and "straw man" solutions are almost always modified after scrutiny by a community of stakeholders, adding to the time required. To maintain a reasonable schedule, the enterprise systems engineer must reduce the possibilities by using his understanding of the new technology, operational concept, mission need, enterprise architecture, performance requirements, and system solution(s). Even if a solution is theoretically possible, the environment that it will see may preclude its successful implementation—the responsibility of the enterprise systems engineer.

Insuring the success of the implementation in hardware and software, including the insertion of new technology, is the second part of the ESE responsibility. This includes assuring that the final product is usable, satisfies the primary mission need, and is compatible with its intended environment (the spirit of AFPD 63-12). Being a part of the implementation design, monitoring its progression, and testing and validation of the final product is the means to the end. Implementation also implies an awareness of several auxiliary areas that are required for the final product, for example, IA, exercise support, and so on.

Thus the enterprise systems engineer bridges several communities of development, monitoring the development to the level of detail necessary to insure an accurate implementation. Being the bridge implies that he or she understands the language spoken in each community, and that they can provide translations between communities where needed. Thus, the ESE understands the technology (algorithm design), the derived system requirements, and the implementation process (S/W), and is able to explain, translate, or produce associated documentation. For instance, from the perspective of S/W implementation, the ESE should know if it is feasible or even possible to translate the required technology into S/W, and even more, should be able to provide an estimate of performance bounds before the development begins. It is important to determine this feasibility before time and resources are spent, and to propose mitigation if appropriate.

The ESE is trying to avoid several things that go wrong with developments, for example:

- Mission need not met
- Enterprise architecture negatively impacted
- Algorithms incorrectly implemented
- Unacceptable performance
- Operational or systems requirements not met

ESE is a crucial bridge that must bring different disciplines together, and all of the disciplines are apparent when translating new technology from theory into practice. The ESE must understand all aspects of the process from theory to implementation—thus he or she deserves the title "multidisciplined." The tendency, however, is to use a narrow interpretation. Thus, a developer of software is apt to use system and software engineering interchangeably, whereas a radar engineer will consider system engineering as the design of a good tracking system. Both are partially correct. However, there exists a mutual underestimation of each others' problems. One should not expect a software developer to appreciate the theory of tracking; nor is it reasonable to expect that a radar engineer will appreciate class diagrams or software interfaces. One should, however, expect the ESE to be familiar with both disciplines, and in addition, to assess the implementation environment, for example, what impacts will communications channels or the host processor(s) have on performance?, or, how can the implementation architecture be "componentized" to improve performance while still being true to the algorithmic design? The ESE is bringing to the table an awareness level that raises the probability of a successful system implementation.

8.13 Summary

We have examined the key facets of enterprise management, paying special attention to the ways in which traditional SM changes in an enterprise environment. No attempt has been made to coherently relate the various elements—in fact the power of the Wiki process is that it gets the wisdom from many sources onto the printed page, the weakness of which is exactly in the organization and coherency of the whole document. That task will be the subject of the next publication series.

List of Acronyms

AETF	Air and Space Expeditionary Task Force
AF	Air Force
AIS	Automated Information System
API	Application Programming Interface
ARS	Action Request System
ATM	Air Traffic Management
BMC	BladeLogic Client Automation
CAC	Common Access Card
CCA	Clinger–Cohen Act of 1996
CDD	Concept Definition Document
CI	Configuration Item
CII	Common Integrated Infrastructure
CIP	Critical Infrastructure Protection
CJCSI	Chairman Joint Chiefs of Staff Instruction
CM	Configuration Management
CMDB	Configuration Management Database
CND	Computer Network Defense
COTS	Commercial Off-the-Shelf
CVS	Concurrent Versioning System
DHS	Department of Homeland Security
DISA	Defense Information Services Agency
DMZ	De-Militarized Zone
DoD	Department of Defense
DoDI	Department of Defense Instruction

DODIIS	Defense Index of Specifications and Standards
DOTMLPF	Doctrine, Organization, Training, Materiel, Leadership and Education, Personnel, and Facilities
DSL	Digital Subscriber Line
DSS	Decision Support System
EBA	Enterprise Business Analyst
EDW	Enterprise Data Warehouse
EII	Enterprise Information Interoperability
EIP	Enterprise Implementation Plan
EIS	Executive Information System
ERP	Enterprise Resource Planning
ESE	Enterprise Systems Engineering
ESM	Enterprise Service Monitoring
ETL	Extract, Transform, and Load
FAR	Federal Acquisition Regulation
FDP	Full-Dimensional Protection
FYDP	Future Years Defense Program
GANS	Global Access, Navigation, and Safety
GES	Global Information Grid Enterprise Services
GIG	Global Information Grid
GPL	General Public License
IA	Information Assurance
I-CRRA	Integrated Capabilities Review and Risk Assessment
IER	Interface Exchange Requirement
IP	Internet protocol
IT	Information Technology
ITIL	Information Technology Infrastructure Library
KMI	Key Management Infrastructure
LAN	Local Area Network
NCES	Net-Centric Enterprise Services
NCW	Network Centric Warfare
OEM	Original Equipment Manufacturer
OMB	Office of Management and Budget
PB	Program Baseline
PKI	Public Key Infrastructure
PM	Program Manager
POC	Point of Contact
POM	Program Objective Memorandum
PPBE	Planning, Programming, Budgeting, and Execution
PPBS	Planning, Programming, and Budgeting System
PSTN	Public Switched Telephone Network
RCS	Revision Control System
SAMP	Single Acquisition Management Plan

SCDR	Shared Cross-Domain Resource
SE	Systems Engineering
SM	Strategic Management
SNMP	Simple Network Management Protocol
SOA	Service-Oriented Architecture
SoS	System of Systems
SPO	System Program Office
SVN	Subversion
SWOT	Strengths, Weaknesses, Opportunities, and Threats
T1	T-carrier 1, Digital Transmission Line, 1.544 Mbps, 24 Voice Channels
TAP	Telocator Alphanumeric (input) Protocol
TRL	Technology Readiness Level
UML	Unified Modeling Language
UPS	Uninterruptible Power Supply
U.S.	United States
USAF	United States Air Force
VPN	Virtual Private Network
WSDL	Web Services Description Language
XML	Extensible Markup Language

References

Ackoff, R., with Elsa Vergara Finnel and Jamshid Gharajedaghi, *A Guide to Controlling Your Corporation's Future*, Wiley, New York, NY, 1984.

Bernard, S., *An Introduction to Enterprise Architecture*, Authorhouse, Bloomington, IN, 2004.

Cummins, F. A., *Enterprise Integration: An Architecture for Enterprise Application and Systems Integration*, Wiley, New York, NY, 2002.

Demarco, T., *Slack: Getting Past Burnout, Busywork, and the Myth of Total Efficiency*, Broadway, New York, NY, 2002.

Haines, S. G., *The Systems Thinking Approach to Strategic Planning and Management*, CRC Press LLC, Boca Raton, FL, 2000.

Kossiakoff, A., and W. Sweet, *Systems Engineering: Principles and Practice*, Wiley, New York, NY, 2003.

Chapter 9

Agile Functionality for Decision Superiority[*]

Kevin A. Cabana, Lindsley G. Boiney, Lewis A. Loren,
Christopher D. Berube, Robert J. Lesch,
Linsey B. O'Brien, Craig A. Bonaceto, and
Harcharanjit Singh

Contents

9.1 Overview 297
    9.1.1 Introduction 297
    9.1.2 Motivation: The Future of Warfare 299
    9.1.3 Response: Military Transformation 300
    9.1.4 Engineering the Command and Control Enterprise 300
    9.1.5 Network-Centricity: The Hallmark of the Next Generation
          Command and Control Enterprise 303
          9.1.5.1 Get Connected.. 304
          9.1.5.2 Expose Data 304
          9.1.5.3 Generate Knowledge... 304

---

[*] The material by Cabana, K. A. et al. (February, 2006) had already been approved for public release. Note from the Principal Author, Kevin Cabana: Each of the internal research and development (Mission-Oriented Investigation and Experimentation [MOIE] and MITRE-Sponsored Research [MSR]) efforts referenced in this chapter has been successfully completed to reflect the agile functionality intended by the design, and is in the process of being transitioned to a group of end users or decision makers.

    9.1.6 Achieving Decision Superiority: The Canonical Enterprise Functions 305
          9.1.6.1 Information Generation 306
          9.1.6.2 Information Management 307
          9.1.6.3 Information Exploitation 307
    9.1.7 Summary: The Remainder of This Chapter 308
9.2 Agile Information Generation (The Right Information) 309
    9.2.1 Introduction 309
          9.2.1.1 A Definition 309
    9.2.2 Models for Information Generation 310
          9.2.2.1 Considerations for an Agile Information Generation Process 312
    9.2.3 The Information Generation Environment 312
          9.2.3.1 Increased Availability of Data 312
          9.2.3.2 Availability of New Data Sources 313
          9.2.3.3 Greater Need for Sharing Data at All Levels 313
          9.2.3.4 Information Needs and Goals Not Always Known in Advance 314
          9.2.3.5 Greater Potential to Task Data Sources to Gather What Is Needed 314
          9.2.3.6 Data Sources Experience the Complexity of an Environment 314
          9.2.3.7 Data with Future Value 315
    9.2.4 Information Generation Challenges 315
          9.2.4.1 Correlating Data from Diverse Sources 315
          9.2.4.2 Managing Information Needs and Goals 316
          9.2.4.3 Exploiting Elements of Complexity within an Environment 316
    9.2.5 Information Generation Tenets and Design Principles 317
          9.2.5.1 Make Explicit the Quality of Data or Generated Information at All Levels 317
          9.2.5.2 Develop, Publish, and Manage Information Needs and Goals 318
          9.2.5.3 Associate/Correlate Data and Generated Information 319
          9.2.5.4 Enable the Fusing of Data or Generated Information 320
          9.2.5.5 Use Visualization to Understand Relationships in Data or Generated Information 320
          9.2.5.6 Enable the Recognition of Complexity in Data and Decision-Making Environments 321
          9.2.5.7 Warehouse Data or Generated Information at All Levels 321
          9.2.5.8 Apply Meaningful Tags to Data or Generated Information 322

- 9.2.6 Enabling Technologies for Information Generation 323
  - 9.2.6.1 Data Fusion 323
  - 9.2.6.2 Data Mining 324
- 9.2.7 Information Generation Use Cases 326
  - 9.2.7.1 Correlating Data from Diverse Sources 326
  - 9.2.7.2 Managing Information Goals 328
  - 9.2.7.3 Exploiting (and Modeling) Complexity 329
- 9.3 Agile Information Management (The Right People and Time) 331
  - 9.3.1 Introduction 331
  - 9.3.2 Information Management Tenets and Design Principles at the Enterprise Level 331
    - 9.3.2.1 Composite Systems and Sources of Complexity 332
    - 9.3.2.2 Managing Complexity in the Enterprise 334
    - 9.3.2.3 The Five W Questions 335
    - 9.3.2.4 How to Be Agile? 336
  - 9.3.3 Information Management Tenets and Design Principles at the Large-Scale Enterprise Level 336
    - 9.3.3.1 Multifocal Organization: Coupling 336
    - 9.3.3.2 Multiscale Organization: Encoding 337
    - 9.3.3.3 Equilibrium and Organization: Modeling 337
  - 9.3.4 What and Where Is the Right Information? 338
    - 9.3.4.1 Focus on Information Exchanges and Communities 340
    - 9.3.4.2 Adaptive Information Access and Distribution Services 341
    - 9.3.4.3 Adaptive Information Discovery through Search Services 342
    - 9.3.4.4 Adaptive Information Discovery through Catalog Services 343
    - 9.3.4.5 Adaptive Information Mediation Services 343
  - 9.3.5 When Is the Right Time? 344
    - 9.3.5.1 Clocks and Time Sources 344
    - 9.3.5.2 Geophysical Time Frames 345
    - 9.3.5.3 Network-Centric Time Frames 345
    - 9.3.5.4 Time Scales and Complexity in Decision Making 345
  - 9.3.6 Who Are the Right People? 346
    - 9.3.6.1 Ex Officio: Roles, Identity, and Clearance 346
    - 9.3.6.2 Identity Information Management Services 347
    - 9.3.6.3 Information Assurance 348
  - 9.3.7 Which Is the Right Budget? 348
    - 9.3.7.1 Information Management Risk–Cost–Benefit Analysis 349
    - 9.3.7.2 The Impact of Increased Information Sharing 350
  - 9.3.8 Robust Information Management 350
    - 9.3.8.1 Redundancy Mechanisms and Issues 350

                9.3.8.2 Adaptability Mechanisms and Issues 352
                9.3.8.3 Information Infrastructures...353
        9.3.9 Information Management Use Cases...355
                9.3.9.1 Dynamic Information Management in Complex
                        Enterprises FY06 Mission-Oriented Investigation and
                        Experimentation Proposal...355
                9.3.9.2 Semantically Enabled Web Services for Effective
                        Decision Making in Network-Centric
                        Environments FYO6 Mission-Oriented Investigation
                        and Experimentation Proposal 355
                9.3.9.3 Managing Infostructure Complexity in Agile Decision
                        Support FY06 Mission-Oriented Investigation and
                        Experimentation Proposal...356
9.4 Agile Information Exploitation (The Right Decisions) 356
        9.4.1 Introduction 356
        9.4.2 The Information Exploitation Environment................................357
                9.4.2.1 A Dynamic and Complex Battlefield 358
                9.4.2.2 Increased Volume and Rate of Information 358
                9.4.2.3 Reduced Manpower and Cost Goals 359
                9.4.2.4 Changing Human Roles...359
                9.4.2.5 High Stakes, Time Pressure, and Uncertainty 360
        9.4.3 Information Exploitation Challenges..361
                9.4.3.1 People Steer, Guide, and Supervise the Engine
                        of Technology...361
                9.4.3.2 Dynamic Information Acquisition, Filtering,
                        and Tailoring..363
                9.4.3.3 Operators' Use of a Multitude of Disjointed Systems
                        and Communications Modalities.................................... 364
                9.4.3.4 Distributed Collaboration and Operations in
                        Complex Environments,...... 366
                9.4.3.5 Trust Formation 367
                9.4.3.6 Coevolution of People and Technology: Performance
                        of the Joint Human-Technological System.....................368
        9.4.4 Enabling Technologies and Methodologies for Information
              Exploitation 370
                9.4.4.1 Cognitive Engineering..370
                9.4.4.2 Intelligent Decision Support Systems and
                        Automation..372
        9.4.5 Information Exploitation Tenets and Design Principles................373
        9.4.6 Information Exploitation Use Cases 377
                9.4.6.1 Mental Models in Naturalistic Decision Making
                        FY01–FY03, FY04–FY06 MITRE-Sponsored
                        Research Project 377

9.4.6.2 Improving Time-Sensitive Team Decision Making FY04–FY05 Mission-Oriented Investigation and Experimentation 379
9.4.6.3 Dynamic Diagrams in Asset Allocation FY06 Mission-Oriented Investigation and Experimentation Proposal....379
9.5 Capabilities-Based Scenarios 380
  9.5.1 Introduction 380
  9.5.2 Operation Anaconda.. 380
  9.5.3 Information Exploitation 381
  9.5.4 Information Management.. 383
  9.5.5 Information Generation 386
  9.5.6 System Acquisition... 387
Acronyms.. 388
References 390

9.1 Overview

9.1.1 Introduction

This chapter lays the groundwork for the Enterprise Systems Engineering (ESE) Research and Development (R&D) paradigm* motivated by a fundamental goal of Network-Centric Operations and Warfare (NCOW), "getting the *right information* to the *right people* at the *right time* to make the *right decisions*" (R4), under the highly complex and dynamic challenges of the twenty-first century warfare. To

---

* Hereafter, denoted as "our paradigm."

contribute to the realization of this goal, we have identified *information generation, management, and exploitation, and their agile combination*, as major building blocks. *The Oxford English Dictionary* defines agile as "having the ability to move quickly and easily" (Simpson and Weiner, 1989). Achieving agility in dynamic military and security environments requires flexibility, that is, the ability to change in response to new and different circumstances (Simpson and Weiner, 1989). Portions of the enterprise, or subsets of people, systems, and processes within which these functions operate, must be able to reconfigure themselves in new and unanticipated ways to rapidly support military objectives under changing and uncertain scenarios. At the highest level,* the enterprise must be robust, or able to withstand change without serious impact to its support of military effectiveness, or its ability to evolve to provide new and better capabilities.

In the twenty-first century security environment, information generation addresses the problem of getting the right information by collecting, fusing, aggregating, and drawing inferences from data gathered from any and all sources on the extended battlefield—and in a network-centric world, any entity that can make observations function as a data-generating sensor. Information management addresses the critical need of constructing sufficient infrastructure to ensure that this information is readily available to "the right people at the right time constrained by the right budget" in the highly distributed and fluid force of the future. The primary focus of information exploitation is the combination of people and technology to process this information and make the right decisions in time-sensitive and unpredictable environments. The agile combination of information generation, management, and exploitation is hypothesized to promote the fluid application of combat power, enabling it to move quickly and easily where it is needed. In our paradigm, information technology (IT) exists to serve the needs of the warfighter, and we acknowledge that a large part of enterprise complexity comes from the fact that its performance and evolution are driven by creative and adaptable people and teams, functioning across organizations, cultures, military echelons, and enterprise scales.

In this section, we begin by addressing the current and future state of warfare and how it has driven the military to transform itself into a network-centric enterprise to establish and maintain a decisive edge over our adversaries. This transformation has broad and deep implications for our R&D paradigm, forcing us to consider how to engineer an agile enterprise in such environments. Our paradigm is based on the integration of leading-edge technologies such as cognitive science, Complex Adaptive Systems (CASs)—including biologically inspired methodology, and ESE, with classical approaches from IT and distributed computing. We go on to consider characteristics of the Command and Control (C2) Enterprise (C2E) and how, within the above integrative framework, our canonical functions can

---

* Alternatively, when thinking of enterprises (or systems) consider multiple scales or views, combinations of scope, granularity, mind-set, and time frame. See White (2007); also, refer to McCarter and White (2008).

serve to enable C2E capabilities. The remainder of this chapter describes characteristics that underpin the agility of information generation, management, exploitation, and their combination in greater detail by highlighting key challenges, engineering design principles, and research needs within and across these areas. The final section of this chapter describes a complex and time-sensitive operational scenario in which all three functions of the paradigm come together as hypothesized to improve military effectiveness and impact.

9.1.2 Motivation: The Future of Warfare

Throughout history, warfare has always been a complex and messy undertaking. Twenty-first century warfare is proving to be even more complex, with multidimensional challenges to the U.S. global interests across a spectrum of conflict never experienced before. There is no clearly defined battlefield—rather, the battlespace is global, stretching to wherever we have interests and vulnerabilities that can be attacked or exploited by our adversaries. There will of course continue to be traditional force-on-force warfighting, but it will be very different and likely to be conducted alongside humanitarian missions, peacekeeping activities, and responses to asymmetrical threats such as the current insurgency in Iraq and the global war on terror, where our adversaries will fight us on their terms with a range of effective and unforeseen techniques, based on complex human and social networks with clear commander's intent. These techniques can often negate our technological advantages and challenge our will to continue (Potts, 2003).

From this emerge three key aspects of the future warfighting environment with significant implications for the U.S. and its coalition partners:

1. We will continue to have global interests and be engaged with a variety of regional actors. The joint force of the future must be prepared to achieve victory across the full range of military operations in any part of the world, to operate with multinational forces, and to coordinate military operations as necessary with government agencies and international organizations.
2. Potential adversaries will have access to the global commercial industrial base, the global commercial infrastructure, and much of the same technology that we possess. We will not necessarily sustain a wide technological advantage over our adversaries in all areas. Our advantages must therefore come through improvements in concepts of operation, doctrine, organizations, training, education, leadership, and force structure that collectively enable us to take full advantage of technology to achieve superior warfighting effectiveness (Alberts, 1996). Lastly, we will need to iterate through these improvement and adaptation cycles more quickly than our adversaries.
3. As our capabilities evolve, we should expect potential adversaries to adapt and make use of asymmetric approaches that avoid U.S. strengths and exploit potential vulnerabilities, such as attacks against U.S. citizens and territory.

9.1.3 Response: Military Transformation

The comprehensive response to this future warfighting environment is the transformation of the military into a joint force that is dominant across the full spectrum of military operations, that is: persuasive in peace, decisive in war, and preeminent in any form of conflict. This is tantamount to an endorsement of a broader and not always conventional vision for how U.S. military power should be applied, which recognizes that the military will likely be involved in humanitarian and peacekeeping missions, while also being called upon to respond quickly to imminent security threats. For the joint force of the future, these goals will be accomplished through a doctrine of *full spectrum dominance*—the ability of the military, operating unilaterally or in conjunction with multinational and interagency partners, to defeat any adversary and control any situation across the full range of military operations.

To make this transformation highly adaptive to the type of asymmetrical threat that characterizes insurgents, terrorists, and other nonstate actors, this vision must be extensible to an emerging concept of maneuver warfare spearheaded by the Marine Corps, *distributed operations*, where a military configured as small, highly skilled, and independently operating units can aggregate quickly to apply sufficient and just-in-time force wherever it is required against a complex and adaptive threat. This threat is fueled by strong social networks and timely, pervasive human intelligence to leverage dispersed attacks that result in chaos, friction, and uncertainty. Without access to equivalent social networks and human intelligence, our response must be based on greater support from technology, collaborative tools, and processes that provide our forces with timely situational awareness to make and execute decisions across military echelons—with the full power of the enterprise behind these decisions. Our R&D paradigm detailed throughout this chapter is a dedicated, albeit modest attempt to address this asymmetrical need while providing greater flexibility for our response to more conventional threat. Capabilities needed to support distributed operations can be summarized as: (1) the ability for dispersed, maneuvering forces to conduct network-enabled operations and make critical, real-time decisions; (2) increased situational awareness for ground forces, based on timely and sufficient (rather than optimal) information; (3) the sustainment of widely dispersed units over extended periods; and (4) enhanced individual and collective mobility of ground forces (McBrien, 2005). Our R&D efforts must therefore be extensible to these capabilities, while serving to provide greater agility to the more centralized and hierarchical demands of time-sensitive team decision making for conventional warfare.

9.1.4 Engineering the Command and Control Enterprise

The primary challenge for this transformation is the engineering of a powerful C2E, characterized by the complex interaction of people (e.g., soldiers, commanders,

Agile Functionality for Decision Superiority ■ 301

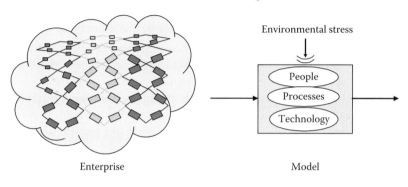

Figure 9.1 A notional enterprise.

and systems engineers), processes (e.g., doctrine, concepts of operation, Rules of Engagement [ROE], and information-sharing practices), and technologies (e.g., information systems, decision support systems, weapons systems, and sensor systems), that interact with each other and their environment to achieve organizational goals such as decision superiority. Figure 9.1 shows the graphical depiction of a notional enterprise that is highly recursive, and comprised of many subenterprises, each constituting an enterprise at its own level. For the C2E, the Airborne Warning and Control System (AWACS) constitutes its own enterprise, and it is likewise comprised of subenterprises of people, processes, and technologies with their own local goals and functions which in turn serve the enterprise-wide goals and functions. This recursive nature* of the C2E adds to its complexity and demands new ways of engineering, shaping, modeling, and understanding these systems and relationships to effectively achieve its organizational goals. The emerging field of ESE embodies the framework necessary to promote this change.

A related and important feature of an enterprise is its dynamism. That is, an enterprise is in a constant state of evolution, reinventing parts of itself through a process of continual innovation and integration. As enterprise systems engineers, we must shape the integration and innovation environment so that the target capabilities are better able to be envisioned and maintained. A particular challenge for researching and developing these capabilities, derived from this characteristic, is the need to work fluidly and transparently across multiple scales of an enterprise.

---

* In general, enterprises (and systems) are not simply nested but rather overlapping as well as nested.

To illustrate the evolutionary nature of an enterprise as it adapts itself to new goals and constraints, consider the military problem of resource allocation, where a limited number of assets (e.g., weapons and tankers) must be paired against a number of objectives (e.g., targets and aircraft in need of refueling) to achieve mission goals and make the best use of these assets. One way to solve this problem would be to develop a system tailored to a specific resource allocation problem. Such a system would be incapable of adapting itself to other and perhaps unforeseen resource allocation problems. As an alternative approach, consider instead, a family of generic resource allocation algorithms and visualization techniques that exist on the enterprise and could be applied to a myriad of resource allocation problems. Such components could be rapidly adapted and reconfigured to meet the demands of any resource allocation problem. Moreover, operators trained in resource allocation decision making could rapidly apply these skills to different areas as the demands of conflicts shift. Such an approach is highly consistent with our vision of an agile enterprise comprised of a variety of adaptable components that can be rapidly reconfigured to meet changing demands, and for which the line between systems acquisition and systems operation is continually blurred by humans who constantly adapt themselves and their systems to meet the emergent challenges of warfare.

Important leading-edge disciplines that serve as foundational elements of ESE include the following:

1. *Complexity theory (CASs)* The C2E is a CAS, comprised of many people, processes, and technologies that interact with each other in ways that continually reshape their collective future. The interactions and connections among these diverse elements lead to emergent features (which may or may not be desirable or predictable), and reflect the enterprise tenet that "the collective whole is greater than the sum of its parts." Because our adversaries also function as CASs (in many instances coevolving with our own), both research and engineering practice need to work to understand where the leverage points are in CASs so that we can harness complexity and use it to our advantage (Axelrod and Cohen, 2000)—and to the disadvantage of our adversaries.
2. *Social science and social network theory* In the C2E people do not work in isolation. Owing to this, we need to understand the dynamics of team collaboration, distributed decision making, information sharing, trust formation, and the development of shared situational awareness. As we move to joint and coalition military and cross-agency venues, we need to understand how to overcome the social and cultural barriers that hinder effective cooperation.
3. *Cognitive science* People are ultimately responsible for making decisions and taking actions in C2. To serve this, we need to understand their decision making and cognitive strategies, their biases, how they build and maintain

situational awareness, and their performance limits. Ultimately, we need to engineer effective support systems and infrastructure to aid them in meeting their decision-making goals.

4. *Information science* Information is the lifeblood of the C2E. We need to understand how to best leverage IT, in conjunction with the above (and other) disciplines, to generate, manage, and exploit information in a manner that provides us with a continual edge over our adversaries.

9.1.5 Network-Centricity: The Hallmark of the Next Generation Command and Control Enterprise

The C2E will continue to mature and evolve as a complex collection of people using processes and technology to achieve their objectives. To a progressively greater extent, these enterprise agents will not be located in only one place. Thus, in order to make the best use of this geographic dispersion, these forces must be able to freely share, exchange, and exploit information to address any level of threat (Alberts et al., 1999). This key principle of NCOW, the Concept of Operations (CONOPS) for the C2E, is summarized in Figure 9.2.

The NCOW tenets from Figure 9.2 organize naturally into three clusters: (1) get connected, (2) expose data, and (3) generate knowledge, which are discussed in the following sections; when fully realized, will promote greater information sharing across the C2E.

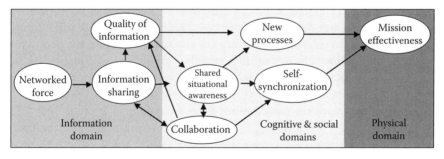

Figure 9.2 Tenets of network-centric warfare. (Modified from Alberts, D. S. et al. [1999]. *Network Centric Warfare: Developing and Leveraging Information Superiority.* CCRP Publication Series.)

9.1.5.1 Get Connected

Get connected includes understanding of *who* and *what* (people and machine agents) are reachable, *where* and *when*, and has as its basis information management systems that facilitate understanding. The major challenges in a network-centric environment center on moving from just physical asset management to parallel management of physical and virtual assets such as services. Leading-edge research on this layer includes areas such as dynamic asset acquisition, deployment and configuration (and their implications for complexity), digital security, dynamic fault tolerance, and performance optimization, as well as the formal analysis of human/system interactions during the generation, management, and exploitation of information.

9.1.5.2 Expose Data

The primary focus of this chapter is on the expose data and generate knowledge clusters. Expose data addresses the challenge of ensuring that any consumer of data and information can discover and access/receive it from any producer in the C2E. Given the number, autonomy, and complex interactions of these groups, as well as the need to perform these functions rapidly in a dynamic environment, this is a daunting challenge that requires, at a minimum, a smooth transfer of information through machine-to-machine (M2M), machine-to-people, and people-to-people transactions. At its most mature level it reflects an intelligent capability to recognize and communicate information patterns across systems and to enable users to discover and retrieve key information—even if they do not know if it exists or how to request it.

9.1.5.3 Generate Knowledge

Generate knowledge addresses the preparation of data and information for exploitation by human decision makers, and includes as its primary focus the fusion and aggregation of data and information for sense-making and situational awareness, as well as the cognitive and collaborative processes necessary to leverage this information for effective decision making in complex environments. From the perspective of our paradigm, warfighters and their missions are the forcing function for the development of technology and collaborative processes that provide strong support for decision making in time-sensitive environments. This view is especially receptive to the increasing trend toward decision making at lower echelons, for scenarios, such as: (1) embedding Close Air Support controllers with the forces they support—while expecting them to remain tightly coordinated with other controllers, (2) enabling local medical assets to launch a coordinated response against multiple and dispersed biological attacks, and (3) supporting distributed operations—Special Operations Forces (SOF) and lower-echelon decision makers dealing with complex insurgencies.

9.1.6 Achieving Decision Superiority: The Canonical Enterprise Functions

At the beginning of this section, we introduced R4 as an overarching goal of the C2E and noted that this would be achieved through the integration of people, processes, and technologies performing the functions of generating, managing, and exploiting information more efficiently than our adversaries (or in rapid response to a natural disaster). This must be supported by a robust and adaptive information infrastructure (infostructure), whose availability under challenging and unpredictable circumstances is enhanced by characterizing and managing its sources of complexity.

Figure 9.3 illustrates the interactions among these canonical functions. In NCOW terms, the battlespace comprises the "physical domain"; information generation and exploitation comprise an overlap of the cognitive and information domains, whereas information management resides in the information domain

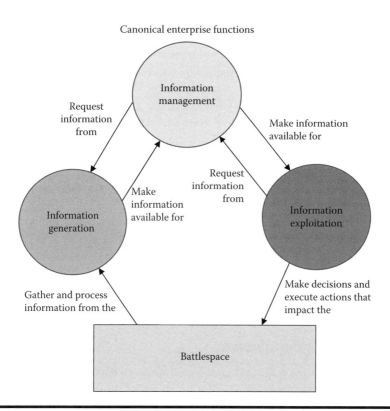

Figure 9.3 Interactions among the canonical enterprise functions and the battlespace.

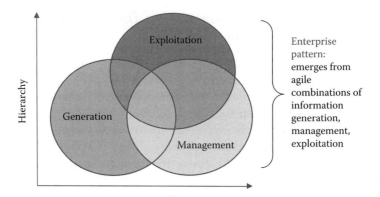

Figure 9.4 The canonical enterprise functions.

(Alberts et al., 1999). For the R&D of enterprise capabilities, this is helpful as guidance in determining what combinations of technologies need to be integrated to produce the target functionality. Within this context, the infostructure can be viewed as an overlap between the cognitive and information domains. Ensuring its availability to warfighters or responders in unpredictable, emergent situations is critical to the achievement of R4.

Figure 9.4 shows the canonical functions in the context of the enterprise pattern they must fluidly form to support R4. The vertical axis (hierarchy) identifies information exploitation (people taking actions in C2) as the forcing function, with the generation and management of information providing strong support.

First-order information flow suggests that information must be generated and managed before it can be exploited. However, different orders and configurations are possible with fully agile functionality (e.g., the iterative aggregation of information and knowledge to support time-sensitive team decision making).

For today's environment, it is necessary to be able to configure these functions in a rapid, on-demand fashion that can leverage the full power of the C2E. We briefly describe each critical function area below. Sections 9.2 through 9.4 provide more in-depth treatment of these functions, enumerating operational goals and objectives, as well as engineering design tenets, principles, and research needs.

9.1.6.1 Information Generation

Information generation is the aggregation and fusion of data and information from sensors and other sources in the C2E for direct use by decision makers. As will be discussed in Section 9.2, information generation utilizes pattern recognition and information fusion technologies that can be fully or partially automated, and focuses directly on contributions to sense-making, situational awareness, and other decision-making activities. An information generation function would heavily leverage the output of netted-sensors and any nonstandard sources of data for better and timely sense-making on objects of interest (and their behaviors).

9.1.6.2 Information Management

To be meaningful, the products of the information generation function must be made available on the enterprise, and either pulled by or pushed to operators, often rapidly, to aid in their decision making. This requires an information management function of commensurate agility to help users discover the relevant information sources for aggregation. Conversely, help discover users who, based on their activity and information request patterns, would benefit from data and information being generated elsewhere in the enterprise (often without knowledge of its existence). This management function is based on dynamic discovery, intelligent agents, and distributed computing technologies (among others), and requires an infostructure that is able to adapt to complexity at different levels and scales of the C2E.

In our paradigm, information generation and management provide strong functional support to the exploitation of this information by people engaged in C2.

9.1.6.3 Information Exploitation

Information exploitation refers to the direct use (explicit or transparent) by people of these functions and underlying technologies as well as the appropriate collaborative processes to make decisions. The exploitation function is based in cognitive science and collaborative information technologies, and pays strict attention to how people best function under complex decision-making situations, and how technology and processes can best be integrated to support them. The net effect of this paradigm, or agile combination of functions, is to add a new forcing function, people, to coevolve with the existing forcing function, technology, to enable them to be more creative and innovative in taking actions in C2. If this is done effectively, it will enable people to better leverage the power of technology and processes, rather than forcing them to compensate for their limitations.

As shown in Figure 9.5, the enterprise pattern comprised of the integration of these functions spans the major process nodes of network-centric information flow in a comprehensive way. Thus, researching and developing capability in the integration of these functions has the potential to yield significant returns-on-investment for decision support and its dependence on the underlying infostructure (i.e., its availability).

To get a clear picture of this, Figure 9.6 shows a hierarchical technology pyramid that underpins our paradigm and much of enterprise technology development and application. A facile mapping to this pyramid would place exploitation in the decision support layer, generation in the boundary between the decision support and IT layers, management, especially when infostructure complexity is considered, and

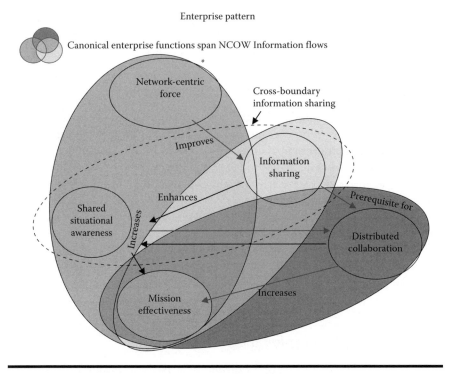

Figure 9.5 How our enterprise research paradigm spans the nodes of network-centric information flow. (Modified from Mitchell, P. J., Cummings, M. L., and Sheridan, T. B. [2004]. *Human Supervisory Control Issues in Network Centric Warfare*. Massachusetts Institute of Technology Humans and Automation Laboratory Research Report No. HAL2004-01. Retrieved December 10, 2005, from http://web.mit.edu/aeroastro/www/labs/halab/papers/HSC_NCW_report.pdf)

firmly connects the IT and network layers of the pyramid. All of this is highly dependent on strong foundational support (sensors and enabling technologies).

9.1.7 Summary: The Remainder of This Chapter

In this section we have briefly discussed the complex challenges the military will face in future warfighting scenarios—from both conventional and asymmetrical threats. We have identified the response to these challenges as the transformation of the military into a network-centric enterprise, within which agile information generation, management, and exploitation, fed by netted, heterogeneous sensors, and supported by a robust infostructure, are necessary conditions for the type and magnitude of the information and decision advantages required. It should be pointed out that the scope of the research landscape associated with this functionality is very large—as is the landscape of the operational and engineering practice venues

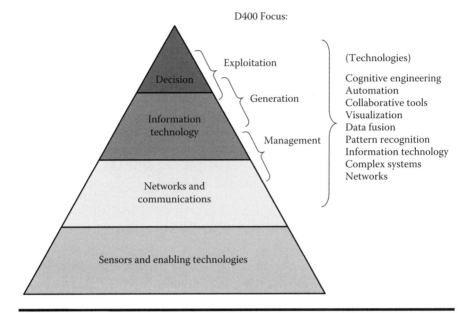

Figure 9.6 Enterprise technology layers.

it targets. The research paradigm introduced here, and treated in more detail in the subsequent sections, covers at best a small, but potentially important part of this landscape, based on the integration of CASs, cognitive science, IT, and ESE.

9.2 Agile Information Generation (The Right Information)

9.2.1 Introduction

9.2.1.1 A Definition

What is information generation? Consider the following as one possible definition:

> Information generation is a process which involves the collecting, processing and displaying of data (i.e., facts of various kinds, informed opinions, wild guesses, etc.) from which one can answer certain relevant questions (who, what, where, when, why) relevant to a given situation. This process may also involve the allocation of resources to gather more data in the event that any one or more questions cannot be answered satisfactorily.

Other equally descriptive definitions are certainly possible, but in the broadest sense, information generation is a process which contributes to or enables the understanding of a dynamic situation. Of course, the degree to which such understanding is needed depends greatly on the problem to be solved (i.e., the context).

For example, it may be sufficient to know where an object (e.g., a person, a vehicle) is located, but not care about its relationship (if any) with other objects. Furthermore, generating information in an *agile* manner considers both the need to be *adaptable* (i.e., modifying a process to accommodate changing circumstances) and *flexible* (i.e., modifying a process to accommodate changing requirements or goals) in one's process.

In the discussions that follow, we will consider data to be the input to a first level of an information generation process, and generated information to be the output. This generated information, in turn, may serve as the input to a second, third, and so on, level of an information generation process. Also, we will sometimes refer to data and generated information collectively as elements.

9.2.2 Models for Information Generation

One model for the information generation process, as it applies to a fairly broad class of problem, the data fusion problem, is the *Joint Directors of Laboratories (JDL) Data Fusion Model*. Informally speaking, data fusion is a process that combines information from multiple sensors and related information sources to achieve better inferences than could otherwise be achieved with just a single sensor or source. This model, originally developed with military systems in mind (but later extended to commercial systems), is a conceptual model which seeks to identify basic processes and functions which may be required to implement a data fusion system. It defines a basic process flow and identifies processes and representative functions. Rather than being a "blueprint" for the development of a data fusion system, the model is intended to be generic and serve as a basis for common understanding and discussion.

Figure 9.7 represents a revised version of the JDL data fusion model. Note that the model is comprised of five levels of data fusion processing and the interaction with sources (of data), a database management system, and a human/computer interface.

Descriptions of each level (provided as part of the model) are provided below.

- Level 0—Subobject data assessment: Estimation and prediction of signal- or object-observable states on the basis of pixel/signal level data association and characterization.
- Level 1—Object assessment: Estimation and prediction of entity states on the basis of inferences from observations.
- Level 2—Situation assessment: Estimation and prediction of entity states on the basis of inferred relations among entities.
- Level 3—Impact assessment: Estimation and prediction of effects on situations of planned or estimated/predicted actions by the participants (e.g., assessing susceptibilities and vulnerabilities to estimated/predicted threat actions given one's own planned actions).

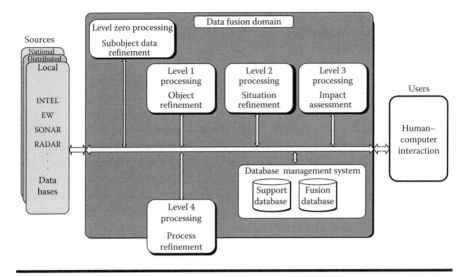

Figure 9.7 The JDL data fusion model. (Adapted from Kessler, O. et al. [1991]. *A Functional Description of the Data Fusion Process*. Office of Naval Technology, Naval Air Development Center, Warminster, PA.)

- Level 4—Process refinement (an element of resource management): Adaptive data acquisition and processing to support mission objectives.

The concept of situational awareness plays a major role in the processes concerned with cognition, decision making, and collaboration in military as well as nonmilitary domains. As defined by Endsley (2000), situational awarenesssimply is knowing what is going on around you. Domains in which situational awareness is studied include aircraft piloting, education, driving, train dispatching, machine and systems maintenance, weather forecasting, and others. Furthermore, Endsley (2000) proposed a *model for situational awareness* that comprised of three levels:

- Level 1—Perception: The perception of cues in an environment.
- Level 2—Comprehension: How people combine, interpret, store, and retain information ... it includes the integration of multiple pieces of information and a determination of their relevance to the person's goals.
- Level 3—Projection: The ability to project from current events and dynamics to anticipate future events (and their implications) allows for timely decision making.

Note the similarity between these two models. For example, the situation assessment level in the JDL model (i.e., Level 2) describes a process in which one attempts to establish relationships among objects, once located and, perhaps, identified. In Endsley's model, the comprehension level (i.e., Level 2) deals with the integration of perceived cues (akin to objects) and determination of the relevance of these cues against one's own goals. At this level, both models are concerned with going beyond the individual collected bits—they are concerned with making sense of the aggregation of information at a higher level.

9.2.2.1 Considerations for an Agile Information Generation Process

Information generation is more than just implementation of algorithms to affect, for example, the various levels of the JDL model. Number crunching by itself cannot hope to bring sufficient understanding that will allow one to answer the questions: who, what, where, when, and why? It is one thing to generate information; it is another to make sense of it (*cognition*, see Section 9.4), act upon it in an appropriate and timely manner (*decision making*, see Section 9.4), and to share data and information with others for their own mutual benefit (*collaboration*, see Sections 9.3 and 9.4). The last observation is an important one. Even within the context of the JDL model (which has a highly automated flavor), the interactions among the human elements of a system are vital to the success of the information generation process—in terms of putting information in the proper context, resolving ambiguities, and sharing this information in an appropriate manner.

9.2.3 The Information Generation Environment

9.2.3.1 Increased Availability of Data

Consider a notional (and scaled-down) information generation environment as depicted in Figure 9.8. Here we see local clusters of information nodes. The localness of a cluster relative to another may be attributed to, for example, physical proximity or the likelihood of affecting some form of interoperability between them. Within each cluster there is some number of nodes with a certain degree of connectedness. Each node has the potential to be a producer as well as a consumer of information. While the number of clusters and nodes within each cluster is small in this representation, in reality, these numbers could be a good deal larger.

Each node has its own needs for information, which may be satisfied based on its own data sources, or in combination with data from other nodes within and outside of its own cluster. Furthermore, it is expected that these information nodes will act in a manner to achieve some mix of well-defined and open-ended goals. To this end, information nodes will engage each other in a collaborative way, sharing data sources, information (at various levels), and information needs and goals throughout the duration of their interaction (which could, in principle, be forever).

Finally, it may very well be the case that any node (or cluster of nodes) will be unaware of, at any given point in time, the existence and, therefore, the potential value of other information nodes. (The single node in the dashed box in Figure 9.8 is meant to represent a node whose existence is yet unknown to the two local information node clusters.)

Agile Functionality for Decision Superiority ■ 313

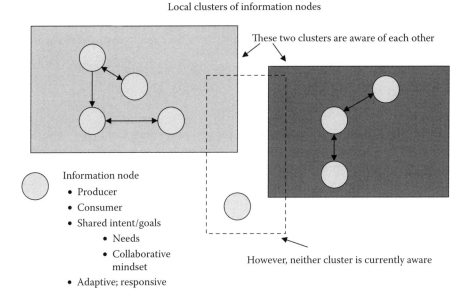

Figure 9.8 Notion of an enterprise from an information node point of view.

9.2.3.2 Availability of New Data Sources

Sensors that exploit various parts of the electromagnetic spectrum (e.g., radar, electro-optic, etc.) constitute important sources of data in many types of military enterprises. In particular, the data that allow for the generation of target attributes such as location, kinematics, and identification almost always originate with such sources. These target attributes are, to be sure, important dimensions of situational awareness (and in some instances, they are all that matters). However, the increased use of internet chat, instant messaging technologies (particularly, in military operations), as well as email and Web logs (blogs) have made possible another dimension of situational awareness—team awareness. Awareness of the changing roles, responsibilities, and knowledge of people (i.e., members of a team) participating in an enterprise can be discerned by exploiting information generated by these nontraditional data sources.

9.2.3.3 Greater Need for Sharing Data at All Levels

Requirements for the effectiveness of any decision-making process within an enterprise are, generally, derived from operational needs. In recent years, the need to reduce fratricide has led to the development of Blue Force Tracking (BFT) technologies providing high levels of situational awareness. Here, the requirement demanded a more comprehensive and dynamic situational awareness that was achieved, in part, by the availability of multiple sources of data and information.

In other decision-making processes, the increased accuracy and timeliness of a reported object attribute (e.g., the location of a high-value target) may be of primary importance. As with the BFT problem, there is a need to share data to achieve required accuracy and timeliness.

9.2.3.4 Information Needs and Goals Not Always Known in Advance

It may very well be the case that for all information nodes within an enterprise, their information needs and goals will always be known, well-defined, and well-articulated. For example, Ground Moving Target Indicator (GMTI) platforms may have associated *a priori* data collection plans from which operators may base their operation of the on-board sensor (collection) assets. In this case, information needs are known and can be mapped into specific resource management actions. Subsequent collection will fulfill the information needs. Beyond the immediate needs embodied in such a collection plan, a node may also possess information goals, motivated by an open-ended or, perhaps, ill-defined problem.

Whether presented as a well-defined need or an open-ended goal, it is assumed that information will be generated based on data that are available at a local node or at other already connected and participating nodes. Such an assumption may impose a limitation on the information generation process—information nodes which exist, but are yet unknown, will not be able to contribute their data to the process. That is, enterprise nodes will not take advantage of the opportunities afforded by the data available at other nodes.

9.2.3.5 Greater Potential to Task Data Sources to Gather What Is Needed

In the future, individual sensor platforms (e.g., Multisensor Command and Control Aircraft [MC2A]) will be capable of (improved) tasking of not only their on board sensor assets but also those of off-board sensors. Far from being a case of selective interoperability, these platforms will be elements of larger sensor networks, each working in a collaborative fashion to achieve information needs and goals. Some primary examples of these networks include the Single Integrated Air Picture (SIAP), Network-Centric Collaborative Targeting (NCCT), and the C2 Constellation.

9.2.3.6 Data Sources Experience the Complexity of an Environment

As discussed in the overview of this chapter (Section 9.1), emergence, as an element of complexity, may occur within and affect elements of an enterprise in a number of ways. In the case of adversary-created emergence, changes in enemy tactics serve to adapt or interact with our efforts to achieve military goals. To the extent that data sources (e.g., sensors, human intelligence, etc.) are able to observe these dynamics within the environment, people and processes within the enterprise will need to understand (at possibly many levels) the implication(s) of this emergent behavior.

9.2.3.7 Data with Future Value

Information has its greatest value in a current context. Arguably, information that is only relevant to a particular situation may lose its value at the end of that situation. However, this does not necessarily mean that the data are without value in the future. The ability to store or warehouse data or generated information may add value to an enterprise in the long run, if one considers retrospective or forensic analyses.

9.2.4 Information Generation Challenges

9.2.4.1 Correlating Data from Diverse Sources

With the increasing availability of data from information nodes, particularly those nodes as yet unknown to other local clusters of nodes, comes the challenge of associating (or correlating, the terms are used interchangeably, here) the data or information from two or more nodes (or even within a single node). In addition, association processes may be carried out over a period of time.

In the data fusion problem, one seeks, for example, to establish whether an observation (e.g., a geo-location on an object) from more than one sensor can reasonably be attributed to the same object—that is, one seeks to correlate the two observations. In doing this, it is then possible to fuse the two observations to obtain a more accurate geo-location. In other settings, the association of data or information may be, for the most part, an end unto itself. For example, the association of intelligence reports from different sources may be all that is required for understanding.

In general, the information generation process is likely to require, at various stages, the association/correlation of data or information. Currently, systems are designed with specific requirements for the types of association/correlation to be carried out. For example, the multisource integration application in the AWACS platform is designed to correlate observations from its on board sensors (e.g., radar and identification friend or foe [IFF]) with observations from off-board sources via the Link-16 tactical data link. Here, the character of the observations is well defined and known *a priori* (i.e., you know the structure of the data), and the algorithms which affect the association are (for the most part) specially tuned to this character. In doing this, one ensures effective (accurate and timely) association in an environment where the particulars of information exchange have been established.

In the broader enterprise, however, the potential exists to interact with information nodes that generate data or information which could be of current or future value to another node. While one does not expect an enterprise to encompass an infinite variety of data or information sources, one can reasonably expect to find some source whose character (i.e., data structure or schema) is not known to you *a priori* but may be useful in the future. The problem is not so much that the data's character is not known (because it can be discovered by its publication), but rather it is the ability of a node to effectively associate/correlate this new data source with existing data sources once the data's structure is known.

9.2.4.2 Managing Information Needs and Goals

Most often, information nodes, as producers of information, work with a collection of well-defined information needs. Here, the problem is known and understood in terms of what information is required to affect a decision or to attain some degree of understanding. However, when a problem is presented in an open-ended way, it may be the case that not all information needs can be articulated. Rather than specific needs, broad information goals may be established.

In an enterprise environment, the availability of data or information from yet unknown information nodes represents both a challenge and an opportunity in terms of developing and refining both information needs and goals. The following assertions, excerpted from Alberts et al. (1999), on the value-creation potential of networks, help shape a view of NCOW in terms of both challenges and opportunities:

Thus, establishing a direct relationship between information and value is at the heart of value creation in the Information Age and is fundamental to understanding the power of network-centric operations. In addition, we believe that:

1. Most potential interactions will never take place.
2. The value of interactions will differ significantly.
3. There will be islands of dense and intense interactions that will dominate the value function.
4. The value of a given interaction will be a function of the content, quality, and timeliness of the interaction.
5. N-way interactions will be the most significant in value creation.

The obvious challenge implied by such a view is to determine which interactions will be of the greatest value, from the perspective of all network nodes. However, value need not be restricted to that which contributes to reducing uncertainty in one's present awareness or understanding; value can also be found in that which reduces future uncertainty. In this sense, an enterprise has value in the form of both present and future opportunity.

Finally, the network-centric view does not directly address the dynamic nature of the network—the necessity or desire to adapt in response to changes in the battlespace environment.

9.2.4.3 Exploiting Elements of Complexity within an Environment

In the emerging concept of NCOW, understanding the behavior and relationships among objects within an environment becomes the focus more so than the understanding of individual objects (Moffat, 2003) In such an enterprise, for example, one concerned with NCOW, elements of complexity, such as self-organization (emergence) and adaptation, can arise. One example is the clustering (i.e., self-organization) of various types of tactical units in a land warfare environment. This clustering (which is part behavior, part relationship) may be observed (most commonly, by sensors) and data produced as a result of these observations. That is, there will be complexity in the data themselves because they mirror a complex system. Thus, certain elements of complexity will be present in the situational awareness of this land warfare environment. However, these elements may not always be readily recognizable. To the extent that such self-organization (here, among tactical land units) is recognizable and results in improved situational awareness, new behaviors may emerge among other objects within the environment, say, from the opposing tactical forces. Here, improved situational awareness produces emergent behaviors within the environment.

In this NCOW enterprise, objects are not only restricted to just those active participants (e.g., objects on the battlefield), but also to those participants who may be observing the environment and making certain decisions as to future courses of action. These decision makers themselves may engineer their own complexity. For example, self-organization (emergence) may be engineered through the collaborative process in which decision makers share data, information, and information goals and needs. That is, there will be complexity within the collaborative decision-making environment. Here, the medium for collaboration may be as simple as verbal communication or it may be of a more complicated form such as internet chat or instant messaging. Regardless of the medium, collaboration can bring about emergent behavior, often driven by the forces of social cohesion and embeddedness (Moody and White, 2003). In such cases, self-organization may directly affect team awareness; that is, an understanding of what others do and know, and what role they have in the larger team. One can assert that improvements in team awareness lead to improvements in situational awareness.

9.2.5 Information Generation Tenets and Design Principles

9.2.5.1 Make Explicit the Quality of Data or Generated Information at All Levels

In almost every situation and at almost every level, to effectively use data or generated information as input to a decision-making process, the quality of these elements must be known. Consider a multidimensional view of quality:

- *The pedigree (origin) of data or generated information:* This can most readily be achieved through the use of standardized tags and nomenclature. For example, the origin of information can be indicated using a digraph/trigraph approach (e.g., AB, DEF, etc.). In instances where new information is being generated, the origin of the constituent elements can be retained.

- *The accuracy of, or uncertainty in, data or generated information:* In general, for continuous information states, accuracy, or uncertainty, can be expressed as a probability density function (PDF) of the error in the information state. Often, however, accuracy is expressed by only the first two moments of the error PDF (e.g., a two-dimensional error ellipse representing the bias and scatter of an estimated information state). In the case of discrete information states, the probability distribution over all possible outcomes of the information state can be used to represent uncertainty. For example, in the case of the identification of an object or an event, the probability of a correct identification can be provided. Of course, assessment of the accuracy of information generated by people (e.g., through observation and analysis, reasoning, or opinion-making) is a much messier endeavor.
- *The latency associated with data or generated information:* Latency is loosely defined as the difference between the time at which information was generated and the time at which the data comprising the information were produced. For example, an intelligence report was made available 2 min after the creation of the constituent intelligence contact. If latency is random rather than deterministic, a statistic (e.g., a 90th percentile value) should be made available with the information.
- *The freshness of data or generated information:* One can represent the degree of freshness in information simply by keeping track of the time since the information was (initially) generated or last updated. From an information producer's point of view, this merely requires placing a time stamp on data.

9.2.5.2 Develop, Publish, and Manage Information Needs and Goals

Development of information needs and goals is a necessary and prudent first step to effective information generation. It places bounds on information from the point of view of both the producer and the consumer. Publication of needs and goals is a powerful idea; it allows for the possibility of one information node to leverage information available at another node. Finally, management of information needs and goals should be considered not only in light of having achieved certain organizational goals (e.g., the successful prosecution of a time-critical target), but also by taking into account the future (projected) value of data.

Underlying this develop, publish, and manage cycle is the ability to express, or represent, information goals and needs. Consider three broad approaches:

- *Specific statement of need:* The statement "I require position and kinematic data on track number ABCD" is explicit and unambiguous. Here, the consumer is providing a specific request for information, thereby strictly limiting the amount of information a potential producer would need to disseminate. Providing a quality metric for position and kinematic accuracy may further limit the amount of disseminated information.

- *General statement of need:* The statement "I require any data relating to ground moving objects between latitudes X and Y and longitudes W and Z" is still explicit, but is now somewhat ambiguous. A potential producer of information would need to decide, based on available bandwidth, just how much information on how many objects could be disseminated. Here, the consumer is hedging against unknown and, possibly, future information needs by being less specific.

    This particular general statement of need can be refined somewhat by ascribing a relative value to classes of ground moving objects. For example, Transport Erectile Launchers (TELs) may be high value, tanks may be medium value, and personnel carriers may be low value.

- *Hierarchy of goals:* Information needs, whether specific or general in nature, are often derived from goals established within an enterprise for a particular mission. Therefore, rather than publishing needs, one can publish a set of goals, which allows an information producer to decide how to best satisfy what they understand to be the needs of a potential information consumer. For example, goal lattices have been used to represent a hierarchy of goals, including a relative apportionment of value among (possibly competing) goals (McIntyre, 1998).

9.2.5.3 Associate/Correlate Data and Generated Information

Association/correlation, either as a first step to fusing data or generated information, or as an end unto itself, is at the center of an information generation process which deals with multiple information sources. Association/correlation should be carried out with the following in mind:

- *Alignment to a common spatial and temporal frame of reference:* This is the first step in any association/correlation process. In the spatial dimension, a transformation of basis vectors from one coordinate system to another is required to affect this alignment. In the case where the spatial data are textual and not numeric, alignment constitutes translation to a common basis language. In the temporal dimension, interpolation or extrapolation to a common time is required.
- *Remove or otherwise account for bias in the constituent data or generated information:* Here, bias is meant as that part of the measurement of an object's attribute (e.g., geo-location) that has a constant offset from the true value of the object attribute, which cannot be eliminated by repeated measurement.

For example, measurements made by a radar sensor contain an azimuth bias when a misalignment is present between a measurement axis and a reference axis (e.g., the direction of truth north).
- *Utilize the accuracy of the data:* When possible, use the accuracy of the data to assess the degree of similarity between objects. Using the known uncertainty that comes along with a measured value of an object attribute (e.g., a geolocation) will allow for a better quality of association/correlation.
- *Allow for the possibility of object attributes at different scales of measurement:* Not all object attributes have values on a numerical (interval or ratio) scale (e.g., value = 1.234). Some attributes will have values on a nominal (e.g., value = yes or no) or an ordinal (e.g., value = 2, which is greater than 1 but less than 3) scale. The association/correlation process should not disregard or under utilize attributes on these nonstandard scales.
- *Allow for the association/correlation of objects over a period of time:* Depending on the threshold(s) used, more than one epoch may be required to obtain a sufficient quality of association/correlation. The usual trade-off will be time versus quality of the result.

9.2.5.4 Enable the Fusing of Data or Generated Information

While not always necessary to attain a desired level of understanding, the fusing of data or generated information requires that certain conditions be met before constituent elements (i.e., data or generated information) are combined:

- *Utilize and then reassess the quality of data or generated information:* The fusion process should use the known quality of the constituent elements and should determine the quality of these combined elements once fusion takes place (see discussion above).
- *Alignment to a common spatial and temporal frame of reference:* See discussion above.
- *Remove or otherwise account for bias in the constituent elements:* See discussion above.
- *Retain links to constituent elements:* Links to the constituent data or generated information are retained as part of the result. At a minimum, the pedigree (origin) of each of the constituents should be retained.

9.2.5.5 Use Visualization to Understand Relationships in Data or Generated Information

Where information is in the form of a relationship (e.g., making the statement that one object is somehow related to another object), visual representations may be the most expedient and reliable means of highlighting such relationships. Often, visualization involves the textual representation of data (e.g., data available in tabular form and across multiple tables). Other visualization approaches involve the construction of a hierarchy (or tree) of data or of generated information. Examples of the use of visualization in the "detection" of relationships in data include the clustering of air traffic control flows and the construction of drug traffic network diagrams (DeArmon, 2000).

9.2.5.6 Enable the Recognition of Complexity in Data and Decision-Making Environments

Recognizing complexity (e.g., self-organization, adaptation) within an enterprise (which includes the data, the decision makers, and the enterprise itself) requires a combination of human and machine processing as well as their interaction. Regardless of whether an element of complexity is engineered or naturally occurring, its recognition will depend on the specific model applied to the CAS under consideration. There are a number of ways to model a CAS, with each approach grounded in a particular discipline (Ahmed et al., 2005). There are, however, some general principles to keep in mind, which may allow one to, at least, initially capture certain elements of complexity:

- *Aggregate data or generated information:* Clustering of multidimensional objects may, at least partially, reveal emergent properties of an enterprise. Here, the less-algorithmic approaches to clustering, such as visualization, should be considered. And, as with association/correlation, object attributes on different scales should be considered.
- *Assess relationships:* To the extent that the selected form of clustering does not adequately capture a relationship, other approaches (which may pick up where clustering left off) should be considered. Again, visualization may get you further than purely algorithmic approaches.
- *Monitor for change:* The time-series behavior of objects (or aggregations of objects) should be examined periodically to detect and characterize underlying dynamics.
- *Allow for push and pull interactions between human and machine:* Often, more meaningful aggregations and relationships can be obtained when a human seeds the process used by the machine. This is one type of pull interaction.

9.2.5.7 Warehouse Data or Generated Information at All Levels

The warehousing of data or generated information from a current context (e.g., a military engagement) will enable their ready analysis in the future. Such retrospective or forensic analyses may facilitate the discovery and modeling of new information and relationships. Beyond the very real issues surrounding what to store and how to store it, other considerations include how to deal with derived data types, such as a network or a pattern expressed as a time series, or perhaps a set of clusters.

9.2.5.8 Apply Meaningful Tags to Data or Generated Information

Data or generated information should be tagged appropriately to facilitate future retrieval and processing. Ideally, the tags should be derived from a vocabulary of terms developed for a specific domain. Here, one can look to the established registries containing such vocabularies. For example, vocabularies and grammars relating to the Department of Defense (DoD) can be found at the DoD XML Gallery (http://diides.ncr.disa.mil/xmlreg/user/index.cfm).

By way of summary, Table 9.1 maps tenets and design principles to the information generation challenges they address. As one might expect, some tenets, for example, applying meaningful tags to data, address all challenges to information generation within an enterprise. In this case, such tagging allows data and generated information to be readily understood by both human and automation within an enterprise.

Table 9.1 Information Generation Challenges and Tenets/Design Principles

Tenets and Design Principles	Challenges		
	Correlate Data from Diverse Sources	Manage Information Goals	Exploit Complexity
Make data quality explicit	☑		
Develop, publish, and manage information goals		☑	☑
Associate/correlate data	☑	☑	☑
Fuse data	☑		☑
Use visualization to assess relationships	☑		☑
Enable recognition of complexity		☑	☑
Warehouse data			☑
Apply meaningful tags to data	☑	☑	☑

9.2.6 Enabling Technologies for Information Generation

There are a number of technologies that enable various aspects of agile information generation within an enterprise. Some of these major technologies and disciplines are listed below:

- Data fusion
- Pattern recognition (in particular, data mining)
- Visualization
- Text and information extraction
- Complexity theory
- Intelligent agents
- Operations research
- Evolutionary and genetic algorithms
- Simulation
- Metadata, schemas, and ontologies
- Data warehouses

Disciplines such as operations research and simulation are well established with wide application and representation in the technical literature. Other technologies, such as evolutionary and genetic algorithms, and intelligent agents, are just coming into their own, relatively speaking. Complexity theory represents somewhat of a hybrid discipline, blending new concepts and technologies, such as the use of agent-based modeling to analyze self-organization and emergence, with traditional technologies such as networks. We briefly discuss two of the most important enabling technologies for agile information generation—data fusion and data mining.

9.2.6.1 Data Fusion

The JDL Data Fusion Model, as described earlier, not only provides a framework for discussing the elements of data fusion processing but also hierarchies of functions and associated enabling technologies (techniques). What follows are some examples of technologies and disciplines used to enable the processing associated with JDL Model Levels 1 through 4, particularly as they relate to the development of data fusion systems for military application (Berube, 2002).

Level 1, Object Assessment, is concerned with the localization and identification of objects. One function within this level of processing deals with estimating the position, kinematics (i.e., dynamics), and other attributes of an object, which is a critical function of almost all military data fusion systems. Figure 9.9 presents a hierarchy of techniques associated with this function. Here, we see that techniques from disciplines such as operations research, evolutionary and genetic algorithms, digital filtering, and stochastic estimation are prominent in their application. In particular, the application of the sequential Kalman filter (both in its linear and nonlinear forms) is nearly ubiquitous in military data fusion systems.

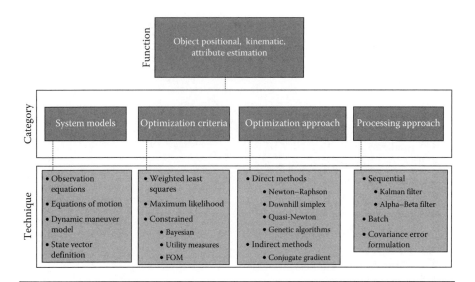

Figure 9.9 A hierarchy of techniques associated with estimating object attributes.

Level 2, Situation Assessment, involves the dynamic generation of a description of current relationships among objects and events in the context of their environment. This processing typically involves aggregating objects, interpreting events and activities, and context-based reasoning. Technologies taken from the fields of artificial intelligence and pattern recognition dominate applications at this level. Examples of these technologies include rule-based and knowledge-based systems employing logical templates, case-based reasoning, fuzzy logic, and clustering.

Level 3, Impact Assessment, involves the projection of current situations into the future to draw inferences about enemy threats, friend and foe vulnerabilities, and opportunities for operations. This processing typically involves aggregate force estimation, intent prediction, multiperspective analysis, and temporal projections. Technologies prevalent in Level 2 are also widely used in Level 3 applications.

Level 4, Process Refinement, involves monitoring the overall data fusion process to assess and improve real-time system performance. Here, processing functions include performance evaluation, process control, determining source requirements, and mission management. Important disciplines at this level include operations research, in particular, optimization and decision theory.

9.2.6.2 Data Mining

Data mining is the art and science of finding meaningful relationships and patterns in large amounts of information (Cabana and Swarz, 2000). Applications of data mining technology are numerous in both commercial and government domains. On the commercial side, data mining has been used for market basket analysis, credit risk assessment, fraud detection, and logistics/inventory modeling.

On the government side, data mining has found application in Internal Revenue Service (IRS) tax compliance, aviation safety, and U.S. Customs Service counter-drug intervention (Cabana et al. [n.d.]).

Problems in data mining may be divided into two classes—predictive and descriptive. In predictive data mining problems, a mathematical model (or set of models) is developed that allows the user to predict a new outcome (or value) based on information about already known (i.e., historical) cases. This is perhaps the most familiar form of data mining and is more commonly referred to as supervised classification or supervised learning. The technologies most commonly applied to this class of problem come from the disciplines of multivariate statistics and artificial intelligence, and include linear and logistic regression, discriminant analysis, neural networks, and decision trees.

In descriptive data mining problems, the goal is to segment information (cases) into some number of distinct groups, where objects in each group are more similar to each other than objects in other groups. This problem is more commonly referred to as unsupervised classification or unsupervised learning. These problems require a definition of similarity, which is usually dependent on the problem being studied. Often, the concept of distance between information (cases) is used as a measure of similarity. As with supervised classification problems, the technologies most commonly applied here come from the disciplines of multivariate statistics and artificial intelligence. One general category of technologies is that of data clustering. Figure 9.10 presents a very high-level sketch of clustering and its

- Find groups of similar data items
- Simplest data mining technique to use/understand
- Requires some definition of "distance"(e.g., between travel profiles)

Uses:

- Demographic analysis

Technologies:

- K-means clustering
- Agglomerative clustering
- Self-organizing maps (Kohonen networks)
- Probability densities

"Groups of squadrons with similar engine failure histories"

- 510FS (Aviano)
- 117FW (ANG) (Atlantic City), 35FW (Kunsan)
- 62FS (Luke)

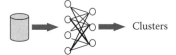

Clusters

Figure 9.10 Clustering and a data mining application. (Adapted from Cabana, K. A. and Swarz, R. S. [2000]. *Applying Data Mining Technology to the CAFC2S Work Program.* Unpublished briefing. The MITRE Corporation.)

application to a data mining problem—determining segments of fighter squadrons with similar engine failure histories. In this very simple example there are three groups of fighter squadrons with similar histories, one of which contains two geographically widely distributed squadrons. Among the technologies that may be applied to such a problem, K-means clustering is one or the more intuitive and easily programmed of those listed in Figure 9.10.

9.2.7 Information Generation Use Cases

There are a variety of research issues relating to the development of agile information generation capabilities in an enterprise. We list a few important issues below, grouped according to a specific challenge (see Table 9.1) to engineering an enterprise:

- Correlate data from diverse sources
  - Ability to correlate sources such as imagery, video, internet chat, and instant messaging with structured sources
  - Ability to do quick, good enough correlation with any data type
- Manage information goals
  - Capabilities-based versus requirements-based publication of information
  - Development of information goals based on opportunities derived from (emerging from) situational awareness
- Exploit complexity
  - Algorithmic and visualization-based approaches to analyzing elements of complexity (e.g., emergence)
  - Real-time data warehousing to support analysis of complexity

Below, we step through a series of example use cases in information generation research.

9.2.7.1 Correlating Data from Diverse Sources

As an example of correlating data from diverse sources, consider the problem of exploiting elements of internet chat and structured data (e.g., GMTI reports) in a time-critical targeting environment. In such environments, operators who rely heavily on chat as part of their sense-making processes have to manually (if they do so at all) establish relationships between chat and structured data, often under time pressure.

To assist in establishing such relationships, an FY06 (fiscal year) Air Force (AF)-sponsored MITRE MOIE has been funded. Figure 9.11 presents a summary of the MOIE, titled "Facilitating Sense Making for Situational Awareness." This research will focus on the development of algorithms and a software prototype for (1) the correlation of chat and structured data types and (2) the aggregation of data for higher levels of situational awareness. The technologies central

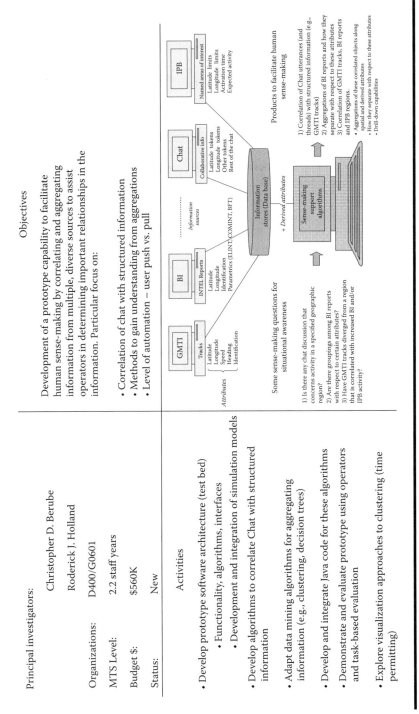

Figure 9.11 Facilitating sense-making for situational awareness, an FY06 AF MOIE.

to this research include data mining, text and information extraction, and data fusion.

9.2.7.2 Managing Information Goals

In the process of developing situational awareness (e.g., in accordance with either the JDL or Endsley models), opportunities to explore as-yet unexplored regions of a battle space may emerge. For example, the aggregation (or clustering) of data and the monitoring of these aggregations over a period of time may suggest hot spots of activity, which, while not of immediate relevance, may be of future interest. In such cases, it may be worthwhile to allocate additional resources, either additional analysis or sensors (or a combination of the two), to a hot spot. The issue of developing a suitable representation (i.e., a model) of a hot spot to support resource/sensor allocation then becomes an issue—that is, how does one develop a goal for information required in the future?

As an example of an opportunity that emerges from exploratory data analysis (which is, essentially, the activity described above), consider a MITRE direct-funded effort to develop data mining techniques to support U.S. Customs counterdrug intervention (Cabana et al. [n.d.]). In this research effort, the goal was to develop a set of smart filters, which when applied to a large volume of air tracks, would allow an operator to focus on a particular subset of these tracks as likely candidates for an airborne drug-traffic-er; that is, targets-of-interest (TOIs). Figure 9.12 illustrates the results of applying data mining techniques to

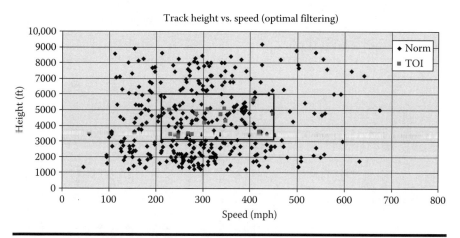

Figure 9.12 Using data mining to discriminate TOIs in counterdrug intervention. (Adapted from Cabana, K. A., Babich, A. M., and Washburn, N. H. [n.d.]. *Counterdrug Surveillance and Control Systems: Using Data Mining Technology to Enhance Detection and Tracking of Targets of Interest*. Unpublished briefing. The MITRE Corporation.)

a set of air tracks. Here, applying data mining reduces the number of tracks for further examination from the 200 produced by the track height-speed filter (black font) down to 40 with a high signal-to-noise ratio (gray font). While the nature of this particular problem allowed for the use of supervised data mining techniques (because historical data sets on airborne drug-traffic-ers were available), one could have worked this problem from an unsupervised viewpoint, simply looking for natural groupings in the volume of track data. In fact, one would not have begun the problem as a problem, but rather as an exploratory data analysis task.

9.2.7.3 Exploiting (and Modeling) Complexity

While the recognition of complexity in certain aspects of military conflicts has been studied by a number of authors (e.g., Moffat, 2003), the recognition and exploitation of elements of complexity within an organization (i.e., an enterprise) appear to have received less attention, at least from a military point of view. Research conducted at the London School of Economics (LSE) in organizational complexity has concentrated on the study of an organization as a CAS. In particular, their research focuses on conducting natural experiments within organizations (e.g., BAE System, Citibank, the World Bank, and AstroZeneca). LSE has developed an integrated research methodology using both qualitative tools (e.g., interviews and narrative analysis) and quantitative techniques (e.g., agent-based models and simulation) to study organizations open to change (Mitleton-Kelly, 2003). Figure 9.13 presents the salient aspects of LSE's research program. Of particular note is the use of the visualization tool NetMap to establish patterns of connectivity using email exchange between elements of an organization.

An important problem in a netted-sensors enterprise is the optimal allocation of sensor resources, in terms of their spatial and temporal distributions. Some recent research from the MITRE Technology Program (Mathieu et al., 2006) explores the application of biologically inspired distributed control algorithms to C2 problems. In particular, this research has focused on the development of a quorum sensing algorithm to support target tracking based on the quorum sensing molecule (QSM), local interaction, and nonlinear dynamics.

Figure 9.14 illustrates an example of the behavior of a QSM tracking algorithm in a netted-sensors environment. The problem under study was one in which 600 sensors were randomly distributed within a spatial field, and whose job it was to detect a single target moving through this field to support a down-stream tracking function. Here, we see that the QSM tracking algorithm affects the activation of sensors (i.e., up-state units) mostly within the "neighborhood" of the TOI, where each sensor has a greater likelihood of target detection.

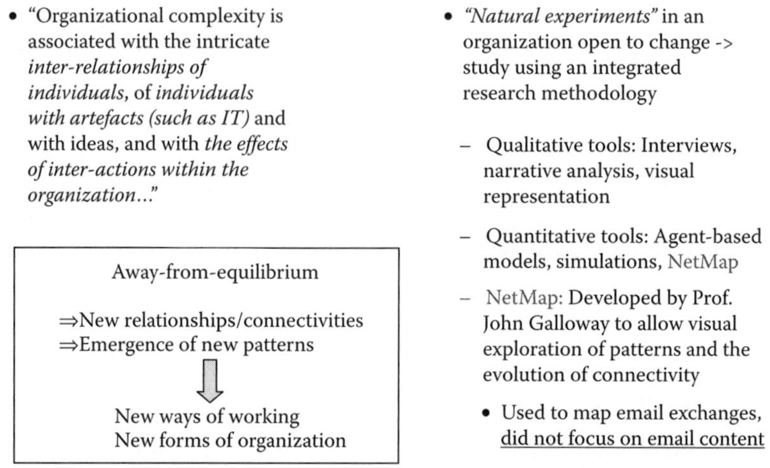

• "Organizational complexity is associated with the intricate *inter-relationships of individuals*, of *individuals with artefacts (such as IT)* and with ideas, and with *the effects of inter-actions within the organization*..."
Away-from-equilibrium
⇒New relationships/connectivities
⇒Emergence of new patterns
New ways of working
New forms of organization
• *"Natural experiments"* in an organization open to change -> study using an integrated research methodology
– Qualitative tools: Interviews, narrative analysis, visual representation
– Quantitative tools: Agent-based models, simulations, NetMap
– NetMap: Developed by Prof. John Galloway to allow visual exploration of patterns and the evolution of connectivity
• Used to map email exchanges, did not focus on email content

Figure 9.13 London School of Economics Research in Organizational Complexity. (Adapted from Mitleton-Kelly, E. [2003]. ***Complexity Research Approaches and Methods: The LSE Complexity Group Integrated Methodology.*** **Retrieved November 10, 2005, from the London School of Economics, Complexity Research Programme Web site: http://www.lse.ac.uk/complexity.)**

QSM tracking (σ_t = 60 m)
y (m)
x (m)
0
100
200
300
400
500
600
700
800
900
1000
Sensors
New sensor hits
Up-state units
Max. neighborhood
Target location

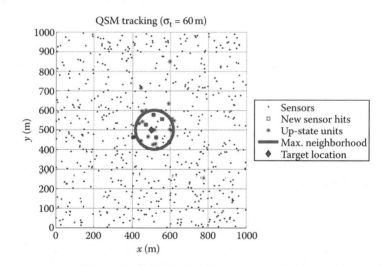

Figure 9.14 An example of QSM tracking of a target in a netted-sensors environment. (Adapted from Mathieu, J., Hwang, G., and Dunyak, J. [2006]. Transferring insights from complex biological systems to the exploitation of netted sensors in command and control enterprise. Manuscript submitted for presentation at the ***2006 Command and Control Research and Technology Symposium,*** **San Diego, CA.)**

9.3 Agile Information Management (The Right People and Time)

9.3.1 Introduction

Information management comprises the designs, techniques, and procedures that enable the right people at the right time to get the right information ... based, of course, on the right budget. Various processing, storage, sensor, and especially network technologies are employed to make information management more agile and better able to both handle stress and heavy load and adapt to potentially unanticipated new demands. This complex and dynamic environment requires rapid, on-demand information access where "right" is defined on the fly and may change unpredictably. Systems must interact in new and unforeseen ways at critical times of need with other information sources discovered on systems whose existence may only have been suspected.

The pull mode CONOPS is no longer sufficient. Designers can no longer presume that operators will know *a priori* which source to utilize to get the necessary information. Push mode support, where the sources generate information based on previous and current operator decisions, exploitation, and usage patterns, must also be added to enterprise architectures and system designs.*

Roles, group affiliation, availability, and connectivity of network participants must also now be presumed even more dynamic than in the past. Flexible authorization and accounting mechanisms that can match dynamic organizations and their funding sources while maintaining attribution and audit trails will also be necessary.

With the increasing reach provided by networks, increasing awareness and analyzed options provided by processing and sensors and the sheer volume of information, now carefully stored, comes the threat of overloaded or misinformed decision makers and unreliable systems at a scale that challenges designers and operators alike. CASs can give warfighters and business managers alike an enduring and keen edge, but they must be handled with care and respect.

9.3.2 Information Management Tenets and Design Principles at the Enterprise Level

The tenets and design principles discussed in the following sections have been found useful in addressing the challenges listed above, but more importantly, they enable

---

* The publish and subscribe paradigm, aided by information brokers, utilizing some aspects of a service-oriented architecture (SOA), is still a viable goal to balance information-sharing needs across enterprises. In some DoD community-of-interest (COIs), at least, there is a greater emphasis on better ways to pull information rather than relying so much on the push.

the enterprise systems engineer to leverage and exploit complex solutions to complex problems.

9.3.2.1 Composite Systems and Sources of Complexity

First and foremost, CASs result when systems are *composed* of distinct, relatively loosely coupled* components or modules. Component-based systems are significantly more robust and adaptable. The more loosely coupled the components, the easier they are to swap in or out on demand, as duly pointed out by enterprise system vendors such as IBM, who find it a compelling selling point, driving the shift from tightly coupled object-oriented components to more loosely coupled Web-service components (Bieberstein et al., 2005; Szyperski, 2003). However, the most prominent example of loose coupling in Command, Control, Communications, Intelligence, Surveillance, and Reconnaissance (C4ISR) systems is that between human operators and machines, as the old joke about the "loose nut behind the wheel" points out. Humans, in turn, are part of a larger CAS, geophysical space; the machines comprise cyberspace. A good part of information management design is dedicated to keeping the engineered systems and the living systems aligned with each other so as to apply resources in concert toward the intentions of the leaders.

The simplest of the complex assemblages are more or less homogenous loosely coupled groups: grains of sand in a dune, a population of some species, and flights of micro-Unmanned Aerial Vehicles (UAVs). Additional complexity occurs when diverse components are grouped—organisms composed of different kinds of cells, circuit boards, modularized operating systems, and airplanes.

The first descriptive mathematical models of complex systems were stochastic. In particular, stochastic descriptions with heavy-tailed distributions (also called power-law distributions) are labeled complex. Stochastic methods are necessarily based on multiple samples. In the case of complex systems, the multiple components or participants that make up the composition are sampled during a finite interval whose length is defined by the lifetime of the target behavior(s) such that sufficient samples are collected to satisfy the modeling constraints. In this way, systems can be captured even if they are short-lived *ad hoc* assemblages.

A number of causative mathematical models can result in heavy-tailed, descriptive distributions. Nonlinear chaotic population models are the most well known; however, there are other models in which the introduction of engineered linear components also causes the characteristic heavy-tailed distributions. Furthermore, some otherwise rigidly specified and engineered systems can alternate between normal and heavy-tailed behavior. What, then, drives a system to be complex? In a phrase: *state space size*. What drives state space definition? The number of variables

---

* Loose coupling occurs when two subsystems only partially share their state spaces—a change in one does not necessarily result in a change in its neighbors. The result is a potentially nonlinear system.

such as inputs and outputs, as well as their processing or relationships and the resulting shift to the next state. The most common way to formally describe a component is as a state machine that operates over a finite state space. The state machine of a component can either be discrete or continuous, depending on the coupling among the components, and the formal specification can range from probabilistic relations to differential equations.

Composite systems are systems that have distinct but interconnected components or systems of systems. They exist and operate at multiple scales, with one model for each component and another for the collection as a whole. Composite state space is an assemblage of all the potential states that are defined by the finite state machines of all the potential participants, whether human beings or engineered components. When the participants can exchange information, their behavior may coordinate. To the extent that their behavior is tightly or loosely coordinated, the components are referred to as *tightly coupled* or *loosely coupled*. In formal descriptions of these interactions, the participants are generally described in terms of their finite state machines, which detail what inputs result in what behavior and outputs. If several participants' state machines shift state in concert over a significant percentage of the combined state space—in one sense, their individual state spaces overlap significantly—they are clearly tightly coupled. The less they act in concert, the smaller the overlap and the looser the coupling.

State spaces can overlap due to coupling in a number of ways, but the most common way to reduce the overlap and decouple is to grow the state space. State space grows when there is an increase in:

1. The number of interactions due to increased number of simultaneous participants
2. The number of types of interactions due to increased diversity of participants
3. The length of the delay in inter-participant interactions beyond a certain threshold

Since composite state space is essentially combinatorial, an increase in any one of these factors can lead to at least exponential growth of the state space. As the state space grows exponentially, the ratio of the overlap to the overall space drops drastically and the coupling shifts from tight to loose. In many systems, there is a threshold, one variant of which is described as period three where this shift is extremely noticeable; to quote the famous article by Li and Yorke (1975): "it implies chaos."

Ironically, the result of such dramatic increases in state space can be an increase in the probability that the interacting participants will find their state machines dealing with second order, unanticipated synchronization. In other words, coupled through unexpected overlaps in a small subset of the state space, which means they are unexpectedly but loosely coupled. This is often a source of unintended side effects in human-designed systems.

Tight-coupling can also be unexpected in engineered systems; it is just much more likely to be detected, either on first inspection of a design or early on in routine observation and thereafter it is an anticipated contingency. Furthermore, loose-coupling is not always bad; margin of error, overengineered, and cutting some slack are all well-known ways to effectively decouple participants and increase the tolerance and reliability in anticipated environments. However, loose-coupling *does* introduce the potential for unexpected complex behavior, especially cascade failures, if the environment goes beyond the anticipated, and previously decoupled participants suddenly find themselves operating at the *bitter end*.*

9.3.2.2 Managing Complexity in the Enterprise

There are numerous ways to manage† complexity in any system, all of which involve reducing the number of states that need to be handled with the intent of *completely* decoupling all behavior except for a carefully specified absolute minimum. In short, while attempting to manage complexity, enterprise systems engineers may introduce it unintentionally due to insufficient decoupling. While the average information systems engineer well understands minimizing unnecessary diversity and may even plan for large-scale numbers, very few understand the implications of introduced delay, especially variable delay. Almost none understand the cumulative effect of all these factors or how to assess and plan for crossing the complexity threshold in a normally well-behaved arrangement.

The upshot is a fair number of systems in which the people explicitly decouple themselves in self-defense from the engineered components, and when that is not possible, decouple the engineered components under their control from others. In such cases, the engineered components are routinely labeled a colossal waste of time and money. This section describes various decoupling techniques when decoupling is appropriate, as well as how to specify and analyze coupling where it is appropriate. Appropriate coupling in enterprise information management enables effective performance by human participants and efficient performance by the rest.

There are a number of challenges introduced by even loosely coupling people and engineered systems into a large-scale enterprise. We list out four important ones:

1. Multiple orders of magnitude increase in potential information inputs—This results in the people in the system trying to "sip from a fire hose."

---

* The very end of a line on a ship is tied to a bitt to keep it from losing it when all the slack is gone, but which when reached results in a great deal of sudden stress load on all concerned.
† An alternative approach is to think about harnessing complexity (Axelrod and Cohen, 2000) in situations that are relatively unstable and where one has very little control; elaboration of this idea is beyond our scope here.

2. Orders of magnitude increase in the inputs to and speed of the decision process, especially those that involve prioritizing and optimizing trade-offs—for example, where and when to bring resources to bear, what information is important, especially when the definition of important is changed by a change in local or enterprise-wide goals.
3. Orders of magnitude decrease in the control of technology due to looser coupling—the humans change from individual tool-user to a competitor or partner with others.
4. Different goals for the same human or system depending on what scope of the enterprise they are operating in—within a scope they may be competitors, but when all the competitors are pushed into the next larger scope, they must cooperate in order to survive in the competition against other coalitions.

In order to minimize overload of the humans in the enterprise, there are a number of critical characteristics that must be present in the system engineering design. They enable the systems to track the humans as the humans adapt to changes. Normally these changes are caused by the humans (and their systems) shifting up and down the scope scale, but changes can also be introduced by factors and forces that originate in systems and processes that may never be integrated into the larger enterprise, such as the natural environment.

9.3.2.3 The Five W Questions

In order to couple any two things, they must share *something* in common. Thus, *orientation* is the second most important tenet for operations involving complex systems. Its importance is emphasized in many analyses that characterize decision making, such as the Observe–Orient–Decide–Act Loop. Much of the complexity in information exchanges about orientation can be summed up in four questions, implemented in the information management processes, procedures, and other formal interfaces such as schemas. A fifth question conveys and adapts the leaders' policies via monetary and political capital control and feedback channels.

1. **What**—type and instance
2. **Where**—location/access routing
3. **Who**—identity and role
4. **When**—time and coordination
5. **Why and Wherewithal**—intent and budget

By categorizing orientation information this way, humans can manage the complex information involved in orientation by organizing it into orientation schemas; the five W questions provide a generic template to be refined by a particular domain.

9.3.2.4 How to Be Agile?

There is a sixth question that goes beyond the day-to-day operations the five "W" questions address. How to ensure that the designs, processes, and procedures of information management enable a vigorous, long-lived enterprise? This requires agility, which we define as a combination of two strategic capabilities, redundancy, and adaptability.

To align two systems at any one time, only a mutually agreed upon common framework is necessary; to align more than two systems or to align two systems across changing circumstances over a period of time requires *standards* for the orientation frameworks. Information management therefore requires a standard for each of the five questions. Five standards are needed for each type of space—living community or engineered. Standards enable *redundancy*: if a component or subsystem fails, then a replacement can be swapped in. Paradoxically, standards also can enable *adaptability* because they clearly delineate not only what is standard but which version of the standard it is and force discussion of migration approaches that allow the enterprise to dynamically evolve and not disintegrate.

This leads to *standard escape mechanisms*. Allowing for ongoing adaptation means covering even the outlier cases where the rest of the standard cannot cope and a replacement standard must be employed. Managing these unexpected cases happens more often in complex systems than random system models predict, and in life-or-death circumstances, having a built-in out can be vital. These escape clauses can be as simple as a free-text block or a placeholder URL that information providers can use to reference additional information. Whenever possible, these on-the-fly extensions should be captured and analyzed so that the schema definitions can catch up to these needs.

9.3.3 Information Management Tenets and Design Principles at the Large-Scale Enterprise Level

People familiar with engineered systems design undoubtedly recognize the above system engineering tenets—composition, orientation and alignment, and standards and standard escape mechanisms. There are other tenets that only come into play when dealing with extremely large-scale enterprises. The complexity that entails from large scales is both a compensatory strategy and the target of constant information management concern.

9.3.3.1 Multifocal Organization: Coupling

The first enterprise tenet to keep in mind is *multifocal organization mechanisms*. Multiple organizational hubs facilitate three classic approaches to organizing large-scale enterprises: subdivisions, councils, and hierarchy. Subdivisions simply institute parallel local domains of control and coordination that enable expedited routine operations within each domain. Subdivisions do not have to be imposed; they often arise from self-organization principles.

Coupling among subdivisions can occur as interactions or communications between sovereign peers if the subdivision of resources is approximately equal, and coupling is relatively loose and coordinated effort only occurs if there is consensus. If the subdivision is not equal, communication occurs between levels of the resource division hierarchy, and coupling is relatively tight, with the superior delegating or loaning resources to and controlling the behavior of the subordinate(s).

However, whenever there is loose coupling, some information may be lost due to differences in resolution, delays in distribution, and so on. Data quality metrics are critical in dynamic environments where information producers and consumers may not have detailed foreknowledge of one another. These metrics allow information consumers to select information that best suits their needs and to make informed decisions. For example, information consumers may select information providers based on these metrics, and put off decisions if information quality is low, or seek out additional information providers to verify or augment information.

Typical data quality metrics include accuracy, precision, completeness, and pedigree. Schemas used to describe data quality should be kept short and simple so as not to overburden information providers.

9.3.3.2 Multiscale Organization: Encoding

Multifocal organizations are complex only to the extent that the various centers are partially independent. It is that combination of independence and interdependence that makes them loosely coupled and the proportion of in- to inter-dependence causes one of the most interesting complex characteristics of all: *multiscaled*. When a collection of subdivisions is sufficiently coordinated in their behavior, they can be treated as a unit whose collective behavior is both more and less predictable.

To the extent that the coordinated behavior can be abstracted, hiding the gory internal details of the individual subdivisions, that behavior is simpler to communicate across space and easier to remember across time because it is cheaper to encode. Encoding can be either in living members of the enterprise or the engineered systems that augment them. *Multiscaled encoding* is the second large-scale enterprise tenet because it prevents complexity costs from overwhelming the resources available for coordinating effort.

9.3.3.3 Equilibrium and Organization: Modeling

Since we are dealing with composite, coupled systems, their components can be perturbed by external forces. Such perturbations can disrupt coordination and force different adaptations within different subdivisions, redistributing the balance of resources and capabilities. Managing a complex enterprise not only requires information management for diverse and dynamic resource and information flows but also *large-scale equilibrium modeling* such that the humans do not feel overwhelmed and the engineering is computationally tractable. This last tenet is one of the hardest to implement because the mathematics is nonlinear, probabilistic, and heavy-tailed. Such mathematical territory is typically avoided by traditional engineering disciplines.

Below, we propose a strategy to address this by looking at how the classic four questions of systems information management, augmented by the why, wherewithal, and how of classic operations information management, are transformed by multifocal organization, multiscale encoding, and equilibrium models to create agile enterprises, suitable for the growing range and scope of situations we face.

9.3.4 What and Where Is the Right Information?

To provide information management that supports the needs of a complex, adaptive enterprise, several key IT enablers must be in place. In dependency order they are:

- *Standardized Loosely Coupled Domain Definition:* the first enterprise prerequisites are the will, authority, and budget necessary to determine a common *why* and *wherewithal*; these in turn define a community or domain both in geophysical space and cyberspace. At extremely large scales, these often will be self-organizing; under certain circumstances the structure will be self-similar (fractal) as well.
- *Standardized Loose Coupling Meta-data for What, Where, When, Who:* at very large scales due to multifocal subdivisions there will be both hierarchical and federated standards processes. Local communities should ensure that they have properly credentialed delegates for all hierarchical standards authority and properly selected representatives to all relevant federations. Procedures whereby the standards can be ignored so that variants can be tested, competed, and evolved are critical.

While developing semantic formal community models, operators and systems should expect to share information with parties that are unknown to them during design and development. This has several important implications for developing and documenting formal semantics models:

- Semantic models must be made accessible. This means they must be captured and published using standard forms that are machine-readable.

Semantic models must be as free as possible from implicit definitions and assumptions; source context information can be used to flag any potential conflicts in definitions and assumptions. A tank from an Armor unit's Lieutenant on the ground might be flagged by software and double-checked by a coalition Airman coordinating fuel deliveries, even though fuel capacity is common to both armored, tracked vehicles and storage facilities. Of course, a General's armored, tracked vehicle might indeed be getting an airlift special delivery.

Semantic models should focus on the meaning of the data and allow for diversity in format. Alternate representations tied to underlying semantic models allow for information to be tailored to the needs of specific operator and communities while maintaining data interoperability.

When developing semantic models, flexible processes should be used to find good-enough solutions. Given the requirements of a complex, adaptive enterprise, it is no longer feasible or advisable to put extensive processes in place that are laden with committees and checks and balances. Searches for optimal solutions should only be undertaken when there is ample time and the environment is stable.*

Many schema management standards delegate responsibility for low-level context services to other standard enterprise-wide services. Name/Authorization mapping is generally handed off to the Lightweight Directory Access Protocol (LDAP) identity service and the Name/Address mapping to the Domain Name Service (DNS).

The understanding of information exchanges can be useful, but real situational awareness comes only with coordinating information generation or discovery across multiple team members. Likewise, real capability comes only with coordinating information exploitation. Coordination requires coordinated real-world context information—clocks must be synchronized, a common translation between geospatial datum agreed upon, and an organizational identity and authorization (clearance and classification) model agreed upon.

Likewise, in the supporting information management universe, agreements must be negotiated for virtual clock/counter synchronization, address assignment management and routing convergence, network identity management, and crypto key exchange methods.

Finally, both the real organizational world coordinated mechanisms and the network virtual world coordinated mechanisms must be aligned through the use of:

- *Adaptable and redundant information distribution* that ensures the distribution of information in a highly dynamic, time-sensitive environment where network connectivity is at risk or network infrastructure is heavily loaded. Build-out and configuration of distribution networks in very large-scale environments is already primarily M2M, but only in response to acquisition and

---

* Generally, it is wise to avoid trying to optimize the performance of individual components of a complex system or enterprise in favor of encouraging a balance among component behaviors for the greater good of the whole, especially when conditions conducive to self-organization in less stable environments exist or can be created.

capital budget life cycles, not decision loops or changing operational situations.* In addition to distribution of operational information, there are a number of shared information infrastructure capabilities that themselves need adaptable information distribution. At very large scales, this means the efficient M2M access to and distribution of advertisements of publication and search engine functions. At large scales, M2M access to and distribution of metadata directories and registries, and data catalogs are also needed.

- *Dynamic information discovery* that ensures user and systems can find the right information not only at design-time, but at run-time as well. This means that at very large scales M2M discovery mechanisms are required in order for the right information to be accessible at the right time.
- *Adaptable and redundant data mediation* that ensures information can be understood and interpreted across the enterprise, even when there are different standards due to different domain focal points. Since the second major source of delay is information access authorization and redaction, this means that at very large scales M2M mechanisms that are well integrated with distribution mechanisms are needed for information assurance. Unfortunately, due to certain complex system self-organizing principles, these integrated facilities often create more complexity in the process of fulfilling other needs. This makes it a fertile area for research, and there are numerous ongoing efforts, primarily in the areas of dynamic mapping among structured data schemas and adaptive redaction and intrusion detection (Lin et al., 2002; Sampada et al., 2004; Tzitzikas and Meghini, 2003).

9.3.4.1 Focus on Information Exchanges and Communities

There were several attempts to create a single common vocabulary across the DoD, the DoD data administration program being just one example. The promise of these efforts was that if everyone used the same data model for representing and storing their information, then data interoperability would be assured. These efforts failed to fulfill this promise for many reasons, including the difficulty in reaching and maintaining agreement of such a large-scale vocabulary involving a multitude of organizations. Two important lessons learned from these efforts are:

- The complexity of reaching and maintaining semantic agreement is a function of the amount of information that must be agreed upon and the number of organizations that must reach agreement.

---

* Consider containerized shipping and telecommunications service; both tend to build out large transport infrastructure networks based on aggregated capital cycles, then retail standard capacity at the edges according to retail capital cycles. Acquisition of on-demand capacity requires first setting up either a retail prepaid or billing account which is periodically rendered according to retail capital cycle standards.

- How organizations and systems represent data internally is of little consequence. What is of importance is the meaning and structure of their information exchanges.

The current DoD strategy states that semantics are to be managed within subdivisions known as COIs. A COI is defined as a group of users and stakeholders "who must exchange information in pursuit of their shared goals, interest, missions, or business processes and who therefore must have a common vocabulary for the information they exchange." COIs simply are a formalization of a subenterprise scope.

9.3.4.2 Adaptive Information Access and Distribution Services

The concept of NCOW envisions an environment where information is posted before it is processed, information is available across the net, and information consumers can discover and access the information in a timely and reliable fashion. The effective M2M sharing of information across a network is essential to the realization of this vision and the ultimate success of our military operations. Without it, decision loops lengthen, situational awareness suffers, and the quality of available information is lower.*

Network-centric drivers and enablers are those network-based architectures and network technologies that improve adaptability to these conditions and enable continuity of operations where previously there would have been a brittle failure. High bandwidth, low-delay communications are technically harder and more expensive, *but* by shifting from point-to-point communications to networked communications the enterprise can take advantage of the *network effect* driver: costs go up only linearly while intelligence sources and assets available for coordinated information exploitation go up exponentially.

The resulting shared higher bandwidth and lower delay communications enable richer potential information exchanges that cover a wider range of contingencies. Planning and configuration values may become minimally compressed information copies, not highly compressed preplanned *references* to information; this enables a wider range of coordination using unplanned information. Remote planning and coordination can also draw on a larger pool of assets now accessible through the net, and there is exponentially wider situational awareness and intelligence.

However, the network-centric M2M sharing of information is made much more difficult by the very dynamic environments in which the AF, as well as its Joint and Coalition partners, operate. Thus, we cannot predict in advance who will need to interact with whom, communications paths may be unreliable, systems and

---

* The problems of breaking down organizational cultures of distrust of others and the dominance of penalties over incentives for sharing information are important topics as well as the enabling technology; see "Which is the Right Budget," and "Information Exploitation Challenges" and "Trust Formation" below.

operators may be unavailable for periods of time during operations, and needs for and uses of information will change.

Several capabilities are critical in enabling adaptable and reliable network-centric M2M distribution of information in the AF range of environments:

- Information providers and consumers must have a range of methods for describing their information capabilities and needs in an environment where a range of network capacities and information assurance capabilities are encountered during information distribution.
- There must be decentralized methods for consumers to identify information providers that best suit their needs, since information selection and prioritization become critical when there is a wealth of potential information providers.
- There must be a reliable information exchange capability that can adapt to changes in network connectivity and availability of participants. In support of this, there must be capabilities to find alternate routes across a network when portions of it have become inaccessible. Furthermore, information must ultimately reach network participants even if those participants were unavailable when the information was originally sent. Also information must be scaled to a variety of resolutions to support the needs of participants at the edge of the enterprise distribution network, including mobile participants operating in limited-bandwidth environments.

Some of the technologies and techniques particularly useful for distribution across loosely coupled components and through domain boundaries are

- Peer-to-peer protocols
- Broadcast and multicast distribution protocols
- Intelligent agents
- Intelligent distribution, for example, content-based or location-based
- Transport-aware and storage-aware data protection services

9.3.4.3 Adaptive Information Discovery through Search Services

Traditionally, information providers and their content are discovered and integrated at design time. A developer identifies the content and information services that their application requires and then constructs the necessary interfaces to access that information. As a result, changes to a consumer's information requirements require that changes be made to the application, the information provider, or both.

To provide true adaptability in the face of changing information requirements, information discovery must shift from being a design-time activity to a system-mediated run-time activity. Systems must be able to search for and access information providers that meet their changing information needs. Before the tighter coupling that defines a particular COI develops, or for discovery outside of and between COIs, the following technologies and techniques are useful:

- Broadcast/multicast/advertise protocols
- Intelligent agents
- *Ad hoc* and natural language search queries and filters
- Protection-aware search access

9.3.4.4 Adaptive Information Discovery through Catalog Services

Once a COI reaches a certain level of maturity, intra-COI discovery can be standardized. This requires that information providers publish rich, machine-readable descriptions of their information and services, and that information consumers have efficient and reliable access to the underlying registries and catalogs. The following technologies and techniques are applicable:

- Service-Oriented Architectures (SOAs)
- Subscription protocols
- Semantics and ontologies
- Orchestration and workflow management
- Protection-aware lookup and browse access

9.3.4.5 Adaptive Information Mediation Services

Mediation services bridge semantic gaps in the enterprise through space, across eras, and between political subdivisions. At very large scales these need to shift from design-time data mediation to true M2M run-time mediation. Like information discovery, these fall into two categories, depending on how mature the community's standards are.

First, in order to communicate with mature communities, the data must be encoded in a common format. Humans must use the same alphabet or characters to represent the concepts; machines likewise must also negotiate the same representation encoding.

In order to communicate meaningfully, the data must be distilled into a common, mutually negotiated vocabulary. Mature COI vocabularies and representation encodings include well-known standards such as defined by the Simple Mail Transfer Protocol (SMTP), those found in unified ontologies such as Web Ontology Language (OWL) and Multipurpose Internet Mail Extensions (MIME), and those referenced in federated (context-sensitive) ontologies such as Resource Description Framework (RDF). Run-time mediation can also support model and schema extensions as long as the *process* of assigning a data type to its referent can be standardized within the community.

Useful technologies and techniques in use among mature COIs include:

- Ontologies
- Semantic Web
- Distributed metadata negotiation and management
- Conceptual modeling techniques
- Language translation services

If there is a mismatch between the structural richness in data encoding or vocabularies due to maturity, legitimate differences in operational resolution, summarization, compression or redaction in response to austere communications resources or protection policies, additional technologies and techniques come into play.

- Emergent pattern analysis based on inferences about What, Where, When, Who obtained from raw sensor-generated information context
- Distributed data quality and stewardship based on export and import policies for process-generated information
- Transport-aware data encoding services
- Protection-aware redaction services

9.3.5 When Is the Right Time?

There are two aspects of time that are important to C2, Intelligence, Surveillance, and Reconnaissance (C2ISR) operations. The first is in establishing one of the foundations of *rate*, especially rate of resource consumption, rate of approach/retreat, and frequency of event occurrence. Monitoring or controlling any process over a period of time requires two types of time specifications or time metadata: the measure or clock and the reference time frame. The first determines the baseline method of sequential changes that establish the units and continuity of all other rates in the system. The second determines the scope and scale to which that baseline applies.

In a large enterprise, especially one that is based on a coalition of peers, there may be multiple time sources or clocks and associated time frames. "Synchronizing watches" and the world time zone standards are two classic examples of methods of aligning time frames.

9.3.5.1 Clocks and Time Sources

There are three main types of clocks: the original diurnal sun-clock, the current atomic-decay clock, and the network-centric virtual clock. The first two are used in geophysical time frames and drive coordination of resource usage and physical interactions based on a locally monotonic sequence of physical events. The third is a logical construct used to drive coordination of discrete event systems based on locally monotonic sequences of virtual event dependencies (Lamport, 1978). In all cases, dynamic systems are by definition rooted in the time frames defined by their clocks.

9.3.5.2 Geophysical Time Frames

Geophysical time frames are important in an enterprise because not only do they have an inescapable direct impact on mass and energy resources but also because they are generally politically defined with the associated direct legal impact. They also have an indirect or second-order impact because they define movement, and movement defines locality in a dynamic system and locality defines the probability of interactions. Assets, entities, and resources that are nearby are more easily and closely coupled since there are more potential communication channels and more potential opportunities for both competition for local resources and collaboration in the face of local problems.

A third-order impact can be created by modulation of a rate of change, which can be used as a means of signaling or communications. An entity that shares resources that increase its consumption may become a potential competitor. One that decreases its consumption decreases the potential for competition. If such increases or decreases mirror neighbors' consumption patterns and follow them closely in time, it may be possible to infer their intentions (Malle and Knobe, 2001).

9.3.5.3 Network-Centric Time Frames

Network-centric time frames are defined by virtual clocks. Although geophysical clocks are used in network-centric operations, they are used exclusively to coordinate geophysical entities and processes through the network. Timing sources are used to synchronize physical layer signaling, generally in the service of human sensory requirements such as the transmission of voice, video, and remote control. Administrative time stamps are used for legal and political coordination in network-centric operations and management. All else may use geophysical clocks but only as a convenient source of ordinal numbers for labeling sequences.

9.3.5.4 Time Scales and Complexity in Decision Making

Every time-based interaction has its scale defined by its native or natural clock. Many of those clocks are second- or third-tier, subsidiary chemical, or engineered measures based on either solar or atomic processes. Therefore, there will be multiple time scales simply because there are multiple, coupled oscillators.*

Delivering information at the right time usually means ensuring that it arrives within an interval which enables decision makers to exploit it effectively and

---

* Note that loosely coupled oscillators exhibit one of the first analyzed forms of complex behavior.

efficiently. Such intervals are defined by combining the availability of resources used in the decision process, any delays in transport and storage, the attention span allocated for processing the information, the availability of resources used in the exploitation, and finally, the availability of exploitation opportunities. Machine multivariable calculations of the right time therefore require metadata that are both timely and time-aligned. As such contributing factors are often cyclic, there are often multiple right time solutions of varying efficacy and efficiency, so computational effort becomes another factor at the right time.

9.3.6 Who Are the Right People?

Sharing information requires that both the provider and the consumer be the right people. *Rightness* has several information management reference frameworks: political, geophysical, and network-centric. In most cases, the geophysical location of the people and equipment defines the political and network-centric reference boundaries and membership identity.

9.3.6.1 Ex Officio: Roles, Identity, and Clearance

The concept of a *role* evolved because individual entities cannot always be on active duty and because humans have capabilities beyond those that they use for a particular job. Roles enable agility in the form of robustness from load-sharing and failover, and adaptability from escalation procedures and differential training. The system shifts from individual to individual among a group while fulfilling a single role; the individuals shift from role to role as they pursue their goals within systems. To the extent that the role can be formally defined, it may benefit from M2M support or even automation. Identity, therefore, exists on three planes: the individual identity on the geophysical plane, the individual identity on the information space or network-centric plane, and role identities that formally define information management capabilities. Clearance is a declaration of fitness of an individual to comply with information assurance requirements while operating in a role; it must map the individual identity to the role identity. Consequently, all identity subsystems in an enterprise information management system must deal with all four kinds of identity data, including:

- *Geophysical identity:* This material identity reference framework is the essence of most traditional, political, and legal systems. Sovereign nations are based on interactions within a geographic scope; human organizations such as corporations are based on interactions among the people that compose it.
- *Network-centric identity:* This information identity reference framework is based on the geophysical framework because the expenditure of energy to encode, process, transport, and store information is directed by a geophysical

entity. Mapping between first-order information space virtual entities to the corresponding geophysical entities constitutes a large part of identity management systems because it enables one first-order entity to locate another.
- *Role identity:* Roles are a way to create an adaptive, robust information space identity. They map a second-level or abstract virtual entity to multiple equivalent first-order virtual entities. Shifts can change and people and material assets can be replaced, but the role's operational capabilities have continuity.
- *Authority to clear, authority to classify:* Authority is the last of the "Who" collection of metadata because it binds the control of resources to identities. At its most simple level, geophysical identity represents a single entity's authority and command of material and energy assets—the warm body, in personnel terms. Network-centric identity at its most simple level represents a single entity's authority and command of information and intelligence resources. The authority to control usage of the material and energy assets or control access to the information assets is a foundational interaction with other entities. That *classification* control information is often implicitly captured in Who identity information—shared assets are often represented by shared identifiers. The authority to declare that another entity is trusted or *cleared* for otherwise restricted access is encoded as identity information, either implicitly as shared identifiers or explicitly in an access control binding like a Public Key Infrastructure (PKI) certificate.

9.3.6.2 Identity Information Management Services

How far to extend trust is simple at a single scale: trust all members of your local scope, do not trust nonmembers. However, due to the principle of subdivision, at enterprise scales both material and information resources must be imported and exported to both peer scopes at the same scale and to superior or subordinate scopes at different scales. Representing how far the trust extends is a key aspect of information management's identity management subsystems. It must be able to represent *delegation* for the export of authority, and it must be able to represent *pedigree* for the importance of authority.

Standard interaction information types or *metadata*, such as identity, delegation, and pedigree metadata, are necessary for two identity management subsystems to support export/import interactions necessary for mission command or coordinated effort. When two identity subsystems interact they also need to align their authorities by exchanging authority delegation and pedigree metadata about their identity information. This process, known as *authentication*, is fundamental. To the extent that information is duly bound to delegation and pedigree metadata there is *transparency*, as each participant knows exactly who has done what to whom within each scope at each scale.

Multiscale authority information represents a chain of trust that crosses scopes and scales. The PKI is an open information management standard for trust chains.

It defines a trust scope as the realm of a server that stores identity/authority information, enables M2M cryptographic authentication methods, and provides a management capability that allows authority information to be created, updated, retired, and distributed. It supports multiscale trust chains by encoding authority as a hierarchy of PKI scopes, where each scale derives its identity (and implicitly, authority) from its parent scope.

9.3.6.3 Information Assurance

However, in going up or down scale, authority information is often deliberately lost. Superiors take credit (and responsibility) for their subordinates' efforts. New information is often generated by aggregating information from other sources, but the organization and exploitation of that source information is not performed under the authority or responsibility of the source entities. Privacy and redaction requirements also require either loss or masking of identity information. Finally, in very large enterprises, the identity, delegation, and pedigree information soon grows to the point that it may completely outweigh the content it describes, and in austere environments this burden may not be justified. All scope boundary information management systems (e.g., next generation firewalls and automated guard cells) must be able to review and prune multiscale identity, delegation, and pedigree metadata, as well as the mission content data, in support of these information import/export capabilities.

9.3.7 Which Is the Right Budget?

Technology solutions are necessary, but not sufficient, for information sharing. To fully realize the benefits of robust information sharing, the socioeconomic barriers to information sharing must also be addressed. Many of those socioeconomic forces are managed through budgets, so information flows in an enterprise are those that are enabled by corresponding financial investments and those whose credibility and reliability are backed by solid funding. We offer the following tenets for building incentives for information enterprise information sharing:

- Acquisition processes must address information sharing by identifying the overarching principles and concepts of operation that rely on sharing and call out to what extent sharing will be implemented through shared infrastructure constructs, including networks, shared processing, storage, and servers. A more detailed review of which information management components constitute infrastructure is presented in Section 9.3.8.3.
- Program managers must be given incentives to build information-sharing capabilities into systems that enable information flows both internally and for sanctioned export and import.

- Information providers must be given incentives to share information. Providers often fail to share information out of fear (of providing the wrong information or making some other mistake) or greed (holding onto information and requiring others to come to them increases their perceived power).

9.3.7.1 Information Management Risk–Cost–Benefit Analysis

Information management should mirror the human organizational structure, which in turn mirrors information generation and exploitation decisions. Information management often uses system architectural constructs to implement policy boundaries and cost/benefit trade-offs. Some of the principles and forces that drive the analysis and the resulting structure are:

- In a network-centric information sharing, the benefits go up as the square of the number of (*constructive*) participants—Metcalfe's Law (O'Brien's extension)
- Costs go down inversely to economies of scale and number of competitor providers—Adam Smith
- Risks go up as the square of the number of *destructive* and *obstructive* participants—O'Brien's Law

The initial investment to share information via a network has several parts:

- Initial threshold or barrier to join:
  - Capital equipment acquisition costs
  - Space and environmental acquisition costs
  - Integration costs
  - Training costs
  - Increased risk of discovery through the network by opponents
- Ongoing costs of operations must also be considered; many sales emphasize low acquisition costs and hide subsequent operational costs
  - Space and environmental leases
  - Power and communications service costs
  - Increased risk of exploitation or attack through the network by opponents

However, in a network-centric environment, the benefits can increase exponentially while the financial costs only increase linearly. Common network effect benefits available immediately include:

- Increased access to larger resource pool
- Improved diversity of resources that can be used to adapt to unexpected situations
- Improved situational awareness due to discovery of others, both friend and foe

Ongoing network-sharing benefits include:

- Shared infrastructure costs
- Continued adaptation capabilities due to diverse resources that enable ongoing continuity of operations
    - Improved on-the-job training due to additional opportunities to observe models and mentors
    - Network-centric automated resource exchanges:
        - Resource market-making
        - Reciprocal service agreements

9.3.7.2 The Impact of Increased Information Sharing

The social dimensions of enterprise information management are not confined to sharing issues. Integrating M2M sharing into an enterprise enables such major improvements in the availability, freshness, and cost of information that it can change concepts of operation and decision-making procedures. Such changes in operation change the acquisition requirements for the system components, with the overall result being enterprise transformation. The areas of social impact—areas that commonly undergo transformation—are those that improve the ability to

- Export appropriate information quickly and cheaply
- Coordinate operations across previously unbridged operational boundaries quickly and cheaply

By improving the speed of interactions and associated information exchanges, and making it easy to reduce the information exposure, trust can be built incrementally and easily between partners.

9.3.8 Robust Information Management

Earlier sections discussed various adaptive approaches. They included a range of information packages based on varying distribution capabilities, decentralized information sources, use of search capabilities to find previously unrecognized information sources, and mediation capabilities to enable their usage. This section covers making information management robust, that is, able to react constructively to change and challenge.

9.3.8.1 Redundancy Mechanisms and Issues

The classic response to a robustness requirement is to add capacity—make it stronger, make it more powerful, and make it larger. Also known as overengineering, redundancy, and margin of error, the immediate issue is to balance the trade-offs between the costs of producing more and the expectation that the more will be needed. In the network-centric information space, adding more translates to *redundant material* components and *replicated information* components.

Whenever there is redundancy or replication, the immediate issue is selection—when and how do the replicated resources come into play? The answer depends on the scope of resource management, which is determined by the coupling between the generic resource and what determines its usage. When a superior subsystem draws on a tightly coupled set of subordinate replica resources, the resource selection process is not visible externally. The only external impact will be that the superior subsystem can cope with a wider range of operating environments or offer a firmer guarantee of service. However, in a network-centric enterprise this is not the only kind of coupling, and looser coupling means the selection mechanism is externally visible. One of the most common network-centric information management capabilities is to provide local replicas of distantly produced information. By decoupling the query process from the distribution process, local consumers receive more reliable access to fresher information with lower delay. This is not only useful for mission information services, it is critical to robustness and performance of enterprise information infrastructure services such as catalogs, directories, and search engines. In short, information management system robustness depends on robust distribution combined with robust storage mechanisms.

Robust distribution primarily depends on redundant network channels—a set of chains of transport resources that can convey information. Selection of a particular channel may be internal if the distribution is within a local scope, but at any time information management must traverse a boundary, the distribution channel needs an individual usage interface that actually conveys the information and an overall management interface used in the selection process. If more than one information flow shares the channels managed by the selection process, the selection process must also manage capacity and arbitrate usage.

Likewise, robust storage depends on redundant information copies. Selection of a copy may be internal or external, and administration of shared storage must manage and arbitrate the usage.

Robust information management also requires multiscale information assurance aspects. If replica selection is externally visible, then an attack on one replica may cause a failover to another in a clearly visible way. However, if the replication is hidden at a lower scale level, attacks of an underlying replica can trigger a silent failover. If the replicas are so similar as to have the same vulnerability, then the next replica also will fail. Eventually, the failover mechanism will come to the end of the chain of dominos, and the fault will propagate up to the next scale. If the next scale up is likewise replicated, attacks on all of the underlying scale components will cause a massive surge of failures to hit the upper scale failover mechanisms almost concurrently, resulting in an unexpected surge that often brings down the whole enterprise in a cascade failure and gives rise to the idea that such engineered systems are brittle.

Such multilevel surges are not planned for primarily because of the way technology investment decisions are made in large-scale enterprises. Accounting domains are replicated to match the subdivision of labor, and the aggregate accounting may or may not exist, depending on the legal organization and accepted rules of accounting. The larger the scale of the enterprise, the less likely there is to be an overall accounting and the more likely that the risk–cost–benefit calculation will only take into account local risk levels. The risk of cascade failure typically falls below the threshold that affects decision making (otherwise, better replica management would be at least considered). The other factor is that until recently most investment decisions did not take into account either the *Law of Large Numbers*, or the fact that the kind of underlying but hidden interdependencies cause failure-risk likelihood distributions to be heavy-tailed. The combination means that such failures will occur much more frequently than traditional analyses would predict.*

9.3.8.2 Adaptability Mechanisms and Issues

True agility requires not just robustness, which can result in brittleness as we have demonstrated above, but also *adaptability*, which is the ability to change strategies by changing the *type* of resources, not just change out the particular instance currently deployed. When applied to information management, this principle results in three main families of tactics:

- *Adaptable information distribution:* Ensure the distribution of information in a highly dynamic, time-sensitive environment.
- *Dynamic information discovery:* Ensure that information about information sources is visible to the right people well before they need it.
- *Adaptable data mediation:* Ensure that visible information is also provided using encodings, structures, and terminology that the local infrastructure can handle, and ensure that local consumers understand how to make sense of the differing semantics.

It also requires corresponding adaptable collateral services:

- *Adaptable and scalable information assurance:* When new information sources are used, command authority and protection mechanisms must automatically adjust accordingly.
- *Adaptable and scalable funding:* When new information sources are used, financial authority and budget transfer mechanisms must automatically adjust accordingly.

---

* For more information on this and other issues surrounding uncertainty, see the excellent book by Taleb (2007).

9.3.8.3 Information Infrastructures

9.3.8.3.1 Sharing Resources (the Case for Infrastructure)

The most important factor in determining an information infrastructure implementation construct is that no single entity can afford the time/money/skills/people to provide that utility alone. Consequently, solving the common problem requires coordinated action and creates stakeholders.

If the service can be broken into component parts and each stakeholder can take responsibility for a part, then they collectively benefit from each other's contribution. This model would not impact the current stand-alone system acquisition model so long as each stand-alone program of record adopts the standards that ensure their part will interoperate with the others. This is the approach advocated by Net-centric Enterprise Solutions for Interoperability (NESI) (see http://nesipublic.spawar.navy.mil/).

However, this "lots of little parallel efforts" approach requires that competition among the stakeholders be managed to minimize the tendency to create parallel standards. More importantly, there are whole areas of infrastructure that cannot be done in parallel but are best done as stand-alone utility company programs. In such cases, the worries come not from standards (the utility company enforces de facto if not de jure standards), but from ensuring that the utility services truly meet the needs of a sufficient number of stakeholders.

9.3.8.3.2 Evolution of Infrastructure

Implementation constructs often evolve from co-ops to corporations quite naturally—that is how the internet has evolved—so any acquisition process should focus on future-proofing the service standards so that they can be governed both ways. As internet operations evolved, the microservices were developed and deployed in parallel under a co-op governance process known as the Internet Engineering Task Force and InterOp. Macroservice bundles that evolved later are based on bundling microservices into corporate business packages and products. Unfortunately for acquisition organizations, the commercial private sector has developed past the microservice level, and they are now primarily concerned with selling macro packages that enable them to at least attempt to monopolize the market. Sometimes there are natural monopolies—distribution networks offering transport services are the most famous example, and Defense Information Systems Agency (DISA) is the DoD utility company organization.

9.3.8.3.3 Enterprise Infrastructure: End-to-End Key Interface Profiles (KIPS)

There is a need to identify key microservice interface profiles that will enable macrobundles that have taken up residence in operational nodes to interoperate.

The following list identifies the ones found to be critical in getting value from a working and workable commercial internet composed of a combination of organic NESI-node-like and utility-company stakeholders.

- Distribution services
- Distributed time service
- Distributed network identity/location mapping and location discovery
- Distributed organizational identity/network identity mapping, and authorization mapping and discovery
- Distributed basic and extended Web services
- Distributed business services

Allowing global end-to-end KIPs also turns out to be critical to future-proofing because it allows new services to evolve naturally as a collection of microservices within an early adopter co-op community and then shifts gears into corporate mode and macroservice packaging.

9.3.8.3.4 Enterprise Service Development and Governance

Global Information Grid-E (GIG-E) microservices will always need parallel co-op development and acquisition because they have to be installed in many nodes that are often termed service access points, mechanisms, clients, or agents. The crypto-based authentication and authorization handshaking mechanisms are classic examples. Network stacks, management agents, and discovery clients are three more (think Winsock, Simple Network Management Protocol (SNMP) agents, LDAP clients, browsers).

GIG-E microservices that are alternating current (AC)/direct current (DC) are power, the original bimodal infrastructure (DC stores and does long-range transport best, AC does retail operations best), information transport (local area network [LAN] vs. wide area network [WAN]) and identity (local vs. global addresses/labels). Note, truly mission-critical utilities are almost forced to be switch-hitters because you cannot always depend on the utility company to deliver.

GIG-E microservices whose macrobundles can be highly optimized (more on this definition another time) are natural utility company services. In this space are long-haul distribution networks (material, power, and information), storage farms (storage area networks, and some server farms), and enterprise-distributed processing (other server farms and grid computing).

Enterprises need both an inclusive infrastructure strategy and a spectrum of governance procedures for all service packaging approaches. At one end are the processes, KIPs, and guidance needed for network-centric node development and at the other the processes, KIPs that focus on internode information distribution. The latter highlights the microservice boundaries between macrobundle-implemented nodes. However, defining that interaction only as a single macrobundle-like the

Enterprise Service Bus (ESB) is insufficient and worse, prone to failure because of too many parts, too many of which are "moving" (active). Even relying on ESB within your node may cause critical information exchanges to fail because of intra-node service dependencies and failures, such as those that occur during distributed attacks from bots and worm. Multilevel defense means each individual system needs to have backup service providers, if only a cache of the remote server's state. In the worst case, a service consumer needs to be able to make progress based at least on local information.

In addition to checking macrobundle interoperability (ESB to ESB), interoperability testing processes need to check that systems gracefully handle failure of each and every underlying microservice (duly identified in the KIPs), plus KIP combinations as identified in standard macroservice implementation constructs.

9.3.9 Information Management Use Cases

The use cases selected for this section focus on the two critical elements of an agile information management function—a sufficient architecture for this capability (and its leveraging of Web services), and the types of patterns needed (ontological; other) to ensure the legitimacy of information moving broadly across the enterprise.

9.3.9.1 Dynamic Information Management in Complex Enterprises FY06 Mission-Oriented Investigation and Experimentation Proposal

Adaptable and reliable information distribution is essential to effective decision-making. Effective information distribution is complicated by the complex, dynamic environment that the AF and its partners operate, for example, network participants may be unavailable at times, COIs and their memberships will be fluid, and communications may be lost. The goal of this MOIE is to investigate an enterprise information distribution capability that combines the strengths of peer-to-peer and publish and subscribe technologies to share information in near-real time while being adaptable to changes in its environment. The prototype resulting from our research will be applicable to a wide variety of areas to improve information sharing including battle management, collateral damage assessment, and BFT.

9.3.9.2 Semantically Enabled Web Services for Effective Decision Making in Network-Centric Environments FY06 Mission-Oriented Investigation and Experimentation Proposal

Given the momentum of Web services and SOAs and the early state of semantic Web technologies and ontology modeling, it is important to evaluate the value of adding semantics to Web services. This MOIE proposal will apply the semantic Web concepts of the Web Services Definition Language (extended for OWL services)

(WSDL-S) approach to a selection of Web services and will evaluate this approach and its effectiveness for enabling and improving the automation of Web services discovery, composition and mediation, reuse of Web services, and the flexibility and reconfigurability challenges of using Web services.

9.3.9.3 Managing Infostructure Complexity in Agile Decision Support FY06 Mission-Oriented Investigation and Experimentation Proposal

The AF MOIE expects to refine a mathematical model based on characteristic parameters of both decision-making information flows and IT systems. Using the model as a base we intend to produce an initial prototype decision aid that takes these classes of parameters as input, generates contours representative of the interface between decision information flows and technologies, and matches requirement contour to capability contour. The model concepts (and the tool that enables rapid "what–if" scenarios) can be used by infostructure designers and managers to rapidly model and evaluate potential infostructure reconfigurations. Such evaluations will not only establish overall acquisition requirements but also to adapt the systems in the field. Use of the model concepts and of the tool enables standardized (and ultimately automated) policy-based reconfiguration specifications whose contents can be coordinated across multiple platforms.

9.4 Agile Information Exploitation (The Right Decisions)

9.4.1 Introduction

The previous two sections discussed the role of agile information generation and management in the C2E. In this section, we address another critical link in the chain: exploiting that information to support enterprise capabilities such as decision making, situation assessment, course-of-action selection, and resource allocation. We define information exploitation as *the interaction of technologies and human cognitive and collaborative processes enabling sufficient sense-making and situational awareness for agile, effective decision making.* Information exploitation refers to the direct use (explicit or transparent) by people of these functions and underlying technologies as well as the appropriate collaborative processes to make decisions. The exploitation function is based on cognitive science and collaborative information technologies, and pays strict attention to how people best function under complex decision-making situations, and how technology and processes can best be integrated to support them.

Effective decisions are the end products of successful information exploitation. And since humans are ultimately responsible for making decisions by exploiting and adapting the capabilities of systems in the C2E, we advocate that systems must be designed with adequate consideration of the role of human operators, teams, and their mission objectives. Moreover, we advocate that systems engineers should construct systems that amplify and augment human cognitive strengths and help overcome human limitations. Technology should support the needs of the warfighter, not vice versa: in effect, people and mission objectives become the "forcing function" for technology. We need to move beyond the design philosophy of "Science Finds, Industry Applies, Man Conforms" to the philosophy of "People Propose, Science Studies, Technology Conforms" (Norman, 1993).

Thus, the central information exploitation challenge is to confront the issue of Human-System Integration (HSI).* HSI is a comprehensive strategy to optimize total system performance (e.g., humans working with machines), minimize total ownership costs, and ensure that the system is built to accommodate the characteristics of the operators (e.g., their strengths, limitations, and biases) who will ultimately operate, maintain, and support it (e.g., the 311 Human Systems Wing, Brooks-City Base, TX). Some of the central issues of HSI include how to determine the information to display to operators, how to format the information display so it is congruent with operator goals and decision-making objectives, how to effectively distribute tasks across team members and technology, and how to determine when a system is usable, effective, and leads to greater performance than either people or technology could achieve working in isolation.

Some would argue that the basic problem in C2 of complex systems is that error-prone humans are simply inadequate to handle the rigors of decision making along with the vast amounts of available information, and "with just a little more automation we can eliminate the 'human error problem' entirely" (Christoffersen and Woods, 2002). But the fact is that an enterprise is a complex interaction of people (e.g., soldiers, commanders, etc.) and processes, as well as hardware and software systems, and people will always be central players in enterprises because of their sense-making skills, creativity, expertise, and adaptability. This is particularly so as the line between systems acquisition and systems operation is blurred by human beings who are constantly adapting themselves and their systems to meet emergent and unexpected challenges of dynamic battlespaces.

In the remainder of this section, we outline a strategy and associated research needs for moving in a direction that will allow us to make the best use of information, people, and technology in large-scale, distributed cognitive systems (enterprise systems).

9.4.2 The Information Exploitation Environment

The increased complexities of future C2 environments pose ever greater demands on the need to quickly and accurately exploit information. There are a variety of

---

* See, for example, Human Systems Integration edition, Fall 2007. Collaborations—Inside Knowledge for MITRE's Systems engineering Community. Volume 5—Number 2. Systems Engineering Process Office (SEPO). The MITRE Corporation. www.mitre.org/work/sepo.

characteristics of the information exploitation environment that complicate effective operation. Next, we relate how these complications affect information exploitation in particular.

9.4.2.1 A Dynamic and Complex Battlefield

The battlefield environment is one where new information is constantly coming in, old information is invalidated, goals are changing as situations evolve, and decisions are made in real-time in response to such evolving conditions. Moreover, decisions are never one shot events. Rather, a series of nonindependent decisions are made that change the state of the battlefield, and it is only possible to judge the quality of a series of decisions in retrospect (Thunholm, 2005). The complexity of the battlefield environment means that the critical function areas of information generation, management, and exploitation must be tightly coupled and integrated. That is, decision makers must have access to the current state of the battlefield to determine the impact of their decisions, a capability provided by information generation. They must also know where to find such information and be able to access it rapidly, which are capabilities provided by information management.

9.4.2.2 Increased Volume and Rate of Information

In the C2E, vast amounts of data and information are available from a multitude of diverse and distributed sources. This increased volume and rate of information available to decision makers has the potential to lead to more informed and therefore better decisions, but it can also lead to information overload. Decision makers may spend an inordinate amount of time determining which information is relevant to their goals, attempting to make sense of that information, searching for additional information, and making a decision based on the totality of that information. Moreover, decision makers must also deal with the trade-off between waiting for near-perfect information before acting, or acting more quickly with good enough information. However, the rapid pace of future battles and the need to stay within adversarial decision loops will punish those who procrastinate (Alberts, 1996). Thus, information systems must support decision makers in developing rapid plans by conveying when the amount of information backing a particular decision or course-of-action is good enough. Information fusion techniques discussed in Section 9.2, including Bayesian updating, can go a long way in extracting the maximum amount of utility from available information. And techniques discussed in Section 9.5 that aid operators in developing flexible and robust plans, rather than potentially brittle optimal plans, can also be of use.

In addition to getting the *right information* to the *right people*, information must also be in the *right format* so people can make the *right decisions*. Determining the proper format and display of information must be based on both decision-making goals and human cognitive and perceptual strengths. People excel at recognizing patterns, matching patterns to past experiences, and selecting courses of action that were successful in the past (Klein, 1999). We shed light on ways to effectively present information to operators in Sections 9.4.4.1 and 9.4.4.2.

9.4.2.3 Reduced Manpower and Cost Goals

Across the Services there is a push to do more with fewer personnel and less cost. Secretary of Defense, Donald Rumsfeld, has advocated making changes to transform the military into a leaner and meaner force capable of winning every single battle that this military is faced with. The Office of Naval Research's Science and Technology Manning and Initiative has set as its goal a fifty-percent personnel reduction while demonstrating operational utility for all functions. Thus, there is a need to design systems that maintain acceptable levels of performance with fewer people.

In order to meet these reduced manpower and cost goals, the role of humans has largely shifted from more narrowly specified missions to multimission tasking, and from manual control of a single system to supervisory control of multiple and possibly automated systems. For example, rather than five operators controlling a single UAV, it is likely that in a future scenario a single operator will supervise a fleet of five or more semi- or fully-autonomous UAVs. However, careful consideration must be made of the roles and interactions of the human and machine elements, and automation should only be introduced when there is a specific need to do so (Mitchell et al., 2004). Section 9.4.2.4 highlights the concerns that must be addressed to ensure effective human–automation collaboration and coordination.

9.4.2.4 Changing Human Roles

Owing to the increased volume and rate of information, as well as the increased operational tempo brought about by NCOW, operators will be increasingly called upon to supervise and utilize automated support systems to assist them in decision-making tasks. They are also confronted with different types of missions and concepts of operations, such as multimission tasking, the trend toward decision making at lower echelons, and distributed operations in which small, highly skilled, and independently operating units must aggregate quickly to provide sufficient capability wherever it is required against a complex and adaptive threat. Given the importance of human insight and adaptability for these new roles, the goal should not be to replace human expertise with automation, but rather to amplify and augment it.

Many current automated systems simply provide optimal decisions without providing insight into the reasoning behind the decisions or leveraging operator expertise. However, several unintended human factors and system performance problems arise from such an approach. For one, it is impossible for designers to predict and model every situation in dynamic and complex battlefields, so the optimal decision may not always be optimal or even correct. This lack of robustness, or brittleness, is a key reason for keeping flexible and creative human operators in the decision-making loop. For another, systems that do not allow operators to rapidly understand how a system came to its advice may cause automation bias, where operators either place too much trust in the system solution when they should not, or place too little trust when they should. Also when operators have little insight into the operation of an automated system, their situational awareness is degraded.

To counter these problems, we believe that automation should allow operators to rapidly understand the reasons and results of automated solutions, adjust system parameters and assumptions as conditions change, and visualize and explore problem spaces so that they can leverage their sense-making skills, creativity, adaptability, and expertise to overcome the problems of automation brittleness and bias.

9.4.2.5 High Stakes, Time Pressure, and Uncertainty

The battlefield is a high-stakes environment where errors may have serious consequences. Moreover, operators are often under extreme pressure to make decisions as quickly as possible where there is much uncertainty and ambiguity. Such high stakes, time pressured, and uncertain environments may lead operators to adopt decision-making shortcuts (heuristics) and biases that information system designers must be aware of. While such strategies are generally effective, they can also hinder decision making in situations where they lead to systematic biases (Mitchell et al., 2004). Relevant heuristics and biases include*:

- *Anchoring and Adjustment Heuristic:* People make an initial guess about the likelihood of a particular event or battlefield state (e.g., I am 80% certain that track is an enemy tank), and they adjust their assessment based on new information (e.g., the track is traveling over rough terrain, so now I am 85% certain that it is an enemy tank). People run into problems because they generally do not make sufficient adjustments to their initial assessment. Systems that employ Bayesian techniques can help people make more optimal adjustments (Burns, 2007).
- *Availability Heuristic:* People judge the likelihood of a particular event based on what has happened in the recent past or what is most readily available in their memory. This can be problematic when the current event differs from any recently available event. Systems that maintain a history of past events and assess their similarity to the current event can help overcome fallible human memory.
- *Representative Heuristic:* People often evaluate probabilities based on the degree to which A resembles B. One side effect of this heuristic is the conjunction fallacy, where specific scenarios appear more likely since they are representatives of how we imagine events. However, specific scenarios are mathematically less likely than more general scenarios.

---

* For a classic reference detailing concepts in this general area, see Tversky and Kahneman (1974).

- *Confirmation Bias:* People tend to seek information that confirms their initial assessment and discount or explain away disconfirming information. Thus, there is a need for information systems to support people in considering multiple possible assessments and selecting the most probable assessment.
- *Automation Bias:* People may place too much or too little trust in solutions provided by automated systems. This is particularly true of automated decision support systems whose reasoning is not well understood by operators.

There is a need for information systems designers to be aware of these heuristics and biases and develop systems that exploit their benefits without performance degradation. For example, systems must display uncertainty so operators can calibrate an appropriate level of trust as contexts change. There is a need for further research in ways to effectively communicate and represent uncertainty in a manner that is coherent to operators and enhances decision making. Bayesian techniques may help people reason under uncertainty and extract the maximum utility from available information (Burns, 2007).

9.4.3 Information Exploitation Challenges

To develop a strategy for effectively leveraging information, people, and technology, we must first identify the major challenges for information exploitation within enterprise systems. An understanding of the underlying challenges will allow us to develop a set of corresponding information exploitation tenets to guide systems engineering efforts.

9.4.3.1 People Steer, Guide, and Supervise the Engine of Technology

In relatively stable decision-making environments, technology has been a powerful driver of human decision processes and functions. For instance, automation of a task that had been time-consuming, tedious, or error prone, can enable people to adjust their roles and associated allocation of time and effort accordingly for better performance.

However in complex enterprise environments where threats and associated priorities change unexpectedly, this technology-driven approach fails. Information exploitation becomes a highly dynamic activity, requiring continual sense-making and adaptability. The human decision makers become the forcing function, since it is people who can apply their expertise, perspective, and contextual knowledge for higher-level sense-making and detection of unanticipated cues in the environment. Just as we would temporarily suspend our car's automated cruise control when we notice highway congestion ahead or the start of inclement weather, dynamic and complex environments require a different type of interaction and feedback between people and technology for high performance.

People provide the experience and perspective to detect cues that a course correction may be needed, for example, instances when seemingly valid and reliable system information no longer makes sense within the current context and thus violates an existing information pattern. They can initiate activities to determine whether the disconnect stems from the system information itself (such as an input error, faulty algorithm, or information *from* the wrong source or sent *to* the wrong recipient), or is due to a perception of the environment that is no longer accurate (e.g., the type of threat, desired effects, or resource constraints have changed). To support this helmsman role, systems must provide feedback on the external environment, in addition to feedback on progress toward specific ongoing tasks. Technologies such as data mining may greatly assist in this role by identifying *potentially* important patterns within vast amounts of data, after which human insight and sense-making are essential for determining those changes in pattern most likely to be truly significant within the operational environment.

Tenet 1: Provide feedback on the surrounding context to facilitate detection of changes in underlying patterns.

Tenet 2: Design for "redirectability" of technological resources.

As discussed in Section 9.1, an important feature of an enterprise is its dynamism. That is, an enterprise is in a constant state of evolution, reinventing parts of itself through a process of continual innovation and integration. Successful navigation through such an environment requires close continual monitoring and characterization of that environment with an eye toward detecting any significant shift in patterns (see also Section 9.2.5.6).

Tenet 3: Enable more *continuous* monitoring/characterization of the decision-making environment.

Moreover, effective human system integration must enable people to readily communicate any significant shifts in environmental patterns via changes in parameters of the supporting systems. On the basis of a detected change in the environment, systems may need to immediately begin monitoring different information sources or scanning for new patterns. Technology can greatly speed information sharing, task completion, and overall performance once the appropriate tasks and direction are identified, but there is a greater need than ever for the human in the loop to perceive when the course of action must be altered: "We are making great progress, but in the wrong direction!" People perform these essential roles of sense making and adaptability, and apply their insights to complement and steer the supporting technologies. Although we continually strive to reduce manpower requirements and the forward footprint, our efforts to automate must focus on tasks that are more stable, rote, time-consuming, and/or error and bias prone—but not those calling out for human sense making, insight, and innovation.

Tenet 4: Leverage and amplify (versus replace) human experience, insight, and adaptability.

In these complex environments, people are also taking on more Human Supervisory Control (HSC) roles over increasingly autonomous systems and vehicles. This involves explicitly supervising and guiding technologies (such as UAVs), intervening when performance is unsatisfactory and then relinquishing control back to the system when its performance is again stable. Systems must be designed with this human–machine interplay and dual feedback in mind; they must provide operators both the system performance information and sufficient contextual information for the necessary sense making we described. This becomes particularly important as we ask people to maintain supervisory control over *multiple* systems and vehicles.

9.4.3.2 Dynamic Information Acquisition, Filtering, and Tailoring

R4 is the goal, but a critical challenge for information exploitation within enterprise systems is that the right information cannot be fully specified in advance; it is situational and dynamically determined. Addressing this challenge requires synergistic efforts between information generation, information management, and information exploitation, as depicted in Figure 9.13.

For example, in confronting asymmetric threats and more agile adversaries, the kind of information that can be exploited to our advantage may be of a new type, in a previously unseen format, or from a new source. This means decision makers must have an infrastructure for information *acquisition* that allows them to modify their information needs in real time (see Section 9.2). The information *management* function must similarly allow them to combine information, filter information, and enable the desired push or pull mechanisms as a need is identified; operators will be unable to fully describe all information needs in advance (see Section 9.3).

Tenet 5: Link information generation, management, and exploitation efforts to enable dynamic extraction of the right information.

Decision makers also contribute to the assessment of the information's source, pedigree, trustworthiness, and reliability before they act on it; systems should provide contextual metadata to facilitate this process. Since ambiguity is the rule rather than the exception in complex enterprises, determining information provenance is particularly important and challenging (see also Section 9.2.5.1).

Tenet 6: Facilitate *human understanding of* and *inputs into* determination of information pedigree, provenance, and trustworthiness.

Moreover, users must be able to locate the required information within their decision cycle or it loses its value. To quote Commander John "Nano" Nankervis, U.S. Naval Reserve (USNR), Operation Iraqi Freedom (OIF) Chief of Air Tasking Order (ATO) Production, "Unknown capabilities are the same as not having any." Since operators are typically bombarded with vast quantities of data, they currently bring informal social networks, as well as information technologies to bear in order to locate the necessary information or expertise. Unfortunately, while experienced teams often develop highly effective social networks and information management processes to assist their exploitation, newer teams without those established relationships, processes, and skill sets may have great difficulty locating and exploiting the relevant information in time. They require both technologies and processes to facilitate this agile composition of needed resources and expertise, as well as mechanisms that permit them to dynamically and adaptively filter, set alerts, and tailor displays without undue effort or delay.

Tenet 7: Value and facilitate information *accessibility*, beyond its existence.

9.4.3.3 Operators' Use of a Multitude of Disjointed Systems and Communications Modalities

Operators must often make use of a variety of systems and communication mechanisms to accomplish their decision-making tasks. As enterprise systems engineers, we must not only consider the impact of a specific system on operator performance in decision-making tasks but also how that system will interact with other systems the operator may be using. The introduction of a new system may overload operators and hinder their ability to work effectively with existing systems to accomplish tasks.

For instance, internet chat has become a pervasive communication modality in military settings, and operators often monitor half a dozen or more chat windows while simultaneously communicating with collocated team members. Operators monitoring a number of communication threads often become cognitively overloaded; people have a limited attention bandwidth and simply cannot attend to each system with the same level of concentration. Thus, careful consideration must be made of how the introduction of a new system will impact operator cognitive load and attention management. A deep understanding of operator tasks is necessary to ensure adequate integration of a new system within an operator's suite of systems. Automated attention cueing strategies, filtering techniques, and the integration of disparate systems or displays into a cohesive system more suited to an operator's overall set of tasks are potential solutions.

Tenet 8: Design and introduce technologies with awareness of the actual work context, including interactions with other systems being used simultaneously.

Not only do operators rely on a multitude of systems, they often also rely on multiple *modalities* (such as phones, radios, headsets, public address systems, email, instant messaging, or face-to-face discussion). Multiple modalities may be needed to communicate with distributed team members and enable the necessary feedback and coordination for real-time team-level exploitation and response. Observation of operators within an Aerospace Operations Center (AOC) (Boiney, 2005) revealed that operators not only have to determine *what* information to focus their attention on and *with whom* to share it but also must make countless secondary judgments about *how* to do so: what means of communicating will provide the desired response most effectively? (see Mathieu et al., 2007).

For example, an operator could post information to a designated database or shared application, in which case it is likely that other team members will know where to pull the information from if it is needed. Alternatively, the operator might push the information via face-to-face discussion or by phone, which guarantees that the intended audience receives the information but provides no persistent record of the communication and does not scale to more than a few people. Instant messaging is another possible means of information sharing, in which case there are decisions to be made about the appropriate chat room(s) to address, or whether a *chat in private* directed at a particular individual would be more appropriate. Yet another option is to use *audio* chat over operator headsets, in which case the operator's comments will be immediately heard by the selected individuals, *if* they are present. Teams employ these various modalities differently, depending on the importance and time sensitivity of their communication and the intended audience.

> *Tenet 9:* Design and introduce technologies to support information exploitation within the multimodal communication environment.

Before any new communication modality is introduced, its utility in meeting operator goals and decision-making tasks should be clearly understood. The efficiency and effectiveness of a particular communication mechanism should be evaluated against other potentially better communication mechanisms, and in terms of the overall impact on performance when used in conjunction with other modes of communication.

> *Tenet 10:* Automation should decrease and not increase the cognitive load in complex enterprises.

Moreover, training should guide operators on how different communication mechanisms should be employed, individually and in combination, to meet different decision-making goals and objectives.

> *Tenet 11:* Train as you Fight; Fight as you Train.

9.4.3.4 Distributed Collaboration and Operations in Complex Environments

Operators in the C2E do not perform their tasks in isolation. Rather, they work in teams that may be collocated or distributed across geographic boundaries. Thus, teamwork and collaborative decision making are critical components of the military's vision of network-centric warfare, which is becoming an integral part of current and future military operations. A basic tenet of networked forces is allowing individuals and/or groups the ability to leverage information both locally and globally to reach effective decisions quickly. Collaboration is both a critical element of networked operations and of military operations in general, and with the increasing trend of automated sensor and weaponry technology being introduced in these environments, successful battlefield operations will involve effective collaboration not only between humans but also humans and automated systems (Scott and Cummings, 2005).

Information exploitation is a collaborative activity, requiring teams of typically dispersed individuals to share information and perspectives, coordinate activities, and reach shared understanding in order to make decisions. Increasingly, collaboration crosses boundaries of service, agency, and country to form highly heterogeneous teams with Joint, Coalition, Cross-Agency, or even Military–Civilian membership. As a result, teams must achieve shared understanding and shared situational awareness in the face of different incentives, cultures, perspectives, and constraints. While a cross-agency team may share unity of *effort*, they may not have unity of *command*. This often means that consensus building, resource allocation, and other decision making must rely on the ability to *influence*, rather than to control. Moreover, in highly complex environments characterized by emergent threats (such as Homeland Defense and Security scenarios), the composition and formation of the team itself becomes dynamic: the right people and corresponding resources are not fully known in advance, but rather must be assembled in real time. This poses additional challenges for information exploitation, since the set of decision makers who must share information, coordinate activities, develop trust, and reach shared awareness may never have worked together before. Complicating matters still further, the team's membership may continue to evolve as new areas of expertise are required.

> *Tenet 12:* The right people in R4 is also dynamic; enable information exploitation across heterogeneous teams with changing membership.

Another challenge of collaborative information exploitation relates to the sheer quantity of connections between people and systems. As we move toward network-centric warfare, we are gaining the ability for anyone, anywhere to communicate with anyone else. Yet effective information sharing necessitates *selective* information sharing to avoid overload. Because there is so much information coming from so many sources in so many forms, and because it is being processed, interpreted, and updated by many systems and people, operators currently report that it takes experience and informal social networks to know where to look and what you are looking for (Boiney, 2005). One of the key findings from observations at Joint Expeditionary Force Experiment (JEFX) 04 was that more links do not equal more or better coordination. There are costs associated with additional links and interconnections that translate into system latency, information overload, and an inability for people to manage their attention effectively across the plethora of systems, communication modalities, and individuals transmitting information.

> *Tenet 13:* Enable *selective* information sharing to balance the need for good judgments with the risk of cognitive overload.
> *Tenet 14:* Value more understanding over more information.

In addition to garnering sufficient situational awareness, teams must develop sufficient awareness of the team itself—including others' identities, roles, activities, and status on tasks. For effective coordination and collaborative information exploitation, they must know something about what others on the team know—and do *not* know—at a given time, whether they are present at their stations at the time of critical communications, and how busy others are. Without these types of team awareness, people may share information inadequately or excessively, misinterpret a response or a delay in receiving one, and lose trust and/or shared situational awareness. Maintaining team awareness is challenging in distributed environments since people cannot readily see or hear one another. It can exacerbate information overload as operators strive to understand *who* they are dealing with and who is doing what, in addition to *what* is going on around them. In such environments, teams currently rely on a mix of technologies, such as tailored coordination displays, phone calls, and instant messages in order to stay abreast of others' activities; no single technology currently addresses this need.

> *Tenet 15:* Provide mechanisms to maintain awareness of others' activities, knowledge, and load (team awareness) to enable effective coordination.

9.4.3.5 Trust Formation

Establishing appropriate trust among decision-making entities is crucial to ensure agile and effective information exploitation. Issues of trust pervade all elements of the information exploitation environment. Trust must be established not only between humans and technology, such as decision support systems and information sources, but also between human team members. Interpersonal trust plays a vital role in people's willingness to share information, as well as in their ability to collectively validate incoming information and resolve differences of interpretation when the information is contradictory or unclear. For example, people are continually vetting information as it arrives, judging its relevance, provenance, pedigree, and trustworthiness. Some of these attributes are inferred based on the *human sources* of information, and their perceived credibility, competence, commitment, and motivation. In time-pressured scenarios, the absence or presence of interpersonal trust and credibility can be the determining factor in which opinions and inputs get full consideration. In short, lack of trust is a potential sociocultural show stopper for effective exploitation and decision making.

To enable appropriate levels of interpersonal trust, team members share much more than directly task-related information. Collaborative information exploitation requires rich interpersonal communication for some issues, and technologies to enable these less obvious but equally important drivers of communication. Different forms and modalities of communication—such as email, chat, video teleconferences (VTC), and M2M posting within systems of record—will be appropriate for certain types and purposes of communication, and potentially detrimental for others. For example, a team may be able to communicate most of the technical information via an online briefing with accompanying audio. But periodic face-to-face or VTC-enhanced discussion may be imperative at key intervals to ensure there is an opportunity to detect social cues, develop relationships, and establish buy-in. This is particularly important if the team is distributed and heterogeneous; it takes significantly more time and effort to establish trust and cohesiveness across organizational boundaries, since corresponding team members will typically not be colocated, not know each other well, and not share all of the same goals, values, priorities, procedures, and social or cultural conventions (Boiney, 2005; McCarter and White, 2007).

Both the development and the *implementation* of these modalities (e.g., training, tactics, techniques, and procedures) must be designed with an awareness of when and how they can aid information exploitation.

> *Tenet 16:* Provide both *systems* to enable social, trust-enhancing interpersonal communication, and *processes* to guide their appropriate use.

9.4.3.6 Coevolution of People and Technology: Performance of the Joint Human-Technological System

Both human cognition and IT are necessary for effective information exploitation; the goal must be complementarity, combining, and augmenting the strengths of each, rather than substitution. Operators *adapt* both their processes and their use of technology to meet emergent, unexpected needs. Many researchers have noted that although systems are typically designed with specific uses and purposes in mind, people will appropriate those tools in whatever way best suits their goals and needs.

> People are "purposive, knowledgeable, adaptive, and inventive agents who engage with technology in a multiplicity of ways to accomplish various and dynamic ends. When the technology does not help them

achieve those ends, they abandon it, or work around it, or change it, or think about changing their ends" (Orlikowski, 2000, p. 423).

Second, the introduction of technology *necessarily* changes human cognition, behaviors, and interactions. This is evident even with relatively simple technological interventions, such as the shift from face-to-face communication to the use of email or chat, in which the sender's attitude and intent must be—sometimes erroneously—inferred. The introduction of technology can result in unintended emergence in which the interdependencies between multiple systems and human processes create undesirable consequences. For example, we discussed earlier how enabling widespread connectivity between individuals and teams who were previously disconnected (or limited to communication up a restricted chain of command) unintentionally lead to information overload. New technologies should be developed, trained for, and assessed with an awareness of these inevitable impacts on human cognition and behaviors, and with the goal of developing appropriate guidance and processes to maximize intended emergence and minimize unintended emergence.

Tenet 17: Emphasize and assess the performance of the joint human-technological cognitive functional system.

In dynamic and complex environments, technologies often force workarounds (or get abandoned) because they were designed for a relatively stable and predictable information exploitation environment. They expertly and optimally implement a specific series of tasks, in a particular order, yet are too brittle to withstand the adaptations that will be needed for fluid information exploitation. One of the reasons that the use of chat has exploded in operational settings is because it is easy to use and designed for general utility. Despite its shortcomings, operators find that they can quickly adapt and appropriate it for emergent needs. In a complex environment, we are better-off doing a set of more general tasks very well, and enabling dynamic recombination of these general capabilities in new ways, than trying to anticipate, optimize, and overspecify all the particular capabilities that will be required.

Tenet 18: Design for general utility; keep technology simple, robust, and flexible.

In the face of advanced automation such as M2M exchange of information, it is easy to temporarily lose sight of human roles, needed capabilities, and contributions. Yet M2M never exists in isolation; all information collection is ultimately human-directed (e.g., tasking selected sensors to collect intelligence of a particular type, at a particular time and location) and is ultimately intended for human consumption, so that it should more properly be labeled *human*-to-machine-to-machine-to-*human* (HMMH). People initiate, contribute to, propagate, and redirect M2M threads. The type and amount of contextual information or metadata that is required for *human* judgments is often different than that required for exchanges between *systems*. A key challenge of HSI is therefore to design information-sharing mechanisms and processes with an eye toward supporting the mixture of, and interplay between, both types of exchanges.

> *Tenet 19:* Facilitate the cocreation and coevolution (between people and technology) of enabling environments.

In short, the enterprise exists to serve the needs of human goals and values. Rather than technology replacing people, people and technology will work together and complement each other in such joint cognitive systems, where cognition is distributed across individuals, teams, and technology (Hutchins, 1995). And people will remain key players in such systems because of their creativity, expertise, and adaptability. Thus, the need to perform HSI in the systems of the future will only grow.

9.4.4 Enabling Technologies and Methodologies for Information Exploitation

9.4.4.1 Cognitive Engineering

Faced with the challenge of implementing HSI in ESE, the question is: How can ESE make the best use of people and systems in large-scale distributed and dynamic enterprises? We believe one potential answer is to augment the practice of systems engineering with the methods of *cognitive engineering*.

Cognitive engineering draws on a variety of disciplines, including human factors engineering, human–computer interaction, decision science, cognitive psychology, computer science, and other related fields. It has roots in task analysis, which identifies the key tasks or functions that are performed in a work domain and then systematically breaks each task into a series of lower-level tasks. Armed with such a task breakdown analysis, it is then possible to make engineering decisions about how to allocate functions between people and systems.

Here it is important to distinguish between *Behavioral* task analysis and *Cognitive* task analysis, since cognitive engineering is most concerned with the latter. Behavioral task analysis is concerned with actions (behavior) that can be directly observed, such as moving a dial or flipping a switch, and it is most often used to measure quantities like time-to-completion or total throughput in a given time. Cognitive task analysis moves beyond observable behavior to measure and model the mental activities (cognition) that drive observable behaviors, and it can be used to assess quantities like throughput as well as quality. For example, cognitive task analysis can be used to assess the potential for human errors in information processing, and thereby serve as a basis for designing decision support systems.

The goal of cognitive engineering is to develop systems, training, and other products that support cognitive functions in decision making, situation assessment, course-of-action selection, resource allocation, and other information processing tasks. Some design questions addressed by cognitive engineering include: What information should be provided to system operators? How should the display be formatted so it is congruent with operator goals and decision-making objectives? How can tasks be effectively distributed across team members and system automation? How can systems support humans so that human–system performance is better than either systems or humans could achieve in isolation?

To answer these design questions, cognitive engineering methods are typically used to construct models that represent the cognitive demands of the domain. Figure 9.15 shows the various targets of analysis of cognitive engineering methods. These may include models of the decisions to be made in a domain and their interrelationships, the information required to support such decisions, cognitive strategies that are employed to process information and make decisions, as well as how decisions and tasks are distributed across agents in the domain, including people and automated support systems.

In order to advance the state of the art of cognitive engineering, we believe it is necessary to tie the analytical products of cognitive engineering more closely to ESE design challenges. Bonaceto (2003) lays the groundwork for integrating cognitive engineering into systems engineering. For further information on cognitive engineering, see Bonaceto and Burns (2006), Bonaceto and Burns (2007),

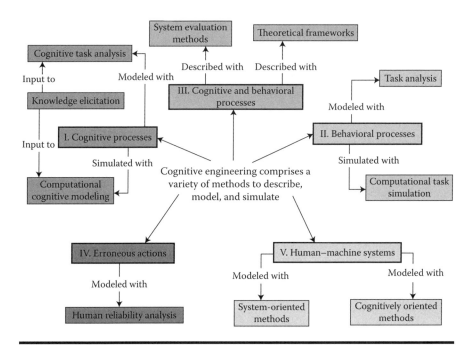

Figure 9.15 Classes of cognitive engineering methods.

Vicente (1999), or visit the Mental Models MITRE-Sponsored Research (MSR) Cognitive Engineering Web site, http://mentalmodels.mitre.org/cog_eng.

9.4.4.2 Intelligent Decision Support Systems and Automation

"Agile, effective decision making" is the crux of information exploitation. With the increased complexities of C2, such as the increased volume and rate of information, diversity of missions, compressed time lines, and unpredictability of adversaries, it is imperative to provide operators with powerful and effective systems to support their decision making objectives. Unless the problem to solve is one that requires no flexibility in decision making and a low probability of system failure, we believe fully automated decision support systems that provide operators with optimal solutions should not be the design goal. Rather, the goal (as stated in Tenet 4) should be to develop decision support systems that complement human decision makers to form an integrated human–machine team capable of solving difficult problems more effectively than either working individually.

To this end, we present an approach to decision support system design called "structure mapping" (Burns and Bonaceto, 2006). The problem with most support systems is that they are not designed with adequate consideration of human cognitive and perceptual strengths. Structure mapping rectifies this shortcoming by constructing visualizations whose features directly map to the structure of the problem at hand. Such displays allow decision makers to reason about complicated situations with intuitive graphical representations. Geospatial maps are a simple example of structure mapping.

However, many decisions in C2 require more robust representations than simple maps. One example is the problem of weapon–target pairing (WTP) in time-sensitive targeting, an important resource allocation problem where operators must task a limited number of assets to destroy high-value targets within a short window of time. Figure 9.16 shows a display called "Pairing pictures" to aid operators in WTP that illustrates the principles of structure mapping. The basic structure of the WTP problem is a matrix of assets paired with targets, which is depicted as the "Problem space" on the left side of Figure 9.16. The cells in the matrix represent the values of various weapon–target pairs, which is computed from a value function that sums the contributions of various factors including kill and loss probabilities, weighted by asset and target priorities The system design maps this structure to a matrix of boxes with color-coded bars that represent the different terms in the value function. This design arose from analysis of both the tasks to be performed in the domain, and the reasoning and representational strategies of human operators.

By designing displays that capture and represent the structure of the problem and mimic representations used by expert human operators, operators can rapidly perceive the problem space and select an appropriate course of action since less effort is spent on understanding the display. Thus, operators are able to devote their cognitive resources to solving the problem rather than figuring out how to use the

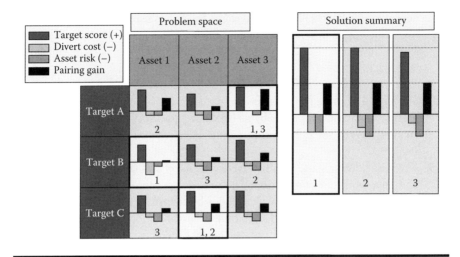

Figure 9.16 "Pairing pictures" support system displays.

tool. Moreover, if an operator's solution is at odds with the automatically computed solutions (shown in the "Solution summary" on the right side of Figure 9.16), they can rapidly understand how the system came to its advice and rectify differences with their preference, which may be based on contextual factors that the system does not account for.

Tenet 20: Engender *appropriate* trust in and reliance on technology.

9.4.5 Information Exploitation Tenets and Design Principles

Information exploitation tenets and design principles have been highlighted and discussed above. Below, we briefly recap and summarize each tenet.

- *Tenet 1:* Provide feedback on surrounding context, as well as on performance. People provide the experience and perspective to detect cues indicating that seemingly valid and reliable system information no longer makes sense within the current context and thus violates an existing information pattern. They can initiate activities to determine whether the disconnect stems from the system information or algorithm itself, or is due to a perception of the environment that is no longer accurate. To support this helmsman role, systems must provide feedback on the external environment, in addition to feedback on progress toward specific ongoing tasks.
- *Tenet 2:* Design for redirectability of technological resources: The battlefield is complex and dynamic, and as situations and workload change, operators should be able to redirect or constrain their technological resources. It is

necessary for decision support systems to make clear their intent, assumptions, and operation so that operators can understand what the system is trying to do and decide whether or not what it is trying to do is congruent with what it should be doing.

- *Tenet 3:* Enable more *continuous* monitoring/characterization of the decision-making environment. Complex and dynamic environments require ongoing human attention, supported by technologies, to detect cues that may herald a change in threat or necessary adjustment in information exploitation strategy.
- *Tenet 4:* Leverage and amplify (versus replace) human experience, insight, and adaptability. Humans are adaptable and creative decision-making entities. The goal of technology should not be to replace humans, but rather to support their strengths and overcome their limitations. Performance of the joint human–technological system should exceed the performance of either working individually.
- *Tenet 5:* Link information generation, management, and exploitation efforts to enable dynamic extraction of the right information. The complexity of the battlefield environment requires that the critical function areas of information generation, management, and exploitation be well linked and integrated. That is, decision makers must have access to the current state of the battlefield to determine the impact of their decisions, a capability provided by information generation. They must also know where to find such information and access it rapidly, which are capabilities provided by information management. Only then can they exploit that information and make timely decisions.
- *Tenet 6:* Facilitate *human understanding of* and *inputs into* determination of information pedigree, provenance, and trustworthiness. Decision makers contribute to the assessment of the source, pedigree, trustworthiness, and reliability of information before they act on it. Thus, systems should provide contextual metadata to facilitate this process. Since ambiguity is the rule rather than the exception in complex enterprises, determining information provenance is particularly important and challenging.
- *Tenet 7:* Value and facilitate information accessibility beyond its (mere) existence. Information exploitation must occur quickly enough to disrupt the enemy's decision cycle. Although the necessary information may exist, it is of no value unless it is identifiable as relevant to the decision at hand, and easily located and assessed by the decision makers in real time.
- *Tenet 8:* Design and introduce technologies with awareness of the actual work context, including interactions with other systems being used simultaneously. How decision makers actually go about their work is often radically different from the contents of documents such as CONOPS and Tactics, Techniques, and Procedures (TTPs). Operators adapt and utilize decision-making strategies that often are not articulated in such formal documents. Moreover, operators

often rely on a multitude of systems and communication modalities. Thus, it is necessary to understand how operators actually perform their work to understand how a new system will impact operators' cognitive load and attention management as they interact with the new system in their work environment.

- *Tenet 9:* Design and introduce technologies to support information exploitation within the multimodal communication environment. Operators often rely on multiple communication modalities (e.g., phones, radios, email, instant messaging, or face-to-face discussion). Multiple modalities may be needed to communicate with distributed team members and enable the necessary feedback and coordination for real-time team-level exploitation and response. Operators not only have to determine *what* information to focus their attention on and *with whom* to share it but also must make countless secondary judgments about selecting a modality that will provide the desired response most effectively.

- *Tenet 10:* Automation should decrease and not increase the cognitive load in complex enterprises: A side effect of many advanced automated systems is that operators are bored most of the time but overloaded during periods of high activity. Another side effect is that operators may either accept potentially incorrect automated solutions at face value or ignore automated solutions because the reasoning strategies of the system are hidden or are not well understood by the operator. Moreover, if an automated solution does not gel with an operator's solution or preference, the operator may spend additional time and energy trying to rectify the systems' advice with his own preferences. Thus, the system has actually made the decision-making task more complex since the operator must decide whether to go with the system's answer or his own. Inserting automation into operator decision-making tasks must be done with careful consideration of human–automation coordination, and automation should only be introduced when there is a specific need to do so.

- *Tenet 11:* Train as you Fight, Fight as you Train. The introduction of new technology changes the nature of people's decision-making tasks and processes. New technology may introduce new types of errors and change the cognitive activities and strategies needed for effective performance. Thus, technology insertion must be an iterative process. We must develop prototypes, assess their utility in realistic scenarios, and refine and adapt them as we uncover additional demands that were not envisioned. This includes emphasizing team-oriented training and assessment for technologies that will be used in collaborative settings.

- *Tenet 12:* The right people in R4 is also dynamic; enable information exploitation across heterogeneous teams with changing membership. Information exploitation is a collaborative activity, requiring teams of typically dispersed individuals to conduct distributed decision making. Teams must achieve shared understanding and shared situational awareness in the face of different incentives, cultures, perspectives, and constraints.

In highly complex environments characterized by emergent threats (such as Homeland Defense and Security scenarios), the composition and formation of the team itself becomes dynamic: the "right" people and corresponding resources are not fully known in advance, but rather must be assembled in real time.

- *Tenet 13:* Enable selective information sharing (to balance the need for good judgments with risk of excessive distractions or cognitive overload). Though it may be tempting to share all information as broadly as possible in a network-centric environment, this quickly overwhelms human decision makers. Information must be posted and/or pushed selectively and appropriately to the right people at the right time.
- *Tenet 14:* Value more understanding over more information. Technological advances make it ever easier to gather and disseminate information. But care must be taken to avoid gathering information for information's sake, potentially making it harder to locate, attend to, and combine the highly relevant pieces of information for improved understanding and more informed decision making.
- *Tenet 15:* Provide mechanisms to maintain awareness of others' activities, knowledge, and load (team awareness) to enable effective coordination. Since most information exploitation is the result of collaboration among people with differing expertise, roles, and resources, they must maintain team awareness in addition to situational awareness in order to coordinate effectively and achieve shared understanding.
- *Tenet 16:* Provide both *systems* to enable social, trust-enhancing interpersonal communication, and *processes* to guide their appropriate use. Establishing appropriate trust among decision-making entities is crucial to ensure agile and effective information exploitation. Issues of trust pervade all elements of the information exploitation environment: trust must be established not only between humans and technology, such as decision support systems and information sources, but also between human team members. Lack of trust is a potential sociocultural show stopper for effective exploitation and decision-making. Collaborative information exploitation requires rich interpersonal communication for some issues, and technologies to enable these less obvious but equally important drivers of communication. Both the development and the *implementation* of these modalities (e.g., training, tactics, techniques, and procedures) must be designed with an awareness of when and how they can aid information exploitation.
- *Tenet 17:* Emphasize and assess the performance of the joint human–technological cognitive functional system. One way to characterize the C2E is as a distributed "cognitive functional system," where cognition in service of decision-making objectives is distributed across individuals, teams, and technology. End-to-end performance of an information exploitation system cannot be assessed in isolation from the people who will use and adapt the system.

Objective criteria must be established to gauge the effectiveness of a combination of people, process, and technology in meeting enterprise goals. And the performance of the combined human–machine cognitive functional system should be greater than the performance of either the humans or machines working in isolation.

- *Tenet 18:* Design for general utility; keep technology simple, robust, and flexible. The right information is dynamic and determined by the current situation. Thus, there is a need for support systems to adapt to changing conditions. Since designers cannot anticipate *a priori*, each possible scenario in dynamic and fluid military environments, people should be able to adapt processes and technology to meet the emergent needs of the battlefield.
- *Tenet 19:* Facilitate the cocreation and coevolution (between people and technology) of enabling environments. In the face of advanced automation such as M2M exchange of information, it is easy to temporarily lose sight of human roles, needed capabilities, and contributions. Yet M2M never exists in isolation; all information collection is ultimately human directed, so it should more properly be labeled HMMH. The type and amount of contextual information or metadata that is required for *human* judgments is often different than that required for exchanges between *systems*. A key challenge is therefore to design information-sharing mechanisms and processes with an eye toward supporting both types of exchanges.
- *Tenet 20:* Engender appropriate trust in and reliance on technology. Operators must be aware of the uncertainty present in the information used by intelligent decision support systems. Moreover, automation should allow operators to rapidly understand the reasons and results of automated solutions, understand and adjust system parameters and assumptions as conditions change, and visualize and explore problem spaces so that they can calibrate an appropriate level of trust in dynamic and uncertain environments.

9.4.6 Information Exploitation Use Cases

In the following sections, we highlight a group of current and proposed research projects in the information exploitation arena.

9.4.6.1 Mental Models in Naturalistic Decision Making FY01–FY03, FY04–FY06 MITRE-Sponsored Research Project

The project's primary goal was to improve human-system performance in complex domains, including military C2, intelligence analysis, and air traffic control. To this end, the project has adopted a blend of behavioral and decision science to understand "how" and "how" well people think, especially in tasks that require making diagnoses and decisions under uncertainty. The project has utilized a Bayesian framework as a normative standard for identifying where people perform well and where they can use help from computerized support systems. To perform experiments on human judgment and decision making that are both scientifically rigorous and relevant to real-world problems, the project has developed a series of synthetic task environments that simulate key cognitive challenges in complex domains. One such synthetic task environment is Pared-down Poker, which is a scaled-down version of poker that simulates key challenges of C2 decision making, including opponent modeling and resource allocation.

As illustrated in Figure 9.17, the project's impacts and products run the gamut from basic to applied research. In the applied research arena, the project performed a decision analysis for the problem of WTP in time critical targeting that resulted in the development of a prototype decision support system ("Pairing pictures" in Figure 9.17). It also participated in a collaborative effort with MITRE's Center for Advanced Aviation System Development (CAASD) to utilize the tools of cognitive engineering to characterize current operations and anticipate future operational challenges that may arise from runway or tower modifications at Chicago's O'Hare International Airport. The project has also performed an extensive survey of cognitive engineering methods (e.g., cognitive task analysis, cognitive work analysis) and has identified how those methods may best be applied to MITRE's

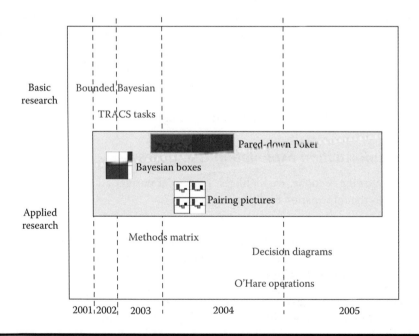

Figure 9.17 Research activities of the mental models in naturalistic decision-making MSR.

systems engineering efforts ("Methods matrix" in Figure 9.17). More information about the project is available at http://mentalmodels.mitre.org.

9.4.6.2 Improving Time-Sensitive Team Decision Making FY04–FY05 Mission-Oriented Investigation and Experimentation

The primary goals of this AF MOIE were to improve time-sensitive collaboration by making it quicker and more accurate in terms of decision outcomes, and with better processes leading up to the resulting human judgments and decisions. To improve time-sensitive collaboration, a critical objective was to identify human-system challenges, adaptations and emergent behaviors in context, in real time. Other objectives were to develop an explicit model of collaborative functions that must be enabled with better automation and processes, and to develop and employ an experimental environment to study team behaviors with additional rigor.

Key accomplishments of this project included the development of a three-part methodology for studying collaboration in context and in real time; application of this methodology to several complex sponsor domains; documentation of operator adaptations and unexpected, emergent behaviors; development of a model of collaboration functions that must be supported by appropriate technology and processes; creation of a realistic experimental team environment for studying key collaborative processes with more control and rigor; developing tailored recommendations for several sponsor domains to improve automation and/or process; and preparing and delivering several research papers and conference presentations on our findings to facilitate knowledge transfer to sponsors, external organizations, and the research community.

9.4.6.3 Dynamic Diagrams in Asset Allocation FY06 Mission-Oriented Investigation and Experimentation Proposal

This AF MOIE, submitted as an FY06 proposal, addresses the increased amount of automation in C2 decision support systems brought about by NCOW. Many current automated systems are black boxes that shed little insight into the answers they provide. Thus, operators have little knowledge of when to trust such systems as contexts change, and such systems do not leverage operators' adaptability and expertise. This MOIE aims to construct and evaluate an asset allocation support system (specifically, asset–target pairing for time-sensitive targeting) that will allow operators to understand, influence, and augment automated answers with their expertise. Its goals are to also identify key visualization and interaction techniques, appropriate levels of automation, and methodologies for developing automated support systems that will be applicable to a wide range of C2 asset allocation systems and automation problems.

9.5 Capabilities-Based Scenarios

9.5.1 Introduction

To demonstrate how the canonical functions of agile information generation, management, and exploitation come together to improve combat effectiveness in the C2E, we describe an actual operational scenario called "Operation Anaconda." Although ultimately successful, Anaconda was poorly executed and resulted in eight U.S. losses. We describe shortcomings in the planning and execution of Anaconda, and demonstrate how the capabilities enabled by the canonical enterprise functions that underpin our paradigm could be leveraged to improve performance in Anaconda and future operations.

9.5.2 Operation Anaconda

On March 1, 2002, an operation code named "Anaconda" was launched against the Al-Qaeda fighters hidden in the Shahi-Kot Valley and Arma Mountains southeast of Zormat in the steep mountainous terrain of southeastern Afghanistan. The operation involved about 2000 soldiers comprised of conventional U.S. forces, SOF, the 10th Mountain Division, the 101st Airborne Division, as well as Afghan forces and coalition forces from Australia, Canada, Denmark, France, Germany, and Norway. Although ultimately successful, Anaconda resulted in eight U.S. losses and dozens more wounded ("Army analyst blames Afghan battle failings on bad command set-up," 2004).

By way of explanation of the casualties, AF, Navy, and Marine Corps officials charged that the Army general responsible for planning the operation had failed to adequately include the other Services and relied on their hastily assembled support only as his battle plan began to fall apart ("Army analyst blames Afghan battle failings on bad command set-up," 2004). When Army forces encountered unexpectedly fierce resistance from Al-Qaeda fighters, their commander issued an emergency call to the AF, Navy, and Marine Corps air and naval fires and logistical assistance ("Left in dark for most Anaconda planning, Air Force opens new probe," 2002). The response to the emergency call for fire was not as prompt or effective as anyone would have liked, contributing to the unnecessary losses and casualties. Figure 9.18 highlights key criticisms of the conduct of the operation.

Army Major Mark Davis asserts that the DoD failed to establish the command structures necessary to integrate the Services, and that the ambiguity in the command structure "created conditions that inadvertently excluded the AF from the planning of Anaconda" (Davis, 2004). A number of officers, however, disagree with Davis' conclusions and point instead to such factors as insufficient planning and poor communication between leaders as more significant contributors to the outcome of the operation ("Army analyst blames Afghan battle failings on bad command set-up," 2004).

> **Operation Anaconda**
> - "Senior Army commanders have been widely criticized by their air and naval counterparts for not coordinating effectively across the services in the days and weeks of ground-force planning that led up to Anaconda [a 2002 battle in Afghanistan]" (Inside the Pentagon [IPT], 29 July 2004, p. 1)
> - "a number of officers ... point to such factors as *insufficient planning*" (ITP, 29 July 2004, p. 1)
> - "Air Force, Navy and Marine Corps officials [charged that] the Army general had failed to adequately include the other components in planning for Operation Anaconda, and instead relied on their hastily assembled support only as his battle plan began to fall apart." (ITP, 28 July 2004, p. 1)
> - "The battle's Army commander ... was forced to issue an emergency appeal for air and naval fires and logistical assistance" (ITP, 3 October 2002, p. 1; and 21 November 2002, p. 1)
> - Result –"a fierce battle against al Qaeda fighters hidden in the steep mountainous terrain of southeastern Afghanistan resulted in eight U.S. losses and dozens more wounded." (ITP, 29 July 2004, p. 1)

Figure 9.18 Key criticisms of the planning and execution of Operation Anaconda.

Overall, the consensus is that the Services were not sufficiently coordinated during the planning phase, and that this lack of coordination contributed to both the need for an emergency call for fires and the difficulty in responding to it. Operation Anaconda posed significant challenges because it involved the use of multiple Joint and Coalition forces in a rapidly changing situation. Although challenging by historical standards, this kind of mission represents the future of warfare, and it is vital that we develop enterprise capabilities that ensure success in such military endeavors. In fact, the operation is often used as an example of the kinds of challenges that pervade planning Joint operations. It is therefore worthwhile to examine how a more effective combination of people, processes, and technology might have better supported the operation and resulted in fewer losses. Below, we examine how information exploitation, management, and generation—the three canonical function areas which underpin our ESE paradigm—may be coordinated and applied to this operation and future conflicts.

9.5.3 Information Exploitation

An information exploitation tenet, discussed previously, is that technology exists to effectively support the processes and decision-making needs of the warfighter. Since finding and exploiting vast amounts of information requires teams of people in distributed environments, a key process in need of technological support and coordination is collaboration. In Operation Anaconda, for example, it is generally acknowledged that the Army general did not adequately coordinate the integration of air assets from the AF and other Services into his battle plans. A collaboration technology that enabled joint operations in the planning phase would have remedied this situation by providing each component commander with a list of each operational activity and its current status. Without adequate procedures, training, and technologies in place, effective collaboration is unlikely to occur, as evidenced in the planning of Operation Anaconda. Thus, adequate collaborative processes and supporting technologies are needed to ensure that all decision-making entities remain informed of the existence and status of tasks to support operational objectives.

While human performance in tasks such as aircraft piloting and tank operation is well understood, human performance in collaborative C2 decision-making processes is less researched and understood. This lack of research permits mistakes and oversights to seep into C2 collaborations at all levels. To remedy this situation, research into Human Factors, and HSI is required in order to develop technologies that enable communication and knowledge-sharing at all echelons of command. If the component commanders involved in planning Anaconda had tools that provided insight into the existence and status of the various operations underway, they would have been more aware of the need for their involvement during the planning phase. Such early awareness creates a collaborative environment that fosters both buy-in and accountability. Early buy-in and accountability would have helped to ensure the availability of air assets and might well have obviated the need for the emergency call for assistance that occurred during the execution of the operation.

Another important aspect of information exploitation is the form and content of information presented to operators. Many systems have been developed without adequate consideration of operators' decision-making needs; they simply display all of the information all of the time regardless of current operator goals (Boiney, 2005). Thus, there is a need for technologies that allow commanders to filter information content and selectively share relevant content with their team. Such technologies would also have applications in joint and coalition environments, permitting operators to share only information that is both relevant and approved. In Operation Anaconda, these technologies could have been used to help coordinate the joint and coalition forces during mission execution, possibly obviating the need to make an emergency call for air support.

Visualization technologies tailored to decision-making needs like those discussed above are clearly needed to eliminate information overload in network-centric environments and to decrease response times. Another enterprise-wide problem that hinders agility is that similar tasks are often supported by separate systems. For example, despite the similarity of resource allocation tasks such as WTP, Intelligence Surveillance and Reconnaissance (ISR) tasking, and Close Air Support mission planning, separate support technologies are being independently developed for each task. A better solution may be a more general resource allocation system for pairing assets to tasks. In addition to cost savings, such a system would enable agility since operators could be trained in the more general task of resource allocation, and thus be able to switch among resource allocation tasks in response to the operational demands of a dynamic conflict. Such a generic system could enable the AOC to alter its internal structure in response to environmental demands and remain agile.

9.5.4 Information Management

Additional difficulties were encountered during the execution phase of Operation Anaconda. Specifically, the other Services had difficulty responding to the emergency call made by the Army Commander. Although it is possible to point to shortfalls in the execution of the operation as the primary cause, a significant contributing factor is the planning process itself and the technologies that support it. Currently, ATOs are developed for each 24-h period. According to operators interviewed at the 9th AF Base, the ATO is most useful during the first 3–4 days of a conflict. Beyond the first 3–4 days, ATO changes become so frequent that the process becomes cumbersome and ineffective. A potential solution is to use wellscripted plans for the first 72 h of a conflict and to then shift to a more dynamic allocation of resources in response to the ebb and flow of the battle. As an example, consider a flexible and dynamic resource allocation strategy where aircraft take-off in response to battlefield needs and are provided with a set of targets in flight. With such a strategy, there is minimal need to replan to perform Close Air Support, prosecute Time-Sensitive Targets (TSTs), or prosecute Dynamic Targets, since there is no fully articulated plan to begin with. Target priorities are formulated based on commanders' intent, ROE, available munitions, and mission dependencies. As a result, a target's priority combines data with pedigree from multiple levels. Once struck, targets would be added to the ISR deck for Battle Damage Assessment. In effect, ISR assets are dynamically tasked in the same manner as strike assets. This contrasts to the current process in which the ISR deck is configured during planning and does not necessarily work in concert with strike plans. Similarly, tanker refueling could also be scheduled dynamically across both the Mobility AF and Combat AF. Although the effectiveness of such dynamic scheduling processes would need to be borne out in simulations or exercises, we have illustrated new capabilities made possible by advances in information management.

As the line between system acquisition and operation continues to blur, creative and adaptable operators should be provided with technologies to build new capabilities on the fly. Currently, operators do indeed implement new capabilities in response to changing demands, typically with Microsoft products such as Excel and Visio. However, such *ad hoc* solutions are rarely documented or implemented in future C2 systems. Thus, there is a need to capture these operator-developed solutions and treat them as software specifications that drive the evolution of the systems of record. Particular attention should be paid to operator-developed solutions that address team decision making across the joint and coalition environment, since these are the most challenging requirements to capture in traditional software engineering requirements. Capturing and exploiting such operator adaptations for future use are essential in enabling a truly agile and robust C2E.

There is also a need to capture and exploit data and information that passes through the AOC to allow warfighters to identify trends in enemy behavior and to automatically generate reports. Little data are warehoused in currently fielded systems, which inhibit commanders from obtaining accurate information about battlefield trends. It is common for commanders to make inquiries about the frequencies of events, such as requests for Close Air Support or diversions from planned targets to TSTs. Operators currently respond to such requests based on potentially flawed recollection and intuition. The lack of warehoused data also inhibits analyses of enemy intent. There is an enormous amount of information flowing through C2 systems that could be used to gain insight into enemy intent, identify trends, and improve our reactions and plans. But such opportunities are lost because the information enabling these insights is not presently captured. In the aftermath of Anaconda, time lines and events were reconstructed through operator interviews and written reports. Had more of the planning data and real-time information been warehoused, it would have been possible to construct such time lines and events more quickly and accurately to better facilitate analysis of the factors behind what went wrong. Furthermore, it is also likely that the existence of warehoused data would have permitted the analysis of more factors at a greater level of detail.

Another information management problem is that information representations tend to be specific to the processing needs of a single system, which inhibits the ability for a wider range of systems in the enterprise to make use of that information (see also Section 9.3). This detrimentally impacts the execution of joint and coalition endeavors such as Operation Anaconda. System designers must focus on the representational needs of the environment in which the system will be fielded, including unanticipated systems that might require the information. A potential solution to providing the kind of interoperability required in joint operations is the concept of "loose couplers," which would make it easier for each of the Services to evolve their systems independently while still maintaining the ability to exchange information (Electronic Systems Center, 2005). Had the joint and coalition forces engaged in Operation Anaconda been able to more easily exchange information during the execution phase, more responsive air support might have been possible.

In addition to the need for increased interoperability among the Services and joint forces, there is a current trend toward greater interoperability among the Services and SOF. In recent conflicts, for example, SOF have both identified and prosecuted TSTs. This close working relationship between SOF and other forces is a new development within the DoD, and has great potential to better deal with asymmetric threats. Previously, SOF worked largely independently of the other forces—a condition that occasionally caused significant problems for Blue Force Deconfliction. Aside from the increased use of SOF, Army Forces like those engaged in Operation Anaconda often work in remote, bandwidth-limited environments. In order to support increased interoperability between the Services and SOF, we need to provide enterprise capabilities that support warfighters in such environments. In Anaconda, lines of communication between Army forces on the ground and supporting AF and Navy assets were circuitous and inefficient. Enterprise services that incorporate remote users could save lives by reducing time lines and increasing efficiency in joint missions. But the systems comprising the C2E are heterogeneous in nature—a fact which requires accommodation when loosely coupling the multitude of subsystems. The U.S. Marine Corps intends to start exploiting smaller units with greater agility and autonomy that can combine into a larger force when more mass is required (McBrien, 2005). Incorporating such units into the C2E will require a much greater emphasis on bandwidth-limited users.

Systems are currently designed with highly specified direct interfaces to their data and computational services. The use of highly specified interfaces inhibits adaptation and agility by preventing systems from being combined in novel and unanticipated ways. Moreover, designers often specify the order of events or activities associated with a system, limiting the ability of operators to adapt their systems in the face of unanticipated events by reordering activities and changing decision-making processes. Systems need to be as flexible as possible by giving operators freedom to adapt their systems to changing contexts. As enterprise systems engineers, we should encourage the design of systems with extensible and generic interfaces, and modules and modular decompositions should be used extensively to provide (re) combinable functionality. Such robust modules and modular decompositions will facilitate adaptation, evolution, and emergence in decision-making environments.

SOAs and ESBs are a potential set of technologies that can address the problems recomposing modules in order to create new capabilities. In the AOC, for example, operators in the planning and ISR cells need to know which planned targets have been prosecuted. Although this information can be gleaned from network messages such as Tactical Data Information Link (TADIL) and Situation Awareness Data Link (SADL), it is not available in a form that makes it easily available to other users whose need for it was not anticipated in advance.

Another information management bottleneck is the tendency to force an artificial distinction between planning and execution. Such a distinction may be appropriate for large-scale force-on-force conflicts, but in the age of asymmetric warfare it limits agility. As an example, consider the problems that arise in performing TST within the constraints of the traditional planning cycle.

In an attempt to remove the distinction between planning and execution, plans are underway to alter the Air Operations Database (AODB) on a continuous basis. The current process employs an artificial distinction between planning and execution in which plans are highly specified and changes are incorporated as updates to the plan. To be truly agile, there is a need to use flexible processes to find "good-enough" solutions to planning and execution problems. Optimization inhibits flexibility and adaptation. In the C2E, deviations are the rule, not the exception. Thus, there is a need to trade optimization for flexibility and adaptation.

It only makes sense to search for optimal solutions when there is ample time and the environment is relatively static.

9.5.5 Information Generation

One of the challenges of working in joint and coalition environments is the fact that there is no common representation of information pedigree across systems and Services. It is important that shortfalls in representations of information pedigree, accuracy, and latency are addressed. Although messages within a specific Service, such as the AF, often include measures of accuracy, there is no common representation for this information across systems within the AF, let alone across the various Services. This lack of standardization hinders the loose coupling of systems across different Services, hindering effective information management and exploitation. Furthermore, information representation standards also need to be developed for chat, UAV video, and human intelligence reports. Because individual systems are part of the larger C2E, it is imperative that issues relevant to enterprise-wide concerns, such as information standardization, be addressed.

Associating chat messages with track data and UAV video data is also problematic, and it is currently being investigated in an FY06 AF MOIE title "Collaborative Data Objects." Maintaining a correlation between a GMTI track and a UAV video track, for example, is unnecessarily difficult because the output from these various systems does not contain enough information to easily support correlation.

The FY05 AF MOIE, titles "Kaleidoscope: MultiSensor GMTI-VMTI (Video Moving Target Indicator) Track Fusion and Visualization," explored ways to correlate color information from video streams with positional information obtained from video or other sources. Such an approach permits the aggregation of various types of information from a variety of sources, enabling more rapid discernment of the attributes of a track. Such an approach can be easily extended to integrate acoustic information as well.

In the C2E, information can be used to support higher-level inferences. For example, information on the current status of a plan execution (e.g., the prosecution status of a target set) can be used to make inferences about the extent to which higher-level operational objectives are being met. Typically there are higher-level combat objectives that are facilitated by a set of tactical objectives, which are in turn facilitated by a set of tactical missions, which may be facilitated by the destruction of individual targets at the lowest level of the hierarchy. This hierarchy can be traversed from the bottom up, so that information about the prosecution of specific targets is used to make inferences about the extent to which the higher-level combat objectives are being met. Figure 9.19 shows an example of objectives hierarchy.

A more complex set of high-level inferences involves Effects-Based Operations (EBO). Typically, missions are intended to achieve some desired effect, such as demoralizing enemy troops or forcing them to abandon certain locations. In EBO, the relationship between the specific tasks and the desired effects is less clear. It is

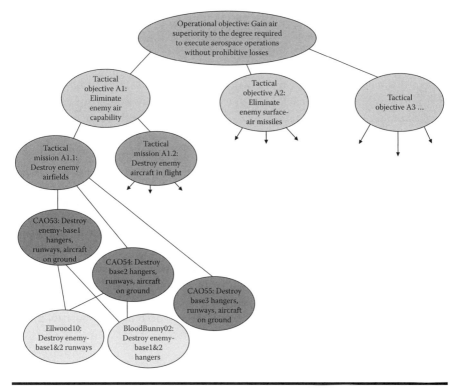

Figure 9.19 An operational objectives hierarchy.

possible, for example, to successfully complete all of the tasks without achieving the desired effects. Similarly, it is also possible to achieve the desired effects without having completed all the tasks designed to achieve that effect. Assessing the status of EBO is just as important as inferring the status of combat objectives based on the prosecution of individual targets, but is far more complicated to determine. Thus, it is an area for further research.

9.5.6 System Acquisition

One problem with the current systems acquisitions process is that in the typical spiral development cycle, the large number of requirements in each spiral has the effect of overfitting the system to the environment that existed when the system was designed, as opposed to the environment in which the system will be fielded upon completion. As the number of requirements increases, the system becomes increasingly sensitive to minor changes in the environment. Fewer requirements and increased iterations over shorter periods of time are one way to keep pace with a changing environment.

Another problem is that there are few established mechanisms for incorporating operational "lessons-learned" gleaned when systems are fielded into the evolution of the system or in the design of new systems. As a result, it is difficult to be agile in responding to emerging operator needs. The only goal should not be to have a system that is well suited to a particular environment, but rather to have a system that is also able to adapt when problems occur or the environment changes. We need to evolve existing capabilities to accommodate change. One way to do this is to use such "lessons learned" to inform the evolution and design of both systems and processes. ESE needs to implement traditions to add resistance to unnecessary change, while still recognizing that immediate action requires a response to severe problems. Rigid tradition and immediate action makes the process self-adjusting.

The acquisition community also needs to develop systems that are less specific and more general. Instead of building a tool that pairs weapons to targets and then modifying it for use in Combat Search and Rescue (CSAR) or ISR cells, it would be better to build a tool that pairs assets to tasks generically. That way, operators trained on the use of the system could perform tasks for TST cells, ISR cells, or CSAR cells. Not only would this save money by building a single, general purpose, tool but it would also make the AOC itself more agile by allowing operators to shift tasks based on the needs at the time. If the CSAR cell becomes very busy for a brief period of time, it would be possible to shift operators to that task until the backlog subsided.

Acronyms

AC	Alternating current
AF	Air Force
AWACS	Airborne Warning and Control System
AOC	Aerospace Operations Center
AODB	Air Operations Database
ATO	Air Tasking Order
BFT	Blue Force Tracking
C2	Command and Control
C2E	C2 Enterprise
C2ISR	Command and Control, Intelligence, Surveillance, and Reconnaissance
C4ISR	Command, Control, Communications, Intelligence, Surveillance, and Reconnaissance
CAASD	Center for Advanced Aviation System Development
CAS	Complex Adaptive System
COI	Community-of-interest
CONOPS	Concept of Operations
CSAR	Combat Search and Rescue

DC	Direct current
DNS	Domain Name Service
DoD	Department of Defense
EBO	Effects Based Operations
ESB	Enterprise Service Bus
ESE	Enterprise Systems Engineering
FY	Fiscal year
GIG-E	Global Information Grid—Ethernet
GMTI	Ground Moving Target Indicator
HMMH	Human-to-machine-to-machine-to-human
HSC	Human Supervisory Control
HSI	Human-System Integration
IFF	Identification Friend or Foe
IRS	Internal Revenue Service
IT	Information technology
JEFX	Joint Expeditionary Force Experiment
JDL	Joint Directors of Laboratories
LDAP	Lightweight Directory Access Protocol
KIPS	Key Interface Profiles
LAN	Local Area Network
LSE	London School of Economics
M2M	Machine-to-machine
MC2A	Multisensor Command and Control Aircraft
MIME	Multipurpose Internet mail extensions
MOIE	Mission-Oriented Investigation and Experimentation
MSR	MITRE-Sponsored Research
NCCT	Network-Centric Collaborative Targeting
NCOW	Network-Centric Operations and Warfare
NESI	Net-Centric Enterprise Solutions for Interoperability
OIF	Operation Iraqi Freedom
OWL	Web Ontology Language
PDF	Probability density function
PKI	Public Key Infrastructure
QSM	Quorum sensing molecule
R4	"getting the *right information* to the *right people* at the *right time* to make the *right decisions*"
R&D	Research and Development
RDF	Resource description framework
ROE	Rules of Engagement
SADL	Situation Awareness Data Link
SIAP	Single Integrated Air Picture
SMTP	Simple Mail Transfer Protocol
SOA	Service-Oriented Architecture

SOF	Special Operations Forces
TADIL	Tactical Data Information Link
TEL	Transport Erectile Launchers
TOI	Target-of-interest
TST	Time-Sensitive Target
TTPs	Tactics, Techniques, and Procedures
UAV	Unmanned Aerial Vehicle
U.S.	United States
USNR	U.S. Naval Reserve
VMTI	Video Moving Target Indicator
VTC	Video teleconferences
WAN	Wide area network
WSDL-S	Web Service Definition Language (extended for OWL services)
WTP	Weapon-Target Pairing

References

Ahmed, E., Elgazzar, A. S., and Hegazi, A. S. (2005). *On Complex Adaptive Systems and Terrorism*. Retrieved November 22, 2005, from the arXiv.org e-Print archive Web site: http://arxiv.org/PS_cache/nlin/pdf/0501/0501032.pdf.

Alberts, D. S. (1996). *The Unintended Consequences of Information Age Technologies*. Washington, DC: National Defense University Press.

Alberts, D. S., Garstka, J. J., and Stein, F. P. (1999). *Network Centric Warfare: Developing and Leveraging Information Superiority*. CCRP Publication Series. http://www.dodccrp.org/files/Alberts_NCW.pdf.

Army analyst blames Afghan battle failings on bad command set-up. (July 29, 2004). *Inside the Pentagon*, 20(31). Retrieved November 16, 2005, from http://www.insidedefense.com/secure/defense_docnum.asp?f=defense_2002.ask&docnum=PENTAGON-20-31-1.

Axelrod, R. and Cohen, M. D. (2000). *Harnessing Complexity: Organizational Implications of a Scientific Frontier*. New York, NY: Basic Books.

Bieberstein, N., Bose, S., Walker, L., and Lynch, A. (2005). Impact of service oriented architecture on enterprise systems, organizational structures and individuals. *IBM Systems Journal*, 44(4), 691.

Berube, C. D. (2002). *Data Fusion and the JDL Model*. Unpublished briefing. The MITRE Corporation.

Boiney, L. (2005). Team decision making in time-sensitive environments. Paper presented at the *10th International Command and Control Research and Technology Symposium*, McLean, VA. Retrieved December 8, 2005, from http://www.dodccrp.org/events/2005/10th/CD/papers/175.pdf.

Bonaceto, C. (2003). *A Survey of Cognitive Engineering Methods and Systems Engineering Uses* (MITRE Product No. MP 03B0000061). Bedford, MA: The MITRE Corporation.

Bonaceto, C. and Burns, K. (2006). *Using Cognitive Engineering to Improve Systems Engineering*. INCOSE Symposium 2006, July 9–13, 2006, Orlando, FL.

Bonaceto, C. and Burns, K. (2007). A survey of the methods and uses of cognitive engineering. In R. Hoffman (ed.), *Expertise Out of Context*. New York: Lawrence Erlbaum Associates.

Burns, K. (2007). Dealing with probabilities: On improving inferences with Bayesian Boxes. In R. Hoffman (ed.), *Expertise Out of Context*. New York: Lawrence Erlbaum Associates.

Burns, K. and Bonaceto, C. (2006). *Structure Mapping in Visual Displays for Decision Support*. Presented at the *2006 Command and Control Research and Technology Symposium*. http://www.dodccrp.org/events/2006_CCRTS/html/papers/182.pdf.

Cabana, K. A., Babich, A. M., and Washburn, N. H. (n.d.). *Counterdrug Surveillance and Control Systems: Using Data Mining Technology to Enhance Detection and Tracking of Targets of Interest*. Unpublished briefing. The MITRE Corporation.

Cabana, K. A., Boiney, L. G., Lesch, R. J., et al. (2006). *Enterprise Systems Engineering Theory and Practice—Volume 9: Enterprise Research and Development (Agile Functionality for Decision Superiority)*. MP05B0000043, MITRE Product. Bedford, MA: The MITRE Corporation.

Cabana, K. A. and Swarz, R. S. (2000). *Applying Data Mining Technology to the CAFC2S Work Program*. Unpublished briefing. The MITRE Corporation.

Christoffersen, K. and Woods, D. D. (2002). How to make automated systems team players. In E. Salas (ed.), *Advances in Human Performance and Cognitive Engineering Research Volume 2*. New York: JAI Press/Elsevier.

Davis, M. G. (2004). *Operation Anaconda: Command and Confusion in Joint Warfare*. Unpublished master's thesis, Air University School of Advanced Air and Space Studies, Maxwell Air Force Base, Montgomery, AL.

DeArmon, J. (2000). Identifying dominant air traffic flows in complex airspace. *The Edge: The MITRE Advanced Technology Newsletter*, 4(2). Retrieved November 16, 2005, from http://www.mitre.org/news/the_edge/august_00/dearmon.html.

Electronic Systems Center. (2005). *Strategic Technical Plan v. 2*. Hanscom Air Force Base: Electronics Systems Center. Retrieved December 8, 2005, from https://enweb.mitre.org/tiles/content/references/documents/stp_guidance/stp/ESC_STP_v2.0_PR.pdf.

Endsley, M. R. (2000). Theoretical underpinnings of situation awareness: A critical review. In M. R. Endsley and D. J. Garland (eds), *Situation Awareness Analysis and Measurement* (pp. 3–32). Mahwah, NJ: Lawrence Erlbaum Associates.

Hutchins, E. (1995). *Cognition in the Wild*. Cambridge, MA: The MIT Press.

Kessler, O., Askin, K., Beck, N., Lynch, J., White, F., Buede, D., Hall, D., and Llinas, J. (1991). *A Functional Description of the Data Fusion Process*. Office of Naval Technology, Naval Air Development Center, Warminster, PA.

Klein, G. (1999). *Sources of Power: How People Make Decisions*. Cambridge, MA: MIT Press.

Lamport, L. (1978). Time, clocks and the ordering of events in a distributed system. *Communications of the ACM*, 21(7), 558–565.

Left in dark for most Anaconda planning, Air Force opens new probe. (2002). *Inside the Pentagon*, 18(40). Retrieved November 16, 2005, from http://www.insidedefense.com/secure/defense_docnum.asp?f=defense_2002.ask&docnum=PENTAGON-18-40-1.

Li, T. Y. and Yorke, J. A. (1975). Period three implies chaos. *American Mathematical Monthly*, 82, 985–992.

Lin, H., Risch, T., and Katchanounov, T. (2002). Adaptive data mediation over XML data. *Journal of Applied System Studies (JASS), Special Issue on Web Information Systems Applications*, 3(2), 393–411.

Malle, B. and Knobe, J. (2001). The distinction between desire and intention: A folk conceptual analysis. In B. F. Malle, L .J. Moses, and D. Baldwin (eds), *Intentions and Intentionality: Foundations of Social Cognition*. Cambridge, MA: MIT Press.

Mathieu, J., Hwang, G., and Dunyak, J. (2006). Transferring insights from complex biological systems to the exploitation of netted sensors in command and control enterprise. Manuscript submitted for presentation at the *2006 Command and Control Research and Technology Symposium*, San Diego, CA.

Mathieu, J., James, J., Mahoney, P. et al. (2007). Hybrid system dynamic, Petri net, and agent-based modeling of the air and space operations center. *INCOSE Symposium 2007*, June 24–28, 2007, San Diego, CA.

McBrien, S. (2005). *United States Marine Corps Distributed Operations Architecture Study*. Unpublished briefing. The MITRE Corporation. Retrieved November 16, 2005, from http://employeeshare.mitre.org/c/cbudde/transfer/Joint%20Lecture%20Series/JLS%2015%20Nov%2005%20Marine%20Corps%20Distributed%20Ops%20Enabled%20by%20C4ISR.ppt

McCarter, B. G. and White, B. E. (2007). Collaboration/cooperation in sharing and utilizing net-centric information. *Conference on Systems Engineering Research (CSER)*. March 14–16, 2007. Stevens Institute of Technology, Hoboken, NJ.

McCarter, B. G. and White, B. E. (2008). *Emergence and SoS, Sociocognitive Aspects*. In M. Jamshidi (ed.), *Systems of Systems Engineering—Principles and Applications*. Atlanta, GA: CRC Press.

McIntyre, G. A. (1998). *A Comprehensive Approach to Sensor Management and Scheduling*. Doctoral thesis, George Mason University, Fairfax, VA (DTIC No. ADA359079).

Mitchell, P. J., Cummings, M. L., and Sheridan, T. B. (2004). *Human Supervisory Control Issues in Network Centric Warfare*. Massachusetts Institute of Technology Humans and Automation Laboratory Research Report No. HAL2004-01. Retrieved December 10, 2005, from http://web.mit.edu/aeroastro/www/labs/halab/papers/HSC_NCW_report.pdf.

Mitleton-Kelly, E. (2003) *Complexity Research Approaches and Methods: The LSE Complexity Group Integrated Methodology*. Retrieved November 10, 2005, from the London School of Economics, Complexity Research Programme Web site: http://www.lse.ac.uk/complexity.

Moffat, J. (2003). *Complexity Theory and Network Centric Warfare*. CCRP Publication Series. http://www.dtic.mil/cgv-bin/GetTRDoc?AD=2d2457288&Location=u2&doc=GetTRDoc.pdf.

Moody, J. and White, D. R. (2003). Social cohesion and embeddedness: A hierarchical conception of social groups. *Sociological Review*, 68(1), 103–127.

Norman, D. A. (1993). *Things That Make Us Smart: Defending Human Attributes in the Age of the Machine*. Jackson, TN: Perseus Books Groups, http://www.perseusbooks.com.

Orlikowski, W. (2000). Using technology and constituting structures: A practice lens for studying technology in organizations. *Organization Science*, 11(4), 404–428.

Potts, D. (2003). Tomorrow's war. In D. Potts (ed.), *The Big Issue: Command and Combat in the Information Age* (pp. 9–28). CCRP Publication Series. http://www.dodccrp.org/files/Potts_Big_Issue.pdf.

Sampada, C., Khusbu S., Neha, D. et al. (2004). Adaptive neuro-fuzzy intrusion detection systems. *International Conference on Information Technology: Coding and Computing (ITCC)*, 1, April 5–7, 2004, Los Vegas, NV.

Scott, S. D. and Cummings, M. L. (2005). *Collaborative Time Sensitive Targeting*. Massachusetts Institute of Technology Humans and Automation Laboratory Research Report No. HAL2005-03.

Simpson, J. and Weiner, E. (eds) (1989). *The Oxford English Dictionary.* 2nd Edition. New York, NY: Oxford University Press.

Szyperski, C. (2003). *Component Software: Beyond Object-Oriented Programming.* 2nd Edition. New York: ACM Press.

Taleb, N. N. (2007). *The Black Swan—Impact of the HIGHLY IMPROBABLE.* New York: Random House.

Thunholm, P. (2005). Planning under time pressure: An attempt toward a prescriptive model of military tactical decision making. In H. Montgomery, R. Lipshitz, and B. Brehmer (eds.), *How Professionals Make Decisions* (pp. 43–56). NJ: Lawrence Erlbaum Associates.

Tversky, A. and Kahneman, D. (1974). Judgment under uncertainty: Heuristics and biases. *Science* (*New Series*) 185(4157), 1124–1131.

Tzitzikas, Y. and Meghini, C. (2003). *Ostensive Automatic Schema Mapping for Taxonomy-based Peer-to-Peer Systems.* Helsinki, Finland: CIA.

White, B. E. (2007). On interpreting scale (or view) and emergence in complex systems engineering. *IEEE Systems Conference* Honolulu, HI, April 9–12, 2007. http://tinyurl.com/6zcda5.

Vicente, K. J. (1999). *Cognitive Work Analysis: Toward Safe, Productive, and Healthy Computer-Based Work.* Mahwah, NJ: Lawrence Erlbaum Associates.

Chapter 10

Enterprise Activities
Evolving toward an Enterprise

Stephen C. Elgass, L. Shayn Hawthorne,
Chris A. Kaprielian, Paul S. Kim, Andrew K. Miller,
Laura R. Ricci, Maura A. Slattery, J. Reid Slaughter,
Peter A. Smyton, and Richard E. Stuebe

Contents

10.1 Introduction 396
10.2 Overview..398
      10.2.1 How Enterprises Form..398
      10.2.2 Evolving toward an Enterprise 400
      10.2.3 Differences in TSE and ESE 400
      10.2.4 Outline...403
10.3 Enterprise Strategic Planning 404
      10.3.1 Capabilities...405
      10.3.2 Capability Analysis, Allocation, and Assessment.......................406
            10.3.2.1 Capabilities-Based Planning Analysis 406
            10.3.2.2 Capability Road Maps...412
            10.3.2.3 Allocation of Capabilities..414
            10.3.2.4 Refinement of Capability Road Maps...........................416
      10.3.3 Mission Threads...419
      10.3.4 Sample Process Description for Creating a Road Map 421
10.4 Framework for the Enterprise..425
      10.4.1 Why a Different Framework Is Needed for ESE? 425

   10.4.2 Service-Oriented Architecture 425
   10.4.3 Technology 428
 10.5 Enterprise Applications Case Studies..429
   10.5.1 Core Attributes 430
   10.5.2 Application Case Study—C2 Constellation............................... 430
   10.5.3 Application Case Study—BFT COI Service..................................431
      10.5.3.1 Capabilities and Analysis..431
      10.5.3.2 Technical Framework 432
      10.5.3.3 Publish and Subscribe Services 432
      10.5.3.4 BFT COI Service Data Schema...433
   10.5.4 Application Case Study: Genetic Algorithms versus Nonlinear
      Optimization for Resource Allocation Tasks 434
Acronyms...435
References 438

10.1 Introduction

Enterprise systems engineering (ESE) is an emerging discipline that requires a holistic view of the people, processes, and technology across a Community of Interest (COI), beyond the system-of-systems (SoS) level [1]. ESE concerns itself with the effective operation of systems within a larger context, such as the Department of Defense Command, Control, Communications, Computers, Intelligence, Surveillance, and Reconnaissance (DoD C4ISR) enterprise, where mission needs are dynamically evolving. In this complex environment, systems possess attributes beyond those normally observed within an SoS, for example, in areas such as flexibility, scaleability, and changeability. In addition, the requirements and system evolution, rather than being tightly controlled as with Traditional Systems Engineering (TSE) methods, must be managed to allow the enterprise capabilities and their interactions to evolve as needed.

  To meet these needs in the present environment, the systems engineer must (1) take a top–down (vice bottom–up) approach with the enterprise, to understand the context and operating space for the system(s) under investigation; (2) accept the need to manage, rather than control, the system design and evolution; and (3) add very different tools and processes to the SE toolkit from what has been used for TSE. These tools and processes involve (1) large-scale visionary planning to support integration of capabilities; (2) complex systems analyses to ensure that the "to-be" enterprise can support changing missions (by allowing even those that are unanticipated to emerge); and (3) the definition of an infrastructure framework that can support the desired enterprise attributes. These three activities are described as enterprise strategic planning, mission thread analysis, and technical framework development, and are briefly discussed in the sections to follow.

To allow systems and SoS to operate effectively in an integrated enterprise, the focus needs to shift from acquiring systems (relatively isolated, inward focused, and rigid) to acquiring easily integrable capabilities needed by the warfighter. As such, the concept of using a capabilities-based strategic planning and analysis construct is proposed.

In this approach, a top–down methodology, from vision and mission, through joint operating and functional concepts, to an enabling concept, is performed. Functional area and needs analyses are conducted to understand the existing and planned system capabilities, and the gaps that must be filled to satisfy the high level vision.

In the United States Air Force (USAF), capability analysis teams and working groups, and the Integrated Capabilities Review and Risk Assessment (I-CRRA) process are currently* used to conduct these analyses. The capabilities road map is a primary tool to guide capability planning by capturing the operational or action plans to achieve objectives. It helps the engineer identify what work needs to be done, what resources are available, and when the work needs to be accomplished. The road map is used to look at alternative approaches to satisfying the capability needs, examine program and technology risks and constraints, and ultimately recommend the allocation of capabilities to systems. Once an initial road map is created, it is refined over time based on operational lessons learned, emergent behavior within the enterprise, and availability of technology.

Once capabilities-based planning and analysis produces a road map, mission thread analysis is performed to ensure the results of the road map activity are robust (i.e., complete and consistent). A mission thread represents a sequence of activities (to include decision making) that supports one or more enterprise-level tasks and makes explicit the relationships among activities and their related products. It can be successfully used to identify operational gaps, determine commonality in system implementations, and support business processes that identify information service orchestration. Threads use operational and system architecture views from the DoD Architecture Framework (DoDAF) to provide a common mechanism for all stakeholders to understand information flow across capabilities and systems to address a particular mission.

Infrastructure framework definition capitalizes on web-based information technology (IT) breakthroughs to fundamentally change the architectural constructs of the building blocks of the enterprise and the way they interact with one another. This technology allows more flexible business processes by applying Service-Oriented Architectures (SOAs) to move toward net-centric, rather than system-centric, operations. At the heart of these frameworks are the concepts of service (chunks of work to serve a specific function demanded by the market) and loose-coupling (defining minimal eXtensible Markup Language (XML)-based data schemas

---

* As noted in Ref. [1], this chapter was first published at The MITRE Corporation in February 2006. The reader should interpret everything in this chapter to have been written in that time frame.

between interfacing systems so that changes to a capability are transparent to interfacing capabilities). The Electronic Systems Center (ESC) Strategic Technical Plan (STP), which advocates a layered architecture framework, is an ideal example of the current thrust in infrastructure frameworks to address ESE objectives.

There are numerous case studies that illustrate the tools and processes described above. The Command and Control (C2) Constellation uses the I-CRRA and mission thread concepts to chart the way forward for the Air Force (AF) components of the C4ISR enterprise. The Blue Force Tracking (BFT) COI service is a specific instantiation of a capability that is derived from the C2 constellation capability analyses, using the SOA infrastructure concepts to inherit enterprise attributes of flexibility, adaptability, scalability, and changeability. Such attributes can even be considered at the application level, where the choice of an algorithm, such as from the class of evolutionary computation algorithms, can more easily meet enterprise requirements than, for example, can traditional nonlinear optimization approaches.

10.2 Overview

10.2.1 How Enterprises Form

The scope of a particular enterprise can be defined by a COI. The enterprise consists of people, processes and technology, and the interrelationships among those constituents. Each enterprise is likely part of a larger enterprise and may subsume or be a superset of other enterprises as defined by a COI with a different view. As such, within any given enterprise some variables can be controlled, some can be influenced, and some can be neither controlled nor influenced but must be understood. The enterprise exhibits different behavior at different scales and potentially from different perceptions by different stakeholders. The proper view is to define each subenterprise in terms of how it serves the goals of the higher-level enterprise.

Enterprises evolve into existence over time as a COI assembles systems, technology, processes, groups, and/or individuals to perform needed or desired services. At some point, there is a realization that affirms the existence of an operationally viable entity that can be identified as an enterprise. This entity could initially be a set of systems or a set of SoS that are recognized as possessing a common or related suite of missions. The portion of the COI that recognizes the potential of an enterprise develops a vision (rather explicitly or implied by their actions) that, initially at least, evolves through common stakeholder purpose, planning, and execution methodologies.

Over time, the environment influences further evolution of the initially conceived enterprise. That is, the enterprise evolves out of inputs from unforeseen or new stakeholders that explore and collaborate in a developing COI or in reaction to forces external to the enterprise (e.g., evolution of related enterprises, a technology breakthrough, etc.) that influences the enterprise direction. Most of the time these stakeholders have minimal understanding of how precisely to implement the necessary enterprise. The how evolves using lessons learned from emerging behavior in the enterprise environment.

Two examples of enterprises can help to illustrate the following:

1. Bill Gates began to create his enterprise from preunderstood visions. Gates saw a computer in every home. He then began to manage the evolution of his enterprise to allow the initial vision to come to fruition. To manage the enterprise's evolution, his "enterprise" was engineered to be highly flexible and adaptive, so that it could evolve when it needed to. This allowed Microsoft to evolve along the paths of "greatest value" and "least resistance." The evolution of the internet, which proved to be a critical enabler for the current Microsoft Enterprise, was not part of the original vision. Rather, unforeseen or new stakeholders of another enterprise (the internet) provided capability and technology that Microsoft evolved to address and exploit. What was unique about Microsoft vice Digital Equipment Corporation, for instance, was that it was able to adapt. The agility of the Microsoft model allowed adaptation of its enterprise to evolve quickly and efficiently; it provides a thoughtful lesson when considering management of other enterprises.
2. The Internet emerged and evolved from a managed development of the Advanced Research Projects Agency Network (ARPANET). That system was created to support the enterprise of research and researchers at various universities and laboratories. The many different stakeholders continually adding capabilities eventually evolved and transformed the initial conceived use into a different form, creating a large Enterprise of its own as a result. COIs used the integrated capabilities in the Microsoft Enterprise to use the World Wide Web (W3) in ways we never could have predicted, over time spinning off a plethora of other new enterprises.

For the purposes of this discussion, an enterprise in our COI is constrained to be (1) a set of DoD mission areas/objectives; (2) the activities and functions that need to be performed to achieve the objectives; (3) the C4ISR systems, services, and components that enable the activities and functions; (4) all the dependencies/interrelationships of these entities; and (5) the people, processes, and technology associated with the DoD C4ISR arena.

The following characteristics define enterprises: (1) constantly adapting to changes in the environment, technology, and processes; (2) continually impacted by uncoordinated organizational, investment, programmatic, and technical decisions controlled by different stakeholders; and (3) sufficiently complex that the engineering team(s) would benefit from formalizing a framework/toolset of engineering and analysis approaches (and supporting tools).

Regardless of whether it is an initial vision or one influenced by environmental factors as it developed and evolved, the confirmation of an enterprise is dependent on a confluence of events and circumstances that conspire to bring together systems, organizations/people, processes, and technologies to provide a set of services, functions, and/or tasks. At some point, the informal confluence of capabilities is recognized as a useful function for and by the communities contributing to and deriving value from the association that henceforth is identified as an enterprise.

10.2.2 Evolving toward an Enterprise

Enterprises are not generally created from a rigorously planned framework to immediately meet a newly identified need; rather, they evolve. Once in existence, the enterprise continues to evolve in response to the changing environment. Initially, the community that causes an enterprise to come into being likely applies an ad hoc approach to assemble processes, technologies, and other components to provide a slowly evolving capability. The engineered and managed evolution of the components contributing to an identified enterprise can be characterized as the skillful planning, design, construction, and extension of an enterprise, that is, evolving toward an enterprise.

Once recognized as such, an attempt can be made to understand, document, and influence or manage the enterprise and its subcomponents. Interactions among the subcomponents of an enterprise, and indeed, interactions between a given enterprise and other enterprises, provide a way for an enterprise's modularization, interdependencies, and current critical paths/threads to be understood, enhancing the opportunity to identify and describe improvements and extensions to the enterprise.

Enterprise evolution encompasses the influence, manipulation, and control of the many different components that are part of its makeup. Expectations, processes, individuals, organizations, interorganizational relationships, interfaces, systems, subsystems, SoS, environments, and even other enterprises are eligible for being used in support of the engineered evolution of the enterprise.

10.2.3 Differences in TSE and ESE

One might conclude that to manage an enterprise requires nothing more than a vision and application of TSE skills, just at a larger scale than is needed for creating a system or an SoS. Although TSE tools and processes are often useful, additional tools and processes are needed. An enterprise possesses a level of complexity, and a need to be flexible, adaptable, and constantly evolving, which can be at odds with the TSE tenets of control and specification. Control and specification are only possible if there is a clear understanding of the specific mission of the entity, such that meeting the specifications allows that entity to perform that mission. But, if there are many missions, some still unknown (and unknowable), overspecification constrains the entity's ability to fulfill those missions.

The key differences between TSE and ESE discussed here are (1) the bottom–up versus top–down analysis and execution model; (2) control versus management of systems evolution; and (3) the suite and use of tools in the practitioner's toolkit.

One of the key considerations in evolving toward an enterprise is that ESE is not so much about finding capabilities in the combinations of various enterprise elements, but about facilitating an environment in which those capabilities can emerge. To that end, all systems engineers must be sure that they do not miss the point: we do not define the elements of an enterprise in order to understand the greater scope and purpose of the enterprise. This is the bottom–up approach. Unlike in a TSE environment, for ESE, we need to start from the top, understanding the boundaries, missions, organizations, and roles and responsibilities.

Using the Joint Operations Concepts (JopSCs) model, for instance, at the Joint Functional Concept (JFC) level, there are concepts such as "Battlespace Awareness" and "C2," However, neither of these is really sufficient for defining, bounding, understanding, or analyzing an enterprise. Conversely, at the Joint Operating Concepts (JOCs) level, we have homeland security and "major combat operations" (among others). Again, while these may be considered enterprises, they are not very well bounded.

However, at the next level down in our example, we analyze the intersection of these two areas (which is a critical discussion within the JopSC environment), and it is here that we start to see the emergence of manageable enterprise constructs, for example, the notion of Battlespace Awareness for Homeland Security (BSAHLS).

For the ESE, this is only the starting point, not sufficient for truly understanding the enterprise. In this example, BSAHLS is really an enterprise of enterprises encompassing the Federal Aviation Administration (FAA), Coast Guard, Cheyenne Mountain Operations Center (CMOC), Customs (border security), and so on. Each of these organizations has a unique mission that already exists and is (reasonably) well defined as an independent enterprise and set of capabilities; the FAA has responsibility for all civilian air traffic, the coast guard for civilian naval traffic, and so forth.

It is only inside each of these independent enterprises that we start to see the emergence of the TSE elements within the ESE construct. For instance, FAA has a Transportation Security Agency (TSA) component responsible for "who is traveling," an airborne component for "what's in the air," a domestic set of processes, an international set of processes, and connections to all of the airlines, all of the airports, and weather elements.

Each of these could reasonably be considered an elemental capability, and very quickly at the next level down, a more classic systems view is encountered. The systems and information that support an evaluation of "who is traveling" and "what's in the air" are clearly evident as distinct entities. The systems at this level are more appropriate for TSE techniques, although elements of ESE can apply, as well.

It becomes apparent that the classic TSE bottoms–up approach to understanding the elements first is insufficient to understanding and analysis within an

ESE construct. The TSE approach is very focused on "what are my pieces?" and "how do I put them together to build a system?"

Decomposition from the top down is far more enlightening from an ESE point of analysis. The key element in differentiating TSE from ESE is to understand that an assessment of the enterprise is not about "building capabilities up" but rather about establishing the boundaries, missions, organizations, and roles and responsibilities from the top–down, and managing (not controlling) the resultant integrated entity. Within our example, ESE would be about helping the FAA enterprise to understand how to support both its well-defined "domestic airspace" mission as well as the more complex and less well-defined BSAHLS enterprise. The TSE building block approach would be very inefficient in this role.

Another key point for the ESE is that enterprise management is less about understanding and anticipating the ways in which these capabilities can be effectively combined and employed, and more about facilitating an environment in which capabilities can emerge. Such facilitation is about affecting/shaping the environment in order to motivate production of a desired outcome. This is one of many effective means of producing desired products/capabilities/outcomes in a short time (e.g., considering consumer demands on industry). TSE is notionally about controlling the environment; ESE is about lessening that control and letting the environment "be all that it can be." Thoughtful ESE introduces to the evolution of the enterprise the virtue of agility to be exercised in the improvement of the enterprise in the face of dynamic situations and conditions.

Too often, systems engineers who are exposed to the ESE environment will want to translate a rigorous control philosophy to the enterprise. This is frequently seen in the attempt to identify potential emergent capabilities. By definition, this does not work because emergent capabilities are not really identified until they emerge. The role of the ESE in emergent capabilities is not that of the magician who commands the capabilities to emerge, but rather as the facilitator who ensures that the environment is sufficiently unconstrained to allow capabilities to emerge and that processes are in place to encourage unanticipated capabilities to emerge.

In short, ESE represents a movement from the philosophy of total control to ensure systems behavior, toward the philosophy of "just enough control" to keep the enterprise from spinning off into chaos. Moreover, it is this shift of emphasis that is a key challenge in moving from TSE to ESE. This shift allows the enterprise to possess the flexibility and adaptability to evolve easily without significant ripple effects throughout the enterprise whenever a new component is added to it. This is not to say that the discipline of TSE is to be abandoned; it can still provide the order it traditionally has to subcomponents that are appropriate for more rigorous control. ESE, however, has joined the engineering toolset to manage the interactions and relationships found at the dimension of the larger, more complex enterprise.

All of our TSE tools and models may still be valid in this type of an environment, such as capabilities identification, architectures, road maps, and libraries, to name a few. However, they will likely all need to be applied differently in support of the enterprise.

For instance, in the ESE world, it is critical to understand the complexities (and subtleties) of disparate enterprises. What are needed are tools that can simplify the presentation of the enterprises without losing access to the fidelity necessary to explore commonality. Our existing TSE tools are generally insufficient to this task given that they tend to represent systems in a manner that is, by design, complex and intended to be understood and applied by those with the various TSE roles. This is completely counter to the ESE goal of simplification.

In the TSE world, more information is better because it provides a more detailed understanding of the interactions necessary to establish control of the system. In the ESE world, the goal is less about obtaining more information but more about attaining greater understanding of that information (i.e., behavioral context, interactions, interdependencies across system boundaries, etc.) in order to determine where control can be loosened. In short, the ESE needs to know why something is being done, not necessarily how it is being done. In addition, it is by managing the enterprise at the understanding level that unanticipated capabilities are allowed to emerge.

ESE is much more about managing necessary aspects of the environment than about identifying solutions. Ultimately, a critical role for the ESE is to look for sharing opportunities, and facilitate their execution. It is not necessarily to solve the problem but rather to ensure that an environment exists in which challenges and issues are visible throughout the enterprise and that the environment supports the emergence of capabilities [2]. The ESE process must influence the engineers to integrate concepts into the enterprise and allow innovation, as shown in Figure 10.1 [3].

Building an enterprise that meets the warfighters' objectives, at the system, SoS, and enterprise levels, requires a collection of strategies and tools. Many of the same tools used for TSE apply to ESE and rely on TSE as their foundation. However, enterprises possess characteristics that, as a result of the complex interaction within its people, processes, and systems, do not lend themselves to the techniques used in TSE. To properly build the components to ensure performance and their interactions address the flexibility, scalability, adaptability, and emergent behavior desired in an enterprise. This requires a suite of fundamentally different approaches to those available to the TSE. A prudent systems engineer will supplement the TSE "tools" in the toolbox when building components that operate effectively in an enterprise.

10.2.4 Outline

The remainder of this chapter focuses on the tools, processes, and considerations for evolving toward the enterprise and continuing its evolution once realized. These concepts involve construction of a larger, holistic view that is at a level consistent

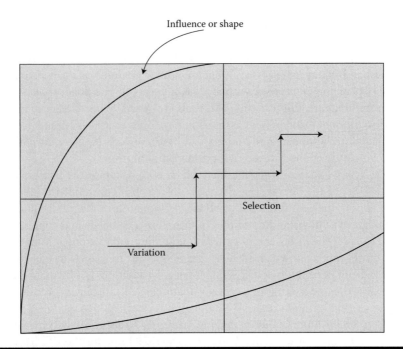

Figure 10.1 Innovation and integration processes 1–6.

with the enterprise scale, and involve (1) large-scale visionary planning in a way that support integration of capabilities; (2) analysis to ensure the "to-be" enterprise can support missions (by allowing even those that are unanticipated to emerge); and (3) the infrastructure framework that can support the desired enterprise attributes. These three activities are described as enterprise strategic planning, mission thread analysis, and technical framework. Finally, case studies of recent activities to evolve enterprises are discussed. This discussion includes describing the need for a new breed of applications, with performance comparable to those developed for systems or SoS, but with enterprise attributes. Examples of such applications and how they address enterprise characteristics are presented.

10.3 Enterprise Strategic Planning

Enterprise strategic planning is the approach to understanding where we are, where we want to go, and what needs to be done (and when and by whom) to evolve the enterprise. This section addresses the key components of enterprise strategic planning:

- Capabilities: why it is important to use an approach to identify the building block capabilities, rather than a system-centric approach, to meet the enterprise needs.

- Capabilities-based planning analysis (CBPA): how to identify the needed capabilities and how to combine them to address mission needs. This subsection will describe the process to flow from the broad, DoD-level vision down to specific capability needs, and how to assess the validity of those capabilities.
- Capability road map: using the results of the CBPA activity to develop the plan to deliver the capabilities as scheduled.
- Capability allocation: using the road map to allocate the capabilities to specific systems.
- Capability road map refinement: updating the road map to reflect lessons learned from operational emergent behavior and incorporating appropriate new technology.
- Mission threads: a process to verify the enterprise strategic plan by stepping through the capabilities needed to perform a specific mission (e.g., time-sensitive targeting) to ensure that capabilities are properly defined, allocated, and interoperable to meet the needs of that mission.

10.3.1 Capabilities

The DoD's progression to a capabilities-based approach for enterprise development and evolution versus a systems-based development has been motivated by a history of well-engineered, unique systems that perform very well on their own but required tremendous efforts to integrate with other systems within an enterprise. In order to realize enterprises that are capable of being easily developed, gracefully evolved, and adapted to changing needs, capabilities-based approaches are superior to more traditional system-based approaches. In a capabilities-based approach, varying sets of capabilities can be brought together to meet changing mission needs and to keep pace with the always evolving goals of the enterprise. This process of integration and innovation can provide the adaptability and scalability desired in today's environment in a way that the narrowly focused systems-based approach that creates systems that are optimized for its original purposes but brittle in meeting other needs does not. As such, capabilities are a key component of the people, processes, and technology that comprise an enterprise. But, they are not by themselves the enterprise. Two national AFs may have a similar set of capabilities, but the people and processes, as well as how they are combined and used, make them each a unique enterprise.

A focus on capabilities, rather than on systems, by the user should provide the developer greater leeway in determining the system (or mix of systems) solution to a problem. This should also afford the user greater flexibility in accomplishing his/her task. Whereas traditional systems-focused approaches allow for high degrees of optimization within defined system boundaries; they can also limit the agility of that system when necessity demands that the system be integrated with others to provide new capabilities, or when the needs of the mission change in unforeseen ways.

In a capabilities-based acquisition methodology, each instantiation of an enterprise must be made up of fundamental sets of capabilities that can be combined and recombined to provide the necessary means to accomplish a desired outcome. In order to facilitate the process of capabilities definition and refinement, it is helpful for the greater scope and purpose of the enterprise to be understood. This is the job of the ESE.

One of the fundamental roles of the ESE is to be able to understand and anticipate the ways in which these capabilities can be effectively combined and employed in concert with the enterprise user. More often than not the entire slate of desired outcomes is not immediately apparent and cannot initially be clearly articulated (in fact, they may never be fully articulated because they continually evolve!). This "give-and-take" approach to capabilities refinement can be a daunting task as it is, by definition, a process filled with ambiguity. The process is not simply about applying TSE principles at a much higher level, but rather represents a fundamental shift in the way business is conducted. Simply put, ESE aims to help define and refine the packages of LEGO® blocks that can be assembled [4], based on desired outcomes, rather than define the LEGO blocks, or even more specifically, the interface between the blocks themselves. As a result, two enterprises with different objectives may share many of the same capabilities (in fact, they may have exactly the same capabilities) as long as they are assembled and used differently.

We will explore the process to apply this capabilities-based approach to developing, analyzing, and allocating capabilities to systems acquisition in the following sections.

10.3.2 Capability Analysis, Allocation, and Assessment

Key to enterprise strategic planning is the CBPA process described in this subsection. Its purpose and the current AF process is described in Section 10.3.2.1, and the remaining subsections address the tools (road maps and threads) used to support and refine the CBPA process.

10.3.2.1 Capabilities-Based Planning Analysis

CBPA is to define and apply a systematic approach for purposeful evolution of the enterprise. It takes a holistic view of the enterprise and defines capabilities in terms of the big picture. In doing so, it defines issues from an enterprise perspective, rather than from a system perspective.

Capability planning is a key part of the strategy to drive the investment decisions that will enable evolution of the enterprise to better meet overall warfighter needs. Its specific purpose is to perform the engineering analysis to identify solutions to meet mission needs. A benefit of capability planning is that it sets into context the myriad defense acquisition community activities and products. Perhaps the easiest way to explain the process is to describe the process used in the DoD as of the publication date of this chapter. (Note that this process will evolve over time and may not represent the current approach at the point in time this chapter is being read; the reader is encouraged to focus on the general top–down philosophy and approach.) The high-level process is shown in Figure 10.2, and the other figures in this section address specific components of the process.

As depicted in Figure 10.3, national security and defense strategies are created; they influence the Joint Vision (e.g., Joint Vision 2020). From the Joint Vision, JOCs and, in turn, JFCs are derived to drive the acquisition process. The JOCs focus on effects, and the JFCs on what functions are needed to provide those effects. To understand JOCs and JFCs, consider, as an analogy, the strategy employed by a commercial hotel corporation. The JOCs are the effects or results that are desired; for example, one obvious JOC may be to make a profit. The JFCs are the functions needed to support the operations necessary to meet the JOCs; in our example, the JFCs may include accomplishing payroll, reservations, guest services, maintenance, and so on.

The JOCs and JFCs give rise to Joint Integrating Concepts (JICs), that is, how to combine or integrate JFCs to satisfy JOCs; in our example, this could include a set of different types of hotels to address different market needs. A high-end hotel and an economy hotel may both be part of the JICs, and they may both need many

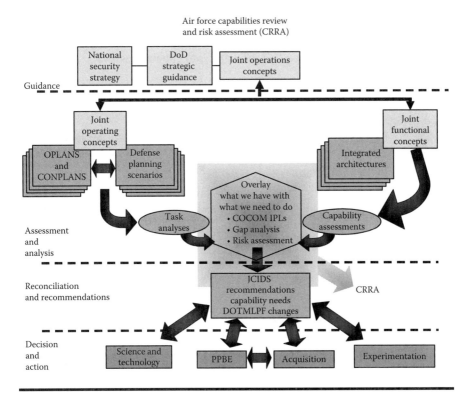

Figure 10.2 Capabilities-based planning analysis processes 2–3.

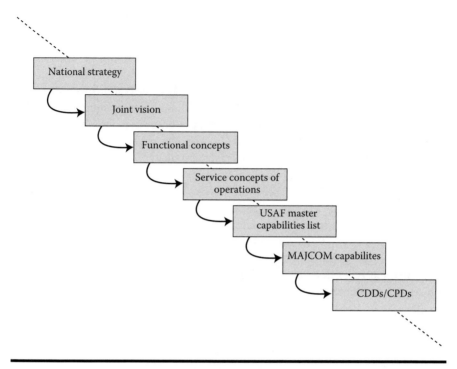

Figure 10.3 Translating vision and priorities into capabilities 2–5.

of the same JFCs (involving payroll, reservations, etc.) but be applied differently to suit different needs.

For the DoD, the JOCs are concepts such as win the war, avoid fratricide, and so on; current JOC categories include Major Combat Operations, Stability Operations, Homeland Security, and Strategic Deterrence. The JFCs are developed by the Combatant Commands, and they currently include the categories of Force Applications, Force Protection, Joint C2 (JC2), Battlespace Awareness, Focused Logistics, NetCentric, Training, and Force Management. JFCs in turn drive service-specific concepts of operations (CONOPS). Currently, a capabilities library approach is used as an organizing construct for the capabilities needed to support the AF CONOPS areas. Figure 10.4 shows an example of how capabilities can be combined and shared for certain service-specific CONOPS (in this case, to satisfy the precision strike JFC). The current AF CONOPS construct is shown in Figure 10.5. The capabilities are listed in a library called the Master Capabilities Library (MCL).

CBPA techniques are used to assess the adequacy of the JICs. CBPAs currently consist of four analysis approaches: functional area analysis (FAA), which addresses the "what's available," functional needs analysis (FNA), which treats the "what's missing," functional solution analysis (FSA), which directs "how to

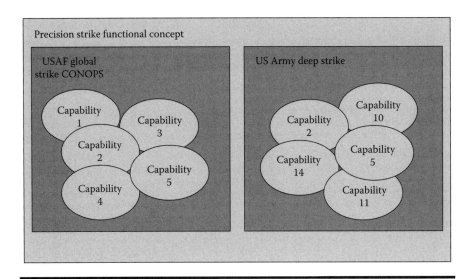

Figure 10.4 DoD capabilities examples 2–5.

fix it," and postindependent analysis (PIA), which aims "how it comes together." The AF process for the FAA, FNA, and FSA is accomplished by the functional teams (such as the C2 Capability Analysis Team for the C2 functional area) and associated working groups. In addition, the I-CRRA process integrates AF CONOPS-based assessments of needs and capability objectives.

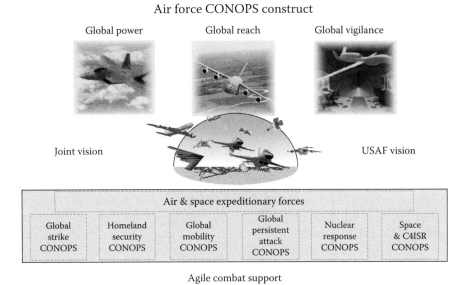

Figure 10.5 AF CONOPS 2–6.

To support the I-CRRA process, seven Task Force-level CONOPS capture (at a high level) user needs, forming a basis for determining capabilities needed to meet these needs, and providing a backdrop against which specific acquisition program products can be assessed and prioritized. The I-CRRA process includes discussions for each CONOPS area (capabilities needed, architectures/systems that support them, and issues) as well as discussions to agree on overall integration CRRA shortfalls across all the CONOPS areas. Distilled down to the service level, the AF CONOPS are developed to provide an understanding of ways in which capabilities outlined in the MCL can be aggregated to provide an operational capability.

Eventually, these MCL capabilities will all be traced (at some level) to the specific Capabilities Development Documents (CDDs) and Capability Product Documents (CPDs) used at the acquisition program level. This tracing will provide information as to which programs are acquiring automated solutions (or parts of solutions) to support specific capabilities, and the time frames and specific program increments intending to deliver specific solutions. The AF is also attempting to capture detailed Measure of Performance-related data associated with specific capabilities. This information will give the AF a view of which programs will, in what time frames, be affecting progress toward necessary levels of performance for specific capability areas.

The AF analysis methodology will continue to evolve, and will hopefully increase emphasis on thread analysis, the detailed role and benefits of which are discussed in Section 10.3.3. In essence, since a combination of specific capabilities (supported by a combination of programs) can enable multiple warfighter activity threads, capability shortfalls and risks must be analyzed in that broader context. Likewise, decisions made with regard to associated capability investments (or cuts) need to reflect (or at least be cognizant of) all those dependencies.

Figure 10.6 depicts a summary of a process whereby user needs, expressed via capability statements and CONOPS, are considered in decision making (such as in the I-CRRA process). Decisions influence SoS functional and performance needs, which in turn influence requirement sets of specific programs positioned to meet those needs. If investments and technical strategies are adequately coordinated across programs, the outcomes (as noted on the right-hand side of the chart) will be integrated SoS solutions and overall integrated enterprise-wide solutions.

Figure 10.7 shows another way to depict the target/objective process for acquisition investment decision making within the AF, from the perspective of one program (mission planning). The current process is called planning, programming, budgeting, and execution, and is documented in Management Initiative Decision (MID) 913 [5]. CBPA information is used in the CRRA process and other Modernization/Capability Planning activities that influence those decisions. These activities leverage an understanding of user-needed capabilities, CONOPS, and system/SoS architecture products as they analyze capability shortfalls/gaps and prioritize investments. Ideally, decisions on investments across programs are related in order to achieve integrated capability road maps, which (eventually) provide improvements in the cross-program delivery of integrated system solutions.

Enterprise Activities ■ 411

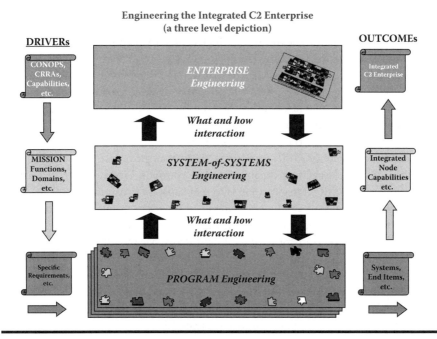

Figure 10.6   Engineering the integrated C2 enterprise 2–7.

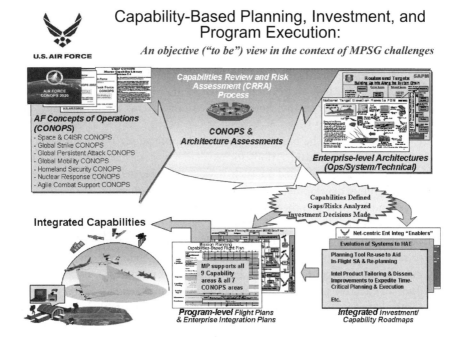

Figure 10.7   Capability-based planning, investment, and program execution 2–8.

10.3.2.2 Capability Road Maps

Capability road maps are the means by which capability planning guides how enterprises evolve. A road map captures the operational or action plans of an organization in achieving its intended objectives. The first step in establishing any road map is to discern the strategic objectives. Before other components of a road map can be determined, a very clear vision of what the organization is planning on achieving must be established. At the enterprise level, enterprise engineers institute a broad vision to serve as the core basis for the organizations under it to develop their respective strategic objectives. Among C4ISR enterprises, these strategic objectives are captured as a means to achieve specified capabilities. The common term to refer to such an enterprise-level C4ISR strategic vision is that of capability road map.

Traditionally, within the DoD, strategic and investment plans have been captured as flight plans. For example, both the Air Force and Air Force Space Command (AFSPC) have Transformational Flight Plans. Meanwhile, the main AFSPC road map is called the Strategic Master Plan, the National Security Space Office's (NSSO's) plan is the National Security Space Plan, the Navy's is the Integrated Strategic Capabilities Plan, and the Joint Chiefs of Staff's (JCS's) is the Joint Strategic Capabilities Plan. The important task at hand as the DoD transforms itself is to delineate these plans in terms of capabilities or express them as capability road maps.

10.3.2.2.1 Key Factors

An effective capability road map comprises two key factors.

The first factor is that the road map must present achievement of capabilities as the main theme. This may seem axiomatic, but in practice, the so-called capability road maps often degenerate into nothing more than systems evolution plans with large categories of system functions offered as capabilities. As discussed in Section 10.3.1, capabilities must be based on operational applications and established top down with no specific considerations for particular system functions. Of course, capability statements developed in such a way must be delineative and cogent enough to facilitate development of other road map components.

The second factor is that the road map must furnish indicators for progress. The hackneyed expression in strategic planning is that of "visions not based on reality are hallucinations" as first averred by Goodstein, Nolan, and Pfeiffer [6]. As C4ISR enterprises develop road maps, they must be aware that each increment along a road map must be practical and achievable. Furthermore, the progress must be measurable. Thus an effective road map must comprise what Goodstein and others referred to as Critical Success Indicators (CSIs). These are those measurable items that specifically indicate the progress along the road map. For example, the extent to which an enterprise adopts a specific and distinct technology is a good CSI, whereas, a more undefinable metric such as the level of net-centricity is a poor one.

10.3.2.2.2 Road Map Cycle

It is a common commercial industry practice to plan the future up to 5 years in advance. This mirrors the DoD Future Years Defense Program (FYDP) cycle. Given this duration as the basis, the actual cycle may be adapted to the needs of each enterprise. Based on measurements of CSIs and other changes, each road map should be updated at least once a quarter.

10.3.2.2.3 General Approach

Generally, there are two approaches to developing a capability road map: top–down and bottom–up.

The prevailing notion seems to be that the bottom–up approach is the preferred methodology. The strategy here is to generate road maps first at the lowest level (with the C4ISR case being at the program level) then aggregating the information incrementally up until the enterprise level road map is established. Thus road maps are "rolled up" from the bottom until the topmost enterprise level road map falls out.

In practice, however, this bottom–up approach must be impractical in that no capability road map seems to have been developed using this methodology (based on the collective experience of the Enterprise Integration Systems Group [EISG] at the ESC). The main reason for this lack of usage appears to be that this approach requires cross-correlating and deconflicting vast amounts of data. Furthermore, different programs tend to build road maps with varying look and feel; hence in some cases, relating respective road maps to each other is infeasible. One way to mitigate this problem may be to dictate a canonical road map format at the enterprise level that all lower-level organizations must abide by, but enforcement of such a paradigm is unwieldy, and it unnecessarily restricts how each organization may want to emphasize a particular aspect of the road map.

Accordingly, the more pragmatic methodology for composing road maps is that of the top down one in which the persons in charge of generating capability road maps retain enough understanding of the lower-level details so that they render the overall strategic vision consistent with the visions of the organizations within the enterprise but are unbounded by the minutia. The overall object remains setting higher-level goals that the whole enterprise may employ to establish more detailed and focused road maps. In developing the top–down approach, the flow down of analysis described in Section 10.3.2.1—Joint Vision to JOCs, JFCs, and JICs, to CONOPS, to I-CRRA results—is used to construct the enterprise road map.

10.3.2.2.4 Critical Information

At the enterprise level, the capability road map should answer the general questions of what work needs to be done to achieve the capability, what resources are available, and when the work can be done. Moreover, contingencies for action when the planned set of priorities becomes inexpedient should be considered.

The major objective of any capability road map is to discern the steps for developing best functionalities possible to deliver the capability while minimizing consumption of resources; an effective capability road map presents a strategy for making the most effective use of given resources and minimizing duplication of effort. Thus an effective capability road map informs each organization within the enterprise of the overall significance of its products as well as how they relate to similar products being built by the other organizations.

10.3.2.3 Allocation of Capabilities

As discussed earlier, ESE is focused on moving from the philosophy of total control toward a philosophy of just enough control to optimize the enterprise to deliver needed capabilities when and where they are required. The capability allocation process is the fulcrum where TSEs' need to control the building blocks of a system in order to deliver assigned capabilities balances the ESE need to reduce or eliminate TSE controls so that new, unplanned capabilities can emerge. This delicate balance requires the creation of an environment where capabilities are loosely allocated; hence new capabilities still emerge instead of an environment where capabilities are rigidly assigned or specified.

10.3.2.3.1 Capability Allocation Process

Given the capability road map(s) created using the methodology described earlier in Section 10.3.2.2, the required capabilities must be analyzed so that they can either be loosely allocated to current and planned systems or unallocated if no current or planned system can provide the required capability. Conversely, all available systems, SoS and family of systems must also be analyzed to identify what capabilities they currently provide and what capabilities they could provide, as the initial step toward identifying emergent capabilities that are not being capitalized. Finally, when capabilities are refined or modified as described in Section 10.3.2.4, the result will be a new capability road map and the capability allocation process will begin again. Capability allocation will be a continually changing, iterative process that never becomes static. The goal will be to reach a steady-state condition where wide oscillations in the allocation of capabilities to systems, or materiel solutions, are damped so that they would eventually reach steady state, if left unchanged for an extended period of time.

To allocate capabilities in a top–down method, the expertise of the entire DoD and other organizations must be engaged to identify initial materiel approaches to provide the capabilities specified in the capability road map(s). The collaborative nature of this effort is meant to develop potential solutions in an integrated fashion that reflect the future requirements of Joint Force Commanders (JFCs). It must be stressed that once a complete capability road map is developed; an almost infinite number of approaches will be available to meet required capabilities. This is the nature of loosely allocating capabilities to current and planned systems—no control is enforced on requiring a desired capability be provided by a specific system. Rather, the control is focused on requiring all systems to provide whichever capability, or part of a capability, they can. This process will necessarily leverage the expertise of all government agencies, as well as industry, in identifying multiple approaches to providing defined capabilities and it should always include existing and future systems that can be modified or integrated to meet the capability need. The integrated doctrine organization training materiel leadership personnel and facility (DOTMLPF) implications of any approach should also be considered throughout the process.

The capability allocation analysis conducted via the I-CRRA will determine multiple approaches to meeting each capability and will identify what is, at a given time, the optimum materiel approach or combination of approaches to provide the desired capability or capabilities. The capability allocation analysis will determine the best way(s) to use the materiel approach(es) to provide a joint capability. Generally, it will not consider which specific systems or system components are the best. As a simplistic example, the capability allocation analysis may determine that a capability is best satisfied by an unmanned aerial vehicle (UAV) employing a bomb instead of approaches employing submarine launched missiles, artillery, or air-launched missiles. The capability allocation analysis will not assess the best UAV or bomb design. This analysis resides in the TSE domain. As stressed earlier, a number of approaches will usually be available to provide the desired capabilities.

The capability allocation analysis will also consider capability gaps, the specified range of military operations, the conditions under which these military operations must be performed, and other factors that are relevant to support JFCs and integrated architectures.

The capability allocation analysis will determine how well each of the identified approaches address identified capability gaps and provide the desired effects. Approaches will include family of systems or SoS that use different ways of filling a capability gap, each addressing operational considerations and compromises in a different way. The materiel approaches identified to provide required capabilities will almost always integrate systems delivered by multiple sponsors and materiel developers.

The product of the capability allocation analysis is a prioritized list of approaches and combinations of approaches ranked by how well each can provide the desired capabilities.

10.3.2.3.2 Assessing Constraints on the Initial Capability Allocation

After the capability approaches are identified, they must be analyzed to identify the technological maturity, technological risk, supportability, and affordability of each approach using the best data available. The capability allocation analysis must also assess the operational risk associated with each approach.

Next, the capability allocation analysis must examine the capability road map again and confirm the nature of the capability or broad-based effect(s) to be provided, when the capability is required, and the applicable operational environment.

Then, the capability allocation analysis must return to the TSE domain and examine the ability of the materiel approaches to provide the desired capability or capabilities under the conditions specified and evaluate the delivery time frame for each approach. For approaches that use existing capabilities or planned capabilities scheduled for delivery, the ESE must examine how a new approach for meeting the proposed capability ties to existing programs. For new materiel approaches, the ESE must evaluate when a useful capability could be delivered using current or planned technology and evaluate the necessity to synchronize the development of systems contributing to SoS, family of systems, and enterprise approaches.

Using these data, the ESE will assess the risks associated with the systems, SoS, and family of systems allocated to the approaches identified that meet the capabilities defined in the capability road map. Then the ESE can assess the programmatic and technical constraints of each approach and create a syllabus of multiple approaches to providing required capabilities as well as a listing of unallocated capabilities where no current or planned system can provide the required capability. Further, this analysis will identify what capabilities current and planned systems, SoS, and family of systems could provide, and the programmatic and technical constraints of each of them, so that emergent capabilities can be capitalized upon. Using these data, the capability road map can be refined as new needs, and capabilities are defined.

10.3.2.4 Refinement of Capability Road Maps

Once an initial capability road map is generated, the road map needs to go through constant refinement to adjust to changes, foreseen and unforeseen as the acquisition programs unfold. A refined capability road map identifies the transition plan to move from the enterprise "as-is" state toward a "to-be" state that evolves to reflect lessons learned from operational emergent behavior and that incorporates appropriate new technology or has been prepared to employ emerging technology. This transition plan includes two products: architecture products that describe the refined capabilities and a refined capability road map that describes how the adjusted framework of capabilities will be realized.

The state of an enterprise reflects the evolution of the enterprise vision, mission, scope, and primary outcomes. The vision, mission, scope, and primary outcomes may remain constant or evolve as emergent behavior is discovered. These changes force a change in state of the enterprise, requiring a refinement of the capability road map.

- Enterprise vision—As stated in Section 10.2, the vision identifies the desired outcome or envisioned "forecasted state" of the enterprise at some point in the future (far enough downstream in time to allow the description to be truly visionary, but near enough in time to allow some reasonable confidence that outcome can occur). The vision statement answers the following critical questions: "who's vision is it?," "what is the forecasted state of the enterprise?," "when is this state desired or expected?," and "what is the context of the enterprise?" The answers constrain the enterprise activities, and focus and provide just enough effort necessary to evolve toward the desired state of the enterprise. Any changes to these answers force a change in state of the enterprise, requiring a refinement of the capability road map. Let us look at a Space Situation Awareness example of an enterprise vision: "By 2010, U.S. AF Space Control Decision Makers have complete knowledge and understanding of the space situation that is necessary to conduct operations by land, sea, and aerospace forces at a given time and place without prohibitive interference by opposing forces (AFDD 2); and to assure friendly use of the space environment, while denying its use to the enemy (AFDD 1)." Note that this example is a good one in that it addresses each of the critical questions needed in the vision: who, what, when, and in what context.

- Enterprise mission—The mission statement describes how the forecasted state is achieved by discussing the high-level activities of the enterprise and answering the following question: "what is the primary activity of the enterprise?" Any changes to how the envisioned state is achieved force a change in state of the enterprise, requiring a refinement of the capability road map. In our space situation awareness example, the mission may be "provide sufficient information to equip warfighting decision makers with the awareness and understanding of the space situation necessary to . . ."
- Enterprise scope/Primary outcomes—The scope of the enterprise is constrained by its stakeholder-expected outcomes. Any stakeholder changes to expected outcomes, capabilities, or products force a change in state of the enterprise, requiring a refinement of the capability road map. Again, let us use our space situation awareness example to make the concept of scope and outcomes more concrete: "situation awareness resulting from having sufficient knowledge about space systems, events, and activities—both current and potential—in, from, or toward space."

In the examples mentioned above, the primary outcome was a state of awareness about the space situation. However, sometimes an expected outcome is a capability to do something with systems. In this case, the product of the enterprise may be processes and systems. These outcomes or capabilities do not constrain the capability provider in methodology. For instance, a needed capability may be to disable the airborne threat. This requirement does not specify gun, missile, or any other type of weapon. The decision is left to the capability developer with performance and stakeholder validation.

In either case, the need to refine the capability road map is determined by monitoring evolution of technology, emergent behavior of the deployment environment, performance of the capability in that environment, and lessons learned and issues feedback from the user and stakeholders of the capability. The frequency that a road map should be refined is dependent on the rate of change of the elements of the enterprise. However, it is fair to suggest that the road map should be reassessed after some major delivery of capabilities or at the start of some major spiral, and so forth. Three interacting roles are required to provide emergent behavior feedback to the enterprise:

- The enterprise architect/ESE incorporates operational lessons learned and applicable technology into the to-be architecture products to explain issues, provide solutions, and apply technology-driven improvements.
- The capability user employs the capability framework in execution of the mission and provides operational feedback to the enterprise architect/systems engineer. The feedback could include the following:
  - User needs: User-defined capability requirements for conduct of the mission, including anticipated capability needs and enabling capability needs that allow adaptable and responsive reaction to urgent and unanticipated changes within the environment.
  - Operational issues: Deficiencies in the user's current capability versus the user's needs.
- The enterprise technologist researches on and develops new technology, and forecasts emergent technology, producing technology updates to the enterprise architect/systems engineer.
  - New technology: New technology available to the enterprise capability developer for realizing the enterprise capabilities.
  - Technology forecast of emerging technologies: Predicts the applicability of newly developed technology or anticipated technology to the enterprise capability needs.

Ultimately, the refinement of the capability road map involves five primary activities:

- Monitor the enterprise user's environment: the ESE performs operational behavior analysis of the as-is system, capability deficiencies, and the needs of the user, and produces operational lessons learned to update architecture products and refine the capability road map.
- Apply technology: Using the capability needs identified by the ESE, the enterprise technologist performs research and development, and recommends application of new technology or an emerging technology forecast to satisfy user capability needs.
- Update architecture products: The ESE collaborates with the enterprise architect to incorporate operational lessons learned and applicable technology into the architecture products that document and explain enterprise

capability issues, solutions, and technology improvements. These products become the to-be architecture products used in refining the capability road map.
- Refine capability road map: The ESE collaborates with the user and the developer to identify development priorities, schedules, and resources to refine the capability road map for the enterprise developer to adjust the capability framework.
- Adjust capability framework: The enterprise developer allocates capability needs to programs or contracts responsible for implementing the capability road map and delivering the capability framework to the user according to the requirements documented in the to-be architecture products.

10.3.3 Mission Threads

Now that we have created and refined our capability road map, how do we know we are applying the results and analysis properly (e.g., allocated capabilities to the correct system[s])? The use of mission threads is a key method. The following is a definition of a thread that was developed in the course of building a Time Critical Targeting C2 Architecture (TC2A) several years ago.

Mission thread: "A sequence of activities (to include decision making) that supports one or more enterprise-level tasks and makes explicit the relationships among activities and their related products. Threads can include multiple (parallel or alternative) activity paths as well as repeating activities (or sets of activities)."

A thread is a means of capturing one view of an operational architecture. A thread can be a very broad end-to-end representation of a set of operational activities such as in the TCT thread that captures the entire kill chain (find, fix, track, target, engage, and assess). A thread can also capture a set of activities for a narrower scope of operations such as combat assessment.

A thread is one of the most useful architecture products in providing a mechanism to convey a common understanding of an operational process to different communities. How are threads commonly used?

- To capture a sequence of operational activities in a more structured fashion than is normally conveyed in a textual description such as training, tactics, and procedures (TTP). This structured representation facilitates analysis.
- As the basis for mapping systems/applications to operational activities to identify gaps, with the aim of identifying potential system improvements (e.g., identifying gaps in system coverage).
- As a means for conveying commonality in operational functionality across C2 platforms in order to identify potential commonality in system implementation. Often the thread may not be at a level of detail adequate to clearly

identify commonality but it can serve as a first cut at identifying potential areas to address commonality.
- As a basis for business process analysis to identify common information services and potential information service orchestration within the C2 enterprise.

Threads offer a construct for different types of analyses by different communities—Joint Expeditionary Force Experiment (JEFX), Space Systems Management Integration Office (SMIO), and Joint Surveillance Target Attack Radar System (JSTARS) Link 16. A thread can cut across multiple enterprises or subenterprises, or be wholly contained within an enterprise of interest. Its use still facilitates analysis of capabilities allocation, gaps, and common implementation and applications. The operational view (OV)-5 and OV-6c are commonly used DoDAF architecture products to capture operational threads. The system view (SV)-5 is the DoDAF product normally used to capture the relationship between system functionality and the operational thread. When we use threads to capture system functionality associated with operational activities, some of the difficulties we have encountered in accurately representing the system relationship to the operational activity include the following:

- Distinguishing which elements (aspects) of the operational activity are automated by a system/system function/application and which are manual.
- Capturing the extent (degree) of automation of an operational activity when a system/system function/application is mapped to it.
- Distinguishing the specific operational activities that a particular system/system function/application automates, as distinguished from other systems.
- Capturing communication or network support to the operational activity.

Difficulties can arise in capturing system functionality associated with operational activities for several reasons, including the following: the methodologies/tools used to capture the threads; the operational threads are not necessarily defined with the aim of using them to capture system functionality; and lack of collaboration between the operational and system communities when the threads are developed.

As we move into the world of information services and orchestration of information services to support operational business processes, better mechanisms are needed to bring the users and systems communities together in the definition of system solutions. Many years ago when there was little or no automation, a great deal of time was spent understanding what the user needed/wanted, working very closely with the user to accommodate their needs in the automated systems. Operational threads, as they are currently being captured, do not appear to bridge the operational/system gap well enough. More effort needs to be invested in a collaborative approach to understanding the future system requirements. Architecture products need to be built that support this close interaction in defining the system capabilities. What methods to recommend are fully understood, but business process analysis constructs to support the definition of SOA solutions should be investigated.

10.3.4 Sample Process Description for Creating a Road Map

The previous subsections in this chapter described the process to perform CBPA and the various tools and methods used to support that process. This subsection uses an example from the mission planning program to illustrate how parts of the process can be applied to a project.

Early in FY05, the mission planning systems group began a high-level capabilities road map. The following summarizes the three basic steps that were used, and highlights parts of that road map.

The first step was to review the latest version of the MCL to determine mission planning's relationship to specific capabilities. Mission planning's Operational Requirements Document (ORD) requirements and the combination of Air Combat Command (ACC) and Air Mobility Command (AMC) CONOPS reflect that mission planning is relevant across all nine capabilities categories. This is true mainly because mission planning supports such a wide range of mission types, and has tools applicable to both preplanning and execution-time frame situational awareness and replanning.

The second step was to create a high-level road map summary as shown in Figure 10.8. As noted in Air Force Materiel Command (AFMC) Interim Guidance 05-001, the summary road map chart is intended to graphically portray current plans and status of future capabilities and technologies required to meet the needs of the warfighter. The mission planning road map summary was based on an example from the Airborne Warning and Control System (AWACS) program, which had been chosen as a standard for use in ESC's battle management systems wing. Note that this standard format is expected to evolve and is not completely aligned with the AFMC-directed format shown in Figure 10.9.

The third step was to compile supporting information to accompany the high-level road map summary. As noted in AFMC Interim Guidance 05-001, a completed materiel capability road map package has not only the road map summary slide, but also a collection of applicable DoDAF architectural views, including (at a minimum) the OV-1, OV-5, SV-5, SV-8, and SV-9.

Groups should also consider supplementing the above-noted information with other materials as needed to effectively tell the story of their responsibilities and products in the context of the "big picture" of the AF warfighter needs, which are framed by the MCL and the set of AF CONOPS. For example, the mission planning systems group compiled a set of charts, including the ones in Figures 10.10 and 10.11, to summarize the nature of mission planning's support to each of the capabilities categories. This information was compiled both as internal reference material internal for the group and also to support discussions with the AF/XOX CONOPS Champs. In addition, information was compiled regarding the SoS challenges facing mission planning and examples of the threads of warfighter activities that relate to those challenges. This information is helpful for cross-program analysis, and explains to AF/XOX some issues that could be addressed via investment in integrated road maps (which would impact across programs).

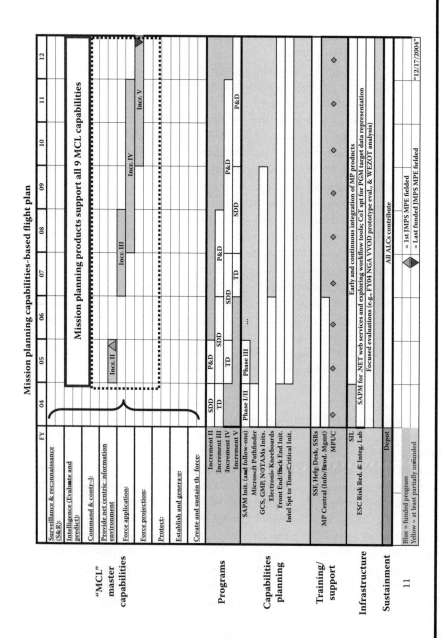

Figure 10.8 Mission planning capability-based flight plan (a.k.a. "Road map") 2–18.

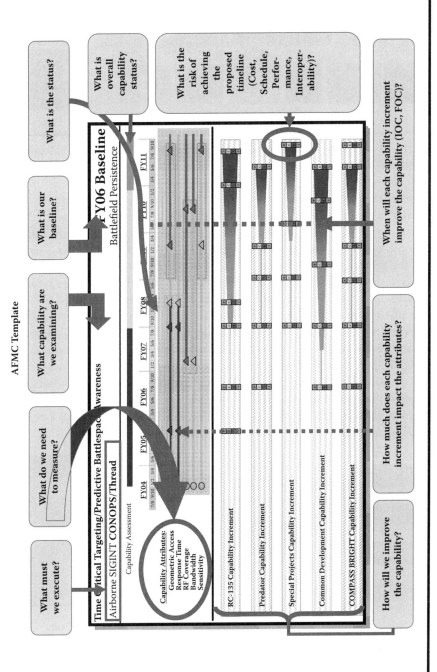

Figure 10.9 AFMC template for road map summaries 2–18.

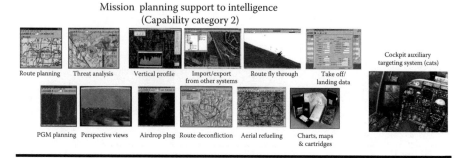

Figure 10.10 Example of mission planning support to MCL capability category 2 (intelligence) 2–19.

Note that, in addition to covering more traditional user need categories such as intelligence and command and control, the MCL includes a capability category (category Number 4) called "provide net centric enterprise information environment." In the mission planning capability road map, this is the category where the mission planning net-centric migration plan is referenced. This plan explains mission planning's net-ready key performance parameter (KPP) approach, the approach to Net-Centric Enterprise Solutions for Interoperability (NESI), and phased plans for leveraging and fielding specific types of web services. All of these plans support evolution toward net-centric operations. They also form a foundation consistent with the ESE objective of facilitating an enterprise environment in which helpful new behaviors could emerge, achieving what is referred to in a recent scientific advisory board (SAB) report [7] as surprise synergy. For example, as mission planning begins to make available to the broader enterprise XML representations of mission plans, subsets of that information may prove useful to currently unanticipated users. Future unplanned loose coupling linkages [7]

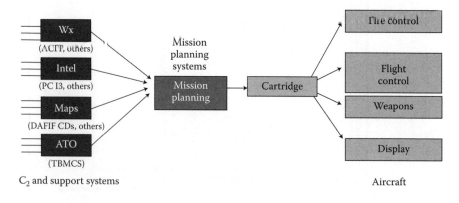

Figure 10.11 Example of mission planning support to MCL capability category 3 (command and control) 2–20.

between mission planning systems and other elements within the broader enterprise may enable helpful new enterprise behavior, resulting in an overall enhanced level of capability for a broader-than-anticipated set of operational users.

10.4 Framework for the Enterprise

This chapter addresses another key element of ESE: the development and use of a common infrastructure, or framework, approach to support the enterprise. Concepts critical to a framework that can support an evolving enterprise, such as an SOA and loose coupling among components, are discussed as enabling concepts useful in achieving the enterprise vision. Applying current IT to these concepts is also described.

10.4.1 Why a Different Framework Is Needed for ESE?

For many people the concept of ESE is nothing more than doing a lot of systems engineering across a large number of systems. However, it is something quite different. It is the idea of focusing on the enterprise-level processes needed for business, defining the necessary components, and engineering their interactions to enable these processes. Although this is a simple concept, it can be difficult to put into practice, as it requires a different approach than TSE—one that has largely not been applied to our enterprise in the past.

While applying ESE efforts, the engineer wants to improve the way the business processes (i.e., the approach used to develop and acquire systems in the enterprise construct) operate. There is a great ongoing demand within the community to provide an enterprise that expands the current set of capabilities by providing more flexibility, scalability, and shared awareness, to name just a few attributes. This is achieved by fundamentally changing the architectural constructs of the building blocks of the enterprise and the way they interact with one another. The thrust is to go from vertical systems that perform large portions of the overall business process (many times these systems have overlapping functionality because they are designed to be self-contained, unique systems) with poor, if any, interactions between them, to smaller, well-defined components of functionality that can be brought together to interact in a larger, more flexible business process. In changing these constructs the enterprise moves from a system- or platform-centric way of doing business toward a net-centric form of operations. In doing so, features will be unleashed, and capabilities wanted in the environment that up to this point, have been elusive.

10.4.2 Service-Oriented Architecture

Initial studying of the movement toward a net-centric way of operations has shown that it will be advantageous to follow the current trends in the commercial sector as they too try to move toward more flexible business processes within this new age of web technologies. The movement of the enterprise from platform-centric to net-centric allows for many improved capabilities and features. The application of an SOA will provide the enterprise with the appropriate framework for achieving this new desired form of operations, as seen in Figure 10.12.

Therefore, what exactly is an SOA? In order to adequately define what an SOA is, we need to first have a common understanding of two terms. The first is service. According to the W3C [8], "a service is an abstract resource that represents a capability of performing tasks that represents a coherent functionality from the point of view of provider entities and requester entities." Essentially, a service is someone or something that performs a defined task or amount of work for another—a car dealership's service department that performs maintenance on an owner's car; restaurants or cafeterias that provide meals for people who are hungry; even movie theaters that provide the latest movies to people who want to be entertained—these are all examples of services people use everyday. There are many other interesting characteristics of services as well:

- The chunks of work or tasks are sized appropriately, given the market's demand
- The service is well defined both by what the service provides and how the requestor interacts with it
- There can be more than one provider of the same service—it just depends on the demand from the market

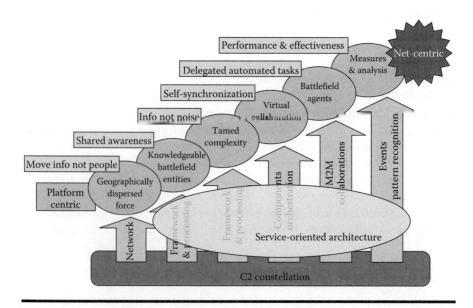

Figure 10.12 SOA framework 3–2.

- In many cases the requester of the service has no idea how the provider actually performs the work, only what the result of that work will be

The second term is loose coupling. Coupling is the dependency between interacting entities—both real and artificial. Real dependencies are ones that always exist and cannot be reduced while artificial dependencies are ones that always exist, but artificial dependencies or their cost can be reduced. Loose coupling describes the configuration in which artificial dependency has been reduced to the minimum. To illustrate this concept, think of an electrical appliance such as a razor. The electric razor has a real dependency on the power plant that provides the electricity. It has an artificial dependency on the outlet. If you were to travel to another part of the world, you would have to use an adaptor to use the razor. If all outlets (and their voltages) were standardized world wide, this would be a loosely coupled configuration. Loose coupling is at the heart of the current ESC STP, which touts the concept of a layered architecture [9]. The loose coupling is tight enough to have a robust, simple, yet standard interface between systems (e.g., the standard electrical outlet), yet sufficiently loose so completely different but similar systems or capabilities on one side can be replaced without ripple effects to the other side (e.g., access to a different power plant).

Combining these two terms we can now define SOA. An SOA is an architectural style whose goal is to achieve loose coupling among interacting services. SOAs are everywhere and in fact, the concept of SOA has been around for more than 20 years. Many years ago, console stereos were the predominate form of home audio equipment—it is a prime example of a closed, platform-centric system (see Figure 10.13). Much like the systems currently fielded within the DoD, each vendor created all the same pieces—amplifiers, speakers, turn-tables, and so on—and shipped (or fielded) them as one monolithic system. Then after many years, RCA developed the RCA connector. This allowed the home audio system to be broken down into components and loosely coupled together, as the owner wanted.

This allows for some very interesting features:

- No longer locked into one company's solution, consumers could construct a home stereo system on a best of breed, component-by-component basis.

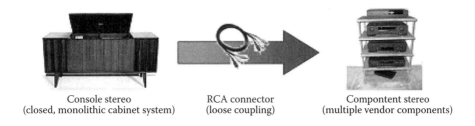

Console stereo
(closed, monolithic cabinet system)

RCA connector
(loose coupling)

Compontent stereo
(multiple vendor components)

Figure 10.13 Example of transformation via a service-oriented architecture 3–3.

- Once in place, the overall stereo could exist indefinitely yet continuously evolve by changing the individual components as needed or desired.
- Enthusiasts could scale a home entertainment system as budget, tastes, and needs dictate.
- New components could be added without changes to other components—many of these new components, such as CD players, did not exist when the configuration was first constructed.

From this example, it is easy to see how the features or capabilities an SOA provides will be very useful within the DoD. As with the stereo example, the goal is to break apart the C4ISR systems and deliver them not as closed stovepipes, but as reusable, functional components. These components need to be sized based on internal market research of how the business processes and their users intend to interact and consume these services. They can then be strung together to form composite applications that can be changed as the needs of the user or as the business environment dictate. In fact, a subset of capabilities can be combined rather than the entire suite, to meet a specific mission. In the stereo example, if a consumer wants, say, a home theater system, he needs some, but perhaps not all, of the same capabilities/components as the stereo aficionado. By developing components, they can be integrated in different systems that serve different missions (but each would require the same capability to meet those different mission needs). Therefore, an SOA provides a methodology appropriate to the basic enterprise attributes (flexibility, scalability, adaptability, allowance of emergent behavior, etc.) and constraints (e.g., standardized interfaces to isolate changes).

10.4.3 Technology

Up to now, this section has addressed moving toward net-centric operations and using an SOA to achieve this. The concept of SOA has been defined, but it has not been set in the context of software and web environment. The technologies used to build out this SOA are as important as the decision to move to SOA itself. Unknowingly, an attempt has been made for many years to implement an SOA framework within the DoD. Client–server technology-based systems and Common Object Request Broker Architecture (CORBA) infrastructures are just a couple of the more recent models implemented in an attempt to build an enterprise that provides some of the features that an SOA can bring to the table. For one reason or the other, each previous attempt has fallen short of the goals. Today, with the improvement in speed of computers, reduction in the cost of hardware, and emergence and growth of web technologies and the internet, there appears to be the real possibility to apply these newer technologies to the enterprise in the form of an SOA and truly achieve the capabilities that it promises.

Focusing on the software aspect of the equation, the trend within the commercial sector is to build out the framework via web services.

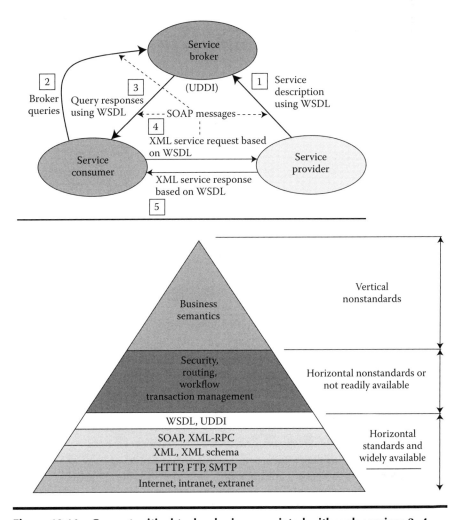

Figure 10.14 Current critical technologies associated with web services 3–4.

Current critical technologies of a web service—such as XML, Simple Object Access Protocol (SOAP), and Universal Description, Discovery, and Integration (UDDI) layered emerging standards (see Figure 10.14)—and where to focus standards implementation—enterprise services versus information services—are addressed fully in the ESC STP.

10.5 Enterprise Applications Case Studies

Previously this chapter has attempted to describe the tools, processes, and technologies critical to ESE. Within this section, we try to make these concepts more real and relevant by discussing several pertinent applications of enterprises. These will be presented in terms of their approaches to the evolution, facilitation, and management of the enterprise examples.

10.5.1 Core Attributes

As previously described, enterprises have certain core attributes such as:

- Flexibility
- Adaptability
- Scalability
- Changeability (encouraging emergent, evolutionary behavior)

Each of these case studies examine some or all of these attributes in terms of how an ESE approach was a necessary facilitator.

10.5.2 Application Case Study—C2 Constellation

As noted in the overview of this chapter, a key point for the ESE is facilitating an environment in which capabilities can emerge. The ESE is a facilitator who ensures that the environment is sufficiently unconstrained to allow capabilities to emerge and that processes are in place to encourage unanticipated capabilities to emerge.

The C2 constellation efforts have attempted to approach the definition of the future constellation architecture in such a way that it relaxes constraints and encourages emergence of new capabilities. A C2 constellation enterprise architecture goal is to develop a midterm target architecture that would capture the mission, data, enterprise infrastructure, and network constructs for air combat operations. Eventually, the intent is to add combat support operations as well.

The current DoDAF architecture, based on the TSE approach, begins with the construct of nodes and derives other architecture products for these nodes. For the C2 constellation future architecture, the starting points are the constructs of air operations function classes (i.e., functionalities of system components) and air operations data classes representative of the entire Constellation. This is consistent with the ESE concepts of establishing boundaries and missions from the top down. The function classes and data classes can be decomposed to whatever extent is necessary to perform architecture analyses. This approach lends itself to defining functional and data classes appropriate for COIs to address SOA definition of information services independent of nodes or physical platforms.

Threads can then provide the specific environment in which the functions or services must operate as well as the context for defining the orchestration of services. This approach supports the constructs of flexibility of operations, minimizing redundancy, and reduction in sustainment and training costs desired in the C2 enterprise of the future.

The BFT COI service study described below extends these enterprise-level constructs to a specific problem.

Initially, thread analysis was the primary approach to analyzing the C2 constellation for gaps/shortfalls that should be addressed by new system functionality. Required operational capabilities were defined for each of the AF CONOPS, but no detailed capability analysis was performed for the C2 constellation. A rudimentary approach to address these capabilities was to map major threads and sub-threads back to capability statements to roughly indicate that the capabilities were being addressed.

Gradually capability planning processes are becoming more formalized to identify capability needs. These capability needs will become the drivers for the evolution of the C2 constellation. ESC/EISG is working with the AFC2ISRC to capture these capability needs derived from the AFC2ISRC FAA, FNA, and HQ USAF/XO/XI I-CRRA process. In addition to supporting the formalized CRRA/I-CRRA process, these and other capability planning products are beginning to be used to drive the C2 constellation implementation and development plans.

10.5.3 Application Case Study—BFT COI Service

BFT COI service is another apt example of the enterprise-level endeavors described in this chapter. The Office of the Secretary of Defense (OSD) sponsored this effort under the ESE initiative to perform a quick proof of concept to deliver a net-centric solution for disseminating blue force information. This reference implementation served as an exploration of the DoD net-centric data strategy that outlines transforming the information environment from the centralized thinking and planning currently reflected in the task, process, exploit, and disseminate (TPED) environment to an edge-focused task, post, process, use (TPPU) approach for information sharing.

10.5.3.1 Capabilities and Analysis

As noted previously, one of the keys to enterprise engineering is to tie functionalities of system components and services directly to capabilities. The reference implementation of the BFT COI service was developed to address a critical capability gap. The FY03 space and C4ISR CONOPS itemized the capability to locate, identify, track, and observe friendly, enemy, nonfriendly and nonaligned forces and actors anywhere and anytime in near real time. Subsequent I-CRRA activities yielded the action item 1B for battle management to develop a roadmap to reduce the number of Combat Identification (CID) programs to "best of type" for inclusion into Family of Interoperable Operational Pictures (FIOP).

To corroborate this action item, US Joint Forces Command outlined in Joint Lessons Learned: Operation Iraqi Freedom Major Combat Operations (FOUO—for office use only) the need to establish a joint force capability using common standards and systems ... which integrate CID and situational awareness to employ the timely use of lethal joint fires, direct and indirect, in all environments with minimal risk of friendly force misidentification. Accordingly, OSD selected this capability gap to demonstrate the applicability of the DoD net-centric data strategy of migrating users and applications from maintaining private data within system-specific storage to making data available by posting it in community- and enterprise-shared spaces.

10.5.3.2 Technical Framework

In order to address the BFT capability gap, the core design principle of the BFT COI service was to de-emphasize the existing hierarchical data-push architecture and streamline warfighter access (pull) to data generated by the legacy applications. This functionality was achieved by employing SOA and implementing information services that loosely coupled content providers using a publish and subscribe methodology.

Within this paradigm, content providers publish BFT data without the knowledge of which clients are subscribing as consumers of the information. Conversely, clients may subscribe to receive the best available information without any foreknowledge of the content providers who are publishing the BFT data.

In other words, the BFT COI service represented a paradigm shift in the routing of information to users. Traditionally, the provider–user relation was one-to-one, or one-to-many, and represented a push of the data. The BFT COI service explored a shift to a many-to-many relationship with a focus on users pulling the best information from any source. The capability allows warfighters easier access to more complete blue force content while eliminating redundancies where consumers received the same data multiple times.

The key enterprise element was that this capability was implemented as an information routing capability that delivers data from many content providers to many clients without either needing knowledge of the other.

This is a classic enterprise solution, wherein the providers and consumers are, in essence, unaware of each other. This could be contrasted with the TSE approach, which would normally insist upon full knowledge and documentation of all of the provider–consumer connections, in order to explicitly meet consumer needs and eliminate redundancy.

10.5.3.3 Publish and Subscribe Services

The BFT COI service development team employed the Publish and Subscribe Services (PASS) of the Army Battle Command System (ABCS) 6.4 System to provide the service functionality. PASS receives commands from content providers and clients and performs the necessary operations. The data input and output of the service utilize the web services standard messaging protocol SOAP. Service descriptions are composed of the standard protocol Web Services Description Language (WSDL).

10.5.3.4 BFT COI Service Data Schema

The BFT COI service data schema utilizing the XML technology establishes the set of syntax and structural data constraints that allows a common understanding of data types across all content providers and clients. The BFT data schema is based on the Air Force Cursor on Target (CoT) object-oriented data definition strategy. The BFT COI service core schema holds the essential and ubiquitous "what, when, where" information. This core schema is extended with subschemas that provide more detailed BFT information such as velocity vector and remarks.

10.5.3.4.1 CoT Strategy

The CoT strategy is particularly interesting from the ESE perspective in that it eschews the common COI-centric view in lieu of a data-centric one.

From the CoT perspective, members of a COI have a common but limited scope of interest. Accordingly, their view of a large dataset is merely a projection of the underlying data. These projections are convenient for COIs, but they may not be isomorphic to the data. A good example is that of many different views or reports being generated from a database with a single schema. While the views and schemas are related, they are not interchangeable.

Thus the COI approach only reveals projections. This is useful but inadequate. Instead the CoT strategy is a data-centric one that seeks to identify the underlying schema or data model. In this approach, the relationships among data elements are intrinsic to the data and have little to do with COI views; the strategy attempts to seek out these natural relationships between key data elements that may not coincide with COI views at all. However, the data may be projected to the needs of respective COIs.

Given the BFT COI service, the CoT "what, when, where" core schema (the base class) along with a set of subschemas (namely, uid, track, remarks, organization, and symbology) synthesized readily with the ground track schema of the ABCS 6.4 System and the Joint Blue Force Situational Awareness (JBFSA) Advanced Concept Technology Demonstration (ACTD) schema to generate the data model of the BFT COI entity schema. In other words, schemas defined from the perspective of intrinsic relationships among data elements were projected straightforwardly to define the data model for the BFT COI.

10.5.4 Application Case Study: Genetic Algorithms versus Nonlinear Optimization for Resource Allocation Tasks

One way to build or influence the direction of activities to meet enterprise objectives is to carefully select or design applications that, themselves, support or possess the attributes desired in the enterprise. These attributes include adaptability, flexibility, and scaleability to support a variety of needs, or even to adapt in real time to support the evolving needs of the current mission. In selecting or designing an application, the ESE (in contrast to the TSE) looks beyond the performance specifications (or only ensuring that key requirements are defined) and assesses application solutions against these enterprise attributes as well. This section considers an example of how to compare two different application solution classes to meet a prevalent mission need.

Many missions require the capability to perform resource allocation. Examples include air campaign planning (allocating weapon resources to targets), deployment planning (where to deploy which resources), sensor scheduling (which sensor to survey which areas or for what targets), logistics planning (how to stage the supply lines), and many, many others. Traditionally, this class of problem is considered an optimization problem, and typically the algorithm of choice for the system developer has been nonlinear optimization methods. Such approaches have the obvious performance benefit of always providing the best solution possible. However, these algorithms also have drawbacks that have been observed operationally:

- Can take too long to return with the optimal result
- Brittle solutions (e.g., a small perturbation to the solution due to an unexpected change in plans collapses the solution and results in an inefficient and ineffective, and perhaps unexecutable, plan)
- Lack of ability to adapt to dynamic real-time changes in relevant factors while the application is processing (i.e., a restart from the initial conditions is required to adapt)

Recent research has identified that evolutionary computation algorithms can provide somewhat degraded, but still acceptable, performance in areas where optimization algorithms excel, and can address the core enterprise attributes that optimization algorithms cannot. Evolutionary computation is used here to refer to a family of algorithms including genetic algorithms, evolutionary algorithms, genetic programming, and evolutionary programming. Evolutionary computation is an iterative two-stage process consisting of random variation (or in an ESE construct and innovation) and selection (integration), though not necessarily in that order. An initial population of solutions is generated randomly or by using domain knowledge to construct reasonable solutions. Each solution in the population is then evaluated using a fitness function, or objective function, to provide an assessment of its quality.

The worst solutions are discarded and the remaining solutions are altered. Alteration can be accomplished by making random changes to a solution, or by exchanging portions of two or more solutions. These new solutions are then evaluated for fitness, and the process iterates until time has elapsed, or an acceptable solution is found.

There are a large number of variations in this general process. Solutions can be encoded in a number of different ways, including vectors, matrices, or lists, among others. The information that is encoded can be real values, integers, binary numbers, or letters. Variation can be introduced (depending in part on the encoding scheme) by swapping sections of the vector, matrix, or list, or by averaging them. Poor solutions can be eliminated deterministically by discarding the worst $x\%$ or probabilistically by randomly selecting N solutions from the population and discarding the M worst solutions (where $M < N$).

Evolutionary computation has the following attributes that support enterprise needs:

- Provides flexibility (multiple solutions, a solution set at any time, etc.)
- Provides adaptability (can change terms in the objective function in midstream, can use prior solution as initial point in replanning, etc.)
- Favors good enough answers (in a timely way) over optimal (but costly and brittle) solutions
- Supports scalability (can accept more resources and/or targets during the process)
- Can be parameter driven to be used for many different purposes (supports changeability)

In short, evolutionary computation can provide the flexibility in many resource allocation problems that nonlinear optimization techniques cannot. Of course, such algorithm classes must be assessed to determine whether they can satisfy the performance specifications identified for that capability; but here again a prudent selection of the key performance requirements, independent of design solutions, needs to be conducted to ensure application solutions that possess enterprise attributes are appropriately considered.

Acronyms

ABCS	Army Battle Command System
ACC	Air Combat Command
ACTD	Advanced Concept Technology Demonstration
AFC2ISRC	Air Force C2 Intelligence, Surveillance, and Reconnaissance Center
AFMC	Air Force Materiel Command

AFSPC	Air Force Space Command
AMC	Air Mobility Command
AWACS	Airborne Warning and Control System
BFT	Blue Force Tracking
BSAHLS	Battlespace Awareness for Homeland Security
C2	Command and Control
C4ISR	Command, Control, Communications, and Computers Intelligence, Surveillance, and Reconnaissance
CBPA	Capability-Based Planning Analysis
CDD	Capabilities Development Document
CID	Combat Identification
CINC21	Commander in Chief 21
CMOC	Cheyenne Mountain Operations Center
COI	Community of Interest
CONOPS	Concepts of Operations
CORBA	Common Object Request Broker Architecture
CoT	Cursor on Target
CPDs	Capability Product Documents
CRRA	Capabilities Review and Risk Assessment
CSI	Critical Success Indicators
DCTS	Defense Collaboration Tool Suite
DFC2	Decision Focus Command and Control
DoD	Department of Defense
DoDAF	Department of Defense Architecture Framework
DOTMLPF	Doctrine Organization Training Materiel Leadership Personnel and Facility
D-SIDE	Defense Strategic Integrated Decision Environment
DST	Decision Support Tool
ESC	Electronic Systems Center
ESE	Enterprise Systems Engineering
EISG	Enterprise Integration Systems Group
FAA	Federal Aviation Administration
FAA	Functional Area Analysis
FIOP	Family of Interoperable Operational Pictures
FNA	Functional Needs Analysis
FSA	Functional Solution Analysis
FTP	File Transfer Protocol
FYDP	Future Years Defense Program
GCCC	Ground Combat Command and Control
HTTP	Hypertext Transfer Protocol
IT	Information Technology
I-CRRA	Integrated Capabilities Review and Risk Assessment
JBFSA	Joint Blue Force Situational Awareness

JC2	Joint C2	
JCS	Joint Chiefs of Staff	
JEFX	Joint Expeditionary Force Experiment	
JFC	Joint Forces Commander	
JFC	Joint Functional Concept	
JIC	Joint Integrating Concept	
JOC	Joint Operating Concept	
JopSCs	Joint Operations Concepts	
JSTARS	Joint Surveillance Target Attack Radar System	
KPP	Key Performance Parameter	
MCL	Master Capabilities Library	
MCU	Multipoint[?] Control Unit	
MID	Management Initiative Decision	
NCES	Net-Centric Enterprise Solutions	
NESI	Net-Centric Enterprise Solutions for Interoperability	
NSSO	National Security Space Office	
ORD	Operational Requirements Document	
OSD	Office of Secretary of Defense	
OV	Operational View	
PASS	Publish and Subscribe Services	
PIA	Postindependent Analysis	
READI	Readiness Evaluation, Assessment and Decision Making Information System	
RPC	Remote Procedure Call	
SFAM	Strategic Force Accounting Module	
SMIO	Systems Management Integration Office	
SMTP	Simple Mail Transfer Protocol	
SOA	Service-Oriented Architecture	
SOAP	Simple Object Access Protocol	
STARS	Standard Terminal Automation Replacement System	
STP	Strategic Technical Plan	
SV	System View	
TC2A	Theater C2 Architecture	
TPED	Task, Process, Exploit, Disseminate	
TPPU	Task, Post, Process, Use	
TSE	Traditional Systems Engineering	
TTP	Training, Tactics, and Procedures	
UAV	Unmanned Aerial Vehicle	
UDDI	Universal Description, Discovery, and Integration	
W3	World Wide Web	
WSDL	Web Services Description Language	
XML	eXtensible Markup Language	

References

1. Smyton, P. A., S. C. Elgass, L. S. Hawthorne, et al. 2006. *Enterprise Systems Engineering Theory and Practice—Volume 6: Enterprise Activities (Evolving Toward an Enterprise)*. MP05B0000043, MITRE Product. The MITRE Corporation, Bedford, MA.
2. Gharajedaghi, J. 1999. *Systems Thinking—Managing Chaos and Complexity*. Butterworth/Heinemann Publishers, Boston, MA.
3. Axelrod, R. and M. D. Cohen. 2000. *Harnessing Complexity*. Basic Books, New York, NY.
4. Norman, D. O. and B. E. White. 2008. Asks the Chief Engineer: "So what do I go do?!" *IEEE Systems Conference*, Montreal, Quebec, Canada, April 7–10.
5. Deputy Secretary of Defense, Management Initiative Decision 913: Implementation of the Planning, Programming, Budgeting, and Execution System (PPBE), May 22, 2003.
6. Goodstein, L. D., T. M. Nolan, and J. W. Pfeiffer. 1993. *Plan or Die*. Pheiffer and Company, San Francisco, CA.
7. USAF SAB. 2005. *System-of-Systems Engineering for Air Force Capability Development*. Executive Summary and Annotated Brief, SAB-TR-05-04.
8. Web Services Architecture Working Group (chairs Chris Ferris, Michael Champoin, and Dave Hollander), World Wide Web Consortium, http://www.w3.org/TR/ws-arch/wsa.pdf, February 2004.
9. ESC/EN, USAF Strategic Technical Plan version 2.1, April 25, 2005.

Index

Note: n indicates footnote.

A

ABCS. *See* Army Battle Command System (ABCS)
AC. *See* Alternating current (AC)
ACC. *See* Air Combat Command (ACC)
Acronyms, 61, 228–229, 289–291, 388–390, 435–437
ACTD. *See* Advanced Concept Technology Demonstration (ACTD)
Action request system (ARS), 282, 289
Adaptability, 336, 342, 352
  adaptable collateral services, 352
  adaptable data mediation, 352
  adaptable information distribution, 352
  dynamic information discovery, 352
  mechanisms, 352
Adaptable, 310
Adaptation, 9, 48n. *See also* Variation
  development through, 10
  pattern, 9
  web browser, 9
Advanced Concept Technology Demonstration (ACTD), 433, 435
Advanced Research Projects Agency Network (ARPANET), 399
Aerospace Operations Center (AOC), 365, 388
AETF. *See* Air and Space Expeditionary Task Force (AETF)
AEW&C systems. *See* Airborne—Early Warning and Control (AEW&C) systems
AF. *See* Air Force (AF)
AF CBP, 148. *See also* Joint Capabilities Integration and Development Systems (JCIDS)

analytical output, 153
analytical process, 150, 151
capability categories, 149
CONOPS-based approach, 148, 149
FAA, 151
FNA, 151
FSA, 153
process flow, 154
risk-assessment output, 152
risk-assessment process, 152
timing, 155
AF/ESC. *See* Air Force—Electronics Systems Center (AF/ESC)
AFI. *See* Air Force—Instruction (AFI)
AFMC. *See* Air Force—Materiel Command (AFMC)
AFPD. *See* Air Force—Planning Document (AFPD)
AFROCC. *See* Air Force—Requirements for Operational Capabilities Council (AFROCC)
AFSPC. *See* Air Force—Space Command (AFSPC)
Agent, 48n, 303
  interactions among, 51, 53
  multiple, 213
  selection, 54–55
Agile, 298, 336, 385
  functionality, 26, 306
  information exploitation, 356
  information generation, 309–310, 323
  information management, 331, 355
Air
  Mobility Command (AMC), 421, 436
  Operations Database (AODB), 385, 388
  Tasking Order (ATO), 364, 383, 388
  traffic management (ATM), 274, 289

439

Air and Space Expeditionary Task Force
    (AETF), 260, 289
Airborne
    Early Warning and Control (AEW&C)
        systems, 35–36, 57–58, 61
    Warning and Control System (AWACS),
        301, 388, 421, 436
Air Combat Command (ACC), 421, 435
Air Force (AF), 100, 148, 261, 289, 326,
        388, 398
    C2 Enterprise, 36–37
    CBP, 148
    CONOPS construct, 408, 409
    EA, 194, 195, 196
    Electronics Systems Center (AF/ESC),
        11, 13
    Instruction (AFI), 148
    Materiel Command (AFMC), 421,
        423, 435
    Planning Document (AFPD), 148
    Requirements for Operational Capabilities
        Council (AFROCC), 148
    RSAF, 198
    Space Command (AFSPC), 412, 436
    United States, 397
Air traffic control, 234, 236–237
    end-use capability, 237
AIS. *See* Automated information system
    (AIS)
Alternating current (AC), 354, 388
AMC. *See* Air—Mobility Command (AMC)
Anaconda. *See* Operation Anaconda
Annual Planning and Programming Guidance
    (APPG), 153
AOC. *See* Aerospace Operations Center
    (AOC)
AODB. *See* Air—Operations Database
    (AODB)
APIs. *See* Application programming interfaces
    (APIs)
APPG. *See* Annual Planning and
    Programming Guidance (APPG)
Application programming interfaces
    (APIs), 280
Architecture(s), 68, 182, 199, 191. *See also*
    Enterprise Architectures (EAs)
    AF C2, 197–198
    continuum, 193
    DCAPES, 199
    E-10 program design activities, 203
    EA, 23
    ESE, 24, 184, 190, 191, 192

FEA, 193–194
framing capability portfolios, 118
JJCIDS, 199
layered, 11
multiyear budgeting, 198
negative reputation, 187–189
procurement transformation, 197
products, 89, 200–202, 416
RSAF, 198
STP, 14
TSE, 191
use, 197
viewgraphs, 185–186
Army Battle Command System (ABCS),
    432, 435
ARPANET. *See* Advanced Research Projects
    Agency Network (ARPANET)
ARS. *See* Action request system (ARS)
Asset allocation FY06 mission, 379
ATM. *See* Air—traffic management (ATM)
ATO. *See* Air—Tasking Order (ATO)
Attributes, 117–118, 191
    enterprise, 428, 430
    estimating object, 324
    evolutionary computation, 435
Authentication, 347
Automated information system (AIS),
    261, 289
AWACS. *See* Airborne—Warning and Control
    System (AWACS)

B

Battlefield, 299, 360, 373
    complexity, 358
Battlespace Awareness for Homeland Security
    (BSAHLS), 401, 436
Behavior, 15n, 21, 39n, 108
    behavioral task analysis, 370
    broad regions, 17
    emergent, 74–75, 190
    incentivize enterprise, 21
BFT. *See* Blue Force Tracking (BFT)
BFT COI service, 431
    capabilities and analysis, 431–432
    CoT strategy, 433
    data schema, 433
    PASS, 432–433
    technical framework, 432
Biennial PPBE, 248
Bilateral, 260
Bitter end, 334

BladeLogic Client Automation (BMC), 284, 289
Blind Men and the Elephant, The, 182
Blue Force Tracking (BFT), 313, 388, 398, 436
    capability gap, 432
    COI service, 398
BMC. *See* BladeLogic Client Automation (BMC)
Boolean networks, 16–17
Bottom–up approach, 413
BSAHLS. *See* Battlespace Awareness for Homeland Security (BSAHLS)
Business planning, 286
    business plan, 286
    military enterprise, 286–287
Butterfly effect, 25

C

C2. *See* Command and Control (C2)
C2, and Intelligence, Surveillance, and Reconnaissance (C2ISR), 197, 344, 388
C2 constellations, 36, 65, 125, 206, 298, 398. *See also* Enterprise Analysis and Assessment (EA&A)
    application, 430
    capability planning processes, 431
    challenges, 128
    DoDAF architecture, 430
    STP architecture, 14
    threads, 430–431
C2E. *See* Command and Control Enterprise (C2E)
C2ISR. *See* C2, and Intelligence, Surveillance, and Reconnaissance (C2ISR)
C4I. *See* Command, Control, Communications, Computer, and Intelligence (C4I)
C4ISR systems. *See* C4I, Surveillance and Reconnaissance (C4ISR) systems
C4I, Surveillance and Reconnaissance (C4ISR) systems, 27, 332
CAASD. *See* Center for Advanced Aviation System Development (CAASD)
CAC. *See* Common—access card (CAC)
CAIG. *See* Cost Analysis Improvement Group (CAIG)
Canonical enterprise functions, 305, 306
    first-order information flow, 306
    information exploitation, 307–308
    information generation, 306–307
    information management, 307
    interactions, 305–306
    R4, 305
    research paradigm, 308
    technology layers, 309
Capabilities-based engineering analysis (CBEA), 22, 67, 69, 97, 103
    AF CBP, 148
    analytical model, 107–112
    capabilities-based perspective, 104
    complex contextual issues, 98–99
    complex system strategies, 105–106
    context of ESE, 110
    DoD, 100
    enterprise, 98
    evolutionary planning, 106, 131, 133
    exploratory analysis, 111, 122, 123
    exploratory modeling, 106
    JCIDS, 144
    modularity, 105–106
    modular representation, 107
    overview, 105
    principles, 108–109
    processes, 109, 111–112
    purposeful formulation, 109, 113
    responsive modernization, 99
    stakeholder analysis, 115
    three phases, 113
    varied implementations, 112
Capabilities-based planning (CBP), 98, 101. *See also* AF CBP
    defining characteristics, 101
    DoD, 100–101
    MCL, 150
    need for, 102
    processes, 103
Capabilities-based planning analysis (CBPA), 405, 406, 407, 411, 436
    AF CONOPS, 409, 410
    analysis approaches, 408, 409
    capability planning, 406–407
    DoD capabilities, 409
    engineering integrated C2 enterprise, 411
    I-CRRA process, 410
    JCIDS approach, 144
    JFCs, 407
    JICs, 407, 408
    JOCs, 407, 408
    MCL capabilities, 410
    "target"/objective process, 410
    Task Force-level CONOP, 410

Capabilities-Based Review and Risk Assessment (CRRA), 100. *See also* Integrated Capabilities Review and Risk Assessment (I-CRRA)
   AF, 101
   integrated, 397
Capabilities Development Documents (CDDs), 410
Capabilities view, the, 98
Capability categories, 149, 424
   functional capability, 149–150
   MCL, 424
   need for, 102
   operational capability, 150
Capability Product Documents (CPDs), 410
Capacity planning, 270
CAS. *See* Complex Adaptive Systems (CAS)
CASE. *See* Complex Adaptive Systems—Engineering (CASE)
CBEA. *See* Capabilities-based engineering analysis (CBEA)
CBEA evolutionary planning, 131
   concept decision, 138
   create vision, 134
   enterprise impacts, 136
   evolution strategies, 132, 134
   partnering, 132
   plan for contingencies, 135
   portfolio plans, 135–136
   portfolio roadmap, 140, 141
   steps, 133
   strategy development, 134–135
CBEA exploratory analysis, 122
   alternatives generation, 129–130
   baseline performance, 125
   bound solution space performance, 126–127
   C2 analysis, 128
   concept development, 130–131
   costs evaluation, 125–126
   exploring concepts, 129, 130
   feasibility, 131
   forecast impact, 131
   gaps and opportunities, 127–128
   modeling taxonomy, 125
   motivations for change, 128
   performance and cost, 122, 124
   steps, 123
CBEA purposeful formulation, 113
   bound relevant structures, 120
   conditions, 116–117
   desired outcomes, 116
   EA, 119
   enterprise context, 121
   frame capability portfolios, 118, 119
   interdependencies, 121
   metrics development, 117–118
   outcome spaces, 113, 115
   reference portfolio, 121
   stakeholder analysis, 115–116
   steps, 114
   structure to effect, 120
   task structure, 117
CBP. *See* Capabilities-based planning (CBP)
CBPA. *See.* Capabilities-based planning analysis (CBPA)
CCA. *See* Clinger-Cohen Act (CCA)
CCs. *See* Commodity Councils (CCs)
CDD. *See* Concept definition document (CDD)
CDDs. *See* Capabilities Development Documents (CDDs)
Center for Advanced Aviation System Development (CAASD), 378, 388
Chairman of the Joint Chiefs of Staff Instruction (CJCSI), 144, 199n, 289
Chairman of the Joint Chiefs of Staff Manual (CJCSM), 144, 199n
Character, 315
Cheyenne Mountain
   Operations Center (CMOC), 401, 436
   Upgrade Program, 5
CID. *See* Combat—Identification (CID)
CII. *See* Common—integrated infrastructure (CII)
CIP. *See* Critical—infrastructure protection (CIP)
CJCSI. *See* Chairman of the Joint Chiefs of Staff Instruction (CJCSI)
CJCSM. *See* Chairman of the Joint Chiefs of Staff Manual (CJCSM)
Client–server architecture, 271
Clinger-Cohen Act (CCA) of (1996), 260, 289
CM. *See* Configuration management (CM)
CMDB. *See* Configuration management—database (CMDB)
CMOC. *See* Cheyenne Mountain—Operations Center (CMOC)
CND. *See* Computer network defense (CND)
Coalition, 260
COAs. *See* Courses of action (COAs)
Cognitive
   functional system, 376–377
   science, 302

Cognitive engineering, 370
    behavioral task analysis, 370
    cognitive task analysis, 370
    design questions, 371
    goal, 370–371
    methods, 371, 378
COI. *See* Community of Interest (COI)
Combat
    Identification (CID), 431, 436
    Search and Rescue (CSAR), 388
Command and Control (C2), 10, 36, 65, 206, 298, 398
Command and Control Enterprise (C2E), 298, 358, 388
    AWACS, 301
    CAS, 302
    cognitive science, 302–303
    information generation, 386
    information science, 303
    military transformation, 300–301
    operators, 366
    recursive nature, 301
    social network theory, 302
    systems, 385
Command, Control, Communications, Computer, and Intelligence (C4I), 198
Commercial off-the-shelf (COTS), 259, 289
Commodity Councils (CCs), 197
Common
    access card (CAC), 262, 289
    integrated infrastructure (CII), 261, 289
    Object Request Broker Architecture (CORBA), 428, 436
Communities of interest (COI), 73, 81, 177, 228, 331n, 341, 388, 396, 436
    BFT, 398, 431, 432
    enterprise, 398–399
    products, 85
    roadmap, 85
    role, 73
    techniques, 344
Competency, 39n
Complementarity, 41, 42–43
Complex Adaptive System (CAS), 10, 48n, 298, 388. *See also* Enterprise
    complexity theory, 302
    Engineering (CASE), 25n
Complexity science, 7, 15
    power laws, 15
    self-organizational constructs, 16
Complexity theory, 33n, 302, 323

Complex-system engineering (CSE), 164
Computer network defense (CND), 240, 289
Concept decision, 138
    analytical information, 139
    need for tools, 139
    possible outcomes, 139
    purposes, 138
Concept definition document (CDD), 235, 236, 289
Concepts of operations (CONOPS), 303, 388, 408, 436
    AF, 148–149, 409
    area, 410
Concurrent versioning system (CVS), 271, 289
    client–server architecture, 271
    limitations, 272
    RCS, 272
Configuration management (CM), 227, 228, 233, 268, 289
    change management, 269–271
    configuration items, 269
    database (CMDB), 283
    ESE, 269
    evolution, 273
    existing structure, 273
    plans, 268
    policies and processes, 269
    software, 271
    steps for evolution, 274
    themes, 273
CONOPS. *See* Concepts of operations (CONOPS)
CORBA. *See* Common—Object Request Broker Architecture (CORBA)
Cost Analysis Improvement Group (CAIG), 126
Cost–benefit analysis, 349
    considering ongoing costs, 349
    initial threshold, 349
    network effect benefits, 349
    network-sharing benefits, 350
    resulting structure, 349
CoT. *See* Cursor on Target (CoT)
COTS. *See* Commercial off-the-shelf (COTS)
Coupling, 427
    loose, 11
    multifocal organization, 336–337
    state space, 333
    tight, 334
Courses of action (COAs), 106, 153
CPDs. *See* Capability Product Documents (CPDs)

Critical
    infrastructure protection (CIP), 240, 289
    Success Indicators (CSIs), 412
Cross-domain resource. *See* Security guard
CRRA. *See* Capabilities-Based Review and Risk Assessment (CRRA)
CSAR. *See* Combat Search and Rescue (CSAR)
CSE. *See* Complex-system engineering (CSE)
CSIs. *See* Critical—Success Indicators (CSIs)
Cursor on Target (CoT), 13, 433, 436
CVS. *See* Concurrent versioning system (CVS)

D

Data, 310
Data fusion, 310, 311, 323
    hierarchy of techniques, 324
    impact assessment, 310, 324
    JDL data fusion model, 323
    object assessment, 310, 323
    problem, 315
    process refinement, 311, 324
    situation assessment, 310, 324
    subobject data assessment, 310
Data mining, 324–325
    application, 325–326, 328
    classes, 325
    descriptive problems, 325
    predictive problems, 325
DC. *See* Direct current (DC)
DCAPES. *See* Deliberate and Crisis Action Planning and Execution Segment (DCAPES)
Decision support system (DSS), 278, 290, 372
    pairing pictures, 373
    problem space, 372
    structure mapping, 372
    tenet (20), 373
    WTP problem, 372
Defense Information Services Agency (DISA), 261, 289
Defense Planning Scenarios (DPS), 145
Deliberate and Crisis Action Planning and Execution Segment (DCAPES), 199
De-militarized zone (DMZ), 281, 289
Department of Defense (DoD), 25, 175, 206, 247, 322
    Architecture Framework (DODAF), 194, 199, 397
    C4I, Surveillance, and Reconnaissance (DoD C4ISR) enterprise, 396
    EA, 194
    Index of Specifications and Standards (DODISS), 274
    Instruction (DoDI), 234, 289
Department of Homeland Security (DHS), 263, 289
Design Patterns, 11
    fundamental, 14
Development, 44
    adaptation, through, 9
    complex systems, 47, 49
    differentiation, 44
    innovation, 44
    integration, 44
    model, 44n
    systems view, 45, 46–47
Developmental/operational test (DT/OT), 216
DHS. *See* Department of Homeland Security (DHS)
Differentiation, 44
Digital subscriber line (DSL), 279, 290
Direct current (DC), 354, 389
DISA. *See* Defense Information Services Agency (DISA)
DMZ. *See* De-militarized zone (DMZ)
DNS. *See* Domain Name Service (DNS)
Doctrine, organization, training, materiel, leadership and education, personnel, and facilities (DOTMLPF), 101, 249, 290, 415, 436
DoD. *See* Department of Defense (DoD)
DODAF. *See* Department of Defense (DoD)—Architecture Framework (DODAF)
DoD C4ISR. *See* Department of Defense (DoD)—C4I, Surveillance, and Reconnaissance (DoD C4ISR) enterprise
DoDI. *See* Department of Defense (DoD)—Instruction (DoDI)
DODISS. *See* Department of Defense (DoD)—Index of Specifications and Standards (DODISS)
Domain Name Service (DNS), 339, 389
DOTMLPF. *See* Doctrine, organization, training, materiel, leadership and education, personnel, and facilities (DOTMLPF)
DPS. *See* Defense Planning Scenarios (DPS)
DSL. *See* Digital subscriber line (DSL)
DSS. *See* Decision support system (DSS)

Index ■ 445

DT/OT. *See* Developmental/operational test (DT/OT)
Dynamic capabilities, 99

E

E-10 program, 203
EA. *See* Enterprise Architecture (EA)
EA&A. *See* Enterprise-level analysis and assessment (EA&A)
EAF. *See* Enterprise Architecture (EA)—framework (EAF)
EAP. *See* Enterprise Architecture (EA)—for Procurement (EAP)
EBAs. *See* Enterprise—business analysts (EBAs)
EBO. *See* Effects-Based Operations (EBO)
EDW. *See* Enterprise—data warehouse (EDW)
Effects-Based Operations (EBO), 386–387
Efficiency–innovation trade, 14
EII. *See* Enterprise information interoperability (EII)
EIP. *See* Enterprise—implementation plan (EIP)
EIS. *See* Executive information system (EIS)
EISG. *See* Enterprise—Integration Systems Group (EISG)
EIT. *See* Enterprise—integration team (EIT)
Electronic Systems Center (ESC), 398
End-use capabilities, 236
  air traffic control, 237
  interfaces, 237
Enterprise, 4, 10, 34–35, 37, 98, 162, 190, 234, 235, 398–399, 400. *See* Complex adaptive systems (CAS)
  application, 55, 429–430
  attributes, 191, 399, 430
  business analysts (EBAs), 249, 250
  C2 constellation, 36–37, 430–431
  capabilities, 51, 98, 236
  COI, 398, 399, 431–433
  complexity, 44, 56, 98–99
  continuum, 193
  control, 37
  crosscutting, 250–251
  data warehouse (EDW), 262
  dynamism, 301
  emergence, 190
  evolution, 52, 302, 400
  framework, 404, 425–429
  implementation plan (EIP), 241
  influence, 37
  Integration Systems Group (EISG), 413
  integration team (EIT), 74
  mission, 417
  model, 35
  modes of thought, 39
  multidimensional, 275
  nestedness, 35n, 36
  operational definition, 37
  parameters, 77
  portfolio management, 57–59
  processes and practices, 59–60
  purposeful questions, 55
  purposes alignment, 36
  recognizing, 233
  resource allocation tasks, 434–435
  resource planning (ERP), 258
  Service Bus (ESB), 355, 385
  service monitoring (ESM), 278, 279, 281–282
  SE to enterprise engineering, 38
  state, 190
  systems, 55–56, 234, 237
  "to-be" enterprise, 404
  TSE and ESE differences, 400–403
  visionary planning, 404
Enterprise-level analysis and assessment (EA&A), 25–26, 69, 90, 206, 207, 212, 220
  C2, 208, 209, 214
  characteristics, 212
  developmental versions, 217
  failure mode impacts, 214
  forensics, 220–221
  government role, 223
  lightweight representations, 216–217
  M&S, OITL, and HWIL, 219
  minimal infrastructure, 218
  multiscale analysis, 213
  need, 208–212
  operational involvement, 215
  practical considerations, 226
  progress key steps, 224–226
  stakeholder roles, 221–223
Enterprise Architecture (EA), 23, 25, 66, 69, 83, 119, 192
  AF, 195–196
  data repository, 119
  framework (EAF), 195
  for Procurement (EAP), 197
Enterprise Engineer, 39n
Enterprise Engineering, 284

Enterprise governance, 8, 17, 247
　approaches to, 18
　budgeting and, 247
　capabilities, 247
　collective knowledge framework, 19
　EBAs, 249
　incentivize enterprise behavior, 21
　individuals and enterprise, 19
　objectives, 18
　opposing incentives, 21
　roles and responsibilities, 248
　shape enterprise success criteria, 20–21
　skill categories, 248–249
　stakeholders, 19
Enterprise impacts, 136
　assessing, 136
　capability impacts, 137
　resource impacts, 137
　technical impacts, 136
Enterprise information interoperability (EII), 262
　association issues, 265, 267
　associations among objects, 266
　hidden costs, 262–264
　synchronization issues, 264–265
Enterprise opportunity and risk, 24–25
　engineering and ESE, 175
　management, 177
　opportunity, 168, 173
　regimen, 173
　risk, 165, 167, 173
　traditional SE, 173
　traditional view, 165
　TSE and SoS, 175
Enterprise SE processes, 21
　analysis and assessment, 23
　CBEA, 22
　enterprise architecture, 23
　stakeholder analysis, 23
　STP, 23
　technology planning, 23
Enterprise strategic planning, 396, 404, 405
　capabilities, 404–406, 414–416
　capability road map, 405, 412–414, 416–419
　CBPA, 405–411
　DoD's progression, 405
　ESE role, 406
　mission threads, 405, 419–420
　road map creation, 421
　systems-based approach, 405

Enterprise systems engineering (ESE), 2, 3, 5–7, 65, 97, 164, 182, 242, 297, 396, 401
　accept need, 396
　agile functionality, 26
　analysis and assessment, 25
　architecture, 25
　authoring process, 23
　capabilities road map, 397
　CAS, 302
　cognitive science, 302–303
　complex environment, 396
　context, 5
　emphasis shift, 402
　engineering analysis, 24
　enterprise activities, 27
　essential elements, 7
　fundamental role, 406
　information science, 303
　infrastructure framework, 397–398
　management, 26
　managing necessary aspects, 403
　mission thread analysis, 397
　opportunity and risk, 24
　organized complexity, 6
　organized simplicity, 6
　pilots and case studies, 24
　preplanning the future, 27
　principles, 73
　rationale, 3–4
　scope, 4
　social network theory, 302
　systems thinking, 23
　technology transition, 287–289
　tools and processes, 396
　top–down approach, 396, 397
　unorganized complexity, 6
ERP. See Enterprise—resource planning (ERP)
ESB. See Enterprise—Service Bus (ESB)
ESC. See Electronic Systems Center (ESC)
ESE. See Enterprise Systems Engineering (ESE)
ESE capability allocation, 405
　analysis, 415, 416
　assessing constraints, 415–416
　assessing ESE, 416
　I-CRRA, 415
　process, 414
　top–down method, 414–415
ESE elements, 7
　development through adaptation, 9–11

Index ■ 447

enterprise governance, 17
enterprise SE processes, 21
technical planning, 11
theory and practice, 8
ESE pilots and case studies, 66
　background, 65
　context, 68
　ESE, 67, 68
　final recommendations, 91–92
　overview, 70
　stakeholder analysis, 67
　summary of findings, 66, 70–80, 81–91
ESE processes, 66, 68
　CBEA, 69
　EA&A, 69
　EA, 69
　interrelationship, 70
　STP, 69
　technology planning, 70
ESE technical planning. *See* Strategic Technical Plan (STP)
ESM. *See* Enterprise—service monitoring (ESM)
ETL. *See* Extract, transform, and load (ETL)
Evolutionary
　biology, 9, 34n
　development, 9, 135. *See also* Adaptation
Evolutionary computation, 434, 435
　attributes, 435
　process, 434
Executive information system (EIS), 278
Exploitation function, 307, 356
eXtensible Markup Language (XML), 13, 397
Extract, transform, and load (ETL), 259

F

FAA. *See* Federal—Aviation Administration (FAA); Functional—Area Analysis (FAA)
Family of Interoperable Operational Pictures (FIOP), 431
FARs. *See* Federal—Acquisition Regulations (FARs)
FCBs. *See* Functional—Capability Boards (FCBs)
FDP. *See* Full-dimensional protection (FDP)
FEA. *See* Federal—Enterprise Architecture (FEA)
Federal
　Acquisition Regulations (FARs), 233
　Aviation Administration (FAA), 401
　Enterprise Architecture (FEA), 193, 194
Federally Funded Research and Development Center (FFRDC), 2
Federated architectures. *See* Enterprise architectures (EAs)
FFRDC. *See* Federally Funded Research and Development Center (FFRDC)
FIOP. *See* Family of Interoperable Operational Pictures (FIOP)
Flexible, 310
FNA. *See* Functional—Needs Analysis (FNA)
Frame capability portfolio, 109, 118–119
FSA. *See* Functional—Solutions Analysis (FSA)
Full-dimensional protection (FDP), 240
Function, 108
Functional
　Area Analysis (FAA), 144, 145, 146, 151, 408
　capability, 149–150
　Capability Boards (FCBs), 147
　Needs Analysis (FNA), 144, 146, 151, 408
　Solutions Analysis (FSA), 144, 146–147, 408
Future Year Defense Programs (FYDPs), 153, 247, 413
FYDP. *See* Future Year Defense Programs (FYDP)

G

GCCS-AF. *See* Global—Command and Control System-AF (GCCS-AF)
GeoRSS, 13
GES. *See* Grid enterprise services (GES)
GIG. *See* Global—information grid (GIG)
GIG-E. *See* Global—Information Grid-E (GIG-E)
Global
　Command and Control System-AF (GCCS-AF), 203
　information grid (GIG), 206, 239
　Information Grid-E (GIG-E), 354
GMTI. *See* Ground Moving Target Indicator (GMTI)
Governance of Commons, 20. *See also* Enterprise governance
Grid enterprise services (GES), 261
Ground Moving Target Indicator (GMTI), 58, 314

H

Hardware-in-the-loop (HWIL), 207, 220
Heavy-tailed distributions, 332
HIS. *See* Human-System Integration (HSI)
HMMH. *See human*-to-machine-to-machine-to-*human* (HMMH)
Homeostasis, 188
HSC. *See* Human Supervisory Control (HSC)
HTTP. *See* Hypertext transfer protocol (HTTP)
Hubs, 15
Human Supervisory Control (HSC), 363
Human-System Integration (HSI), 357
   implementing in ESE, 370
   key challenge, 370
Human-to-machine-to-machine-to-*human* (HMMH), 369, 377
HWIL. *See* Hardware-in-the-loop (HWIL)
Hypertext transfer protocol (HTTP), 12

I

IA. *See* Information—assurance (IA)
IAB. *See* Internet—Architecture Board (IAB)
ICANN. *See* Internet—Corporation for Assigned Names and Numbers (ICANN)
I-CRRA. *See* Integrated Capabilities Review and Risk Assessment (I-CRRA)
Identification friend or foe (IFF), 315
IERs. *See* Information—exchange requirements (IERs)
IETF. *See* Internet—Engineering Task Force (IETF)
IFF. *See* Identification friend or foe (IFF)
Ill-formed problem, 10
IMS. *See* Integrated—master schedule (IMS)
INCOSE. *See* International—Council on Systems Engineering (INCOSE)
Information
   assurance (IA), 240, 348
   exchange requirements (IERs), 235
   science, 303
Information exploitation, 298, 306, 307, 381–383. *See also* Cognitive—science
   battlefield complexity, 358
   C2 environments, 357–361
   cognitive engineering, 370–371
   collaboration and operations, 366–367, 368, 376
   cross-agency team, 366, 367
   data mining, 362
   disjointed systems, 364–365
   DSS, 372–373
   dynamism, 362
   exploitation function, 356
   human roles, 357, 359–360, 362, 363
   increased volume, 358
   information acquisition, 363
   information rate, 358–359
   making decisions, 356, 361, 363
   people and technology, 361, 368–370
   reduced cost goals, 359
   reduced manpower, 359
   sheer quantity, 366–367
   stakes, 360, 361
   team awareness, 367
   time pressure, 360, 361
   trust formation, 367–368
   uncertainty, 360, 361
Information exploitation tenets, 373–377
   automated system, 375
   cognitive functional system, 376–377
   decision making environment, 374
   establishing trust, 376, 377
   feedback on environment, 373
   human adaptability, 374
   information accessibility, 374
   information extraction, 374
   information pedigree, 374
   information sharing, 376, 377
   team awareness, 376
   technological resources redirectability, 373–374
   technology design, 374–375, 377
   technology insertion 375
   understanding the information, 376
Information generation, 298, 309–310, 386–387
   application, 315
   assertions, 316
   complexity, 314–315, 316–317, 326
   considerations, 312
   data availability, 312
   data correlation, 315–316, 326–328
   data fusion, 310–311, 323–324
   data mining, 324–326, 328
   data sharing, 313–314
   data sources, 313, 314
   disciplines, 323

facilitating sense making, 326–327
GMTI platforms, 314
LSE's research, 329, 330
needs and goals, 314, 316, 326, 328–329
operational objectives hierarchy, 387
"optimal" allocation, 329
QSM tracking algorithm, 329, 330
situational awareness, 311–312
technologies, 313, 323, 329
TOI, 328
Information generation tenets, 317
   assess relationships, 321
   bias elimination, 319–320, 320
   challenges, 322
   complexity recognition, 321
   data aggregation, 321
   data correlation, 319
   data fusion, 320
   data links retaining, 320
   data needs and goals, 318
   data quality, 317, 320
   data tags, 322
   data warehousing, 321–322
   "freshness" of data, 318
   general statement of need, 319
   goals hierarchy, 319
   information accuracy, 318, 320
   information pedigree, 317–318
   latency, 318
   monitor for change, 321
   object attributes, 320
   object correlation, 320
   pull interactions, 321
   relationships in data, 320–321
   spatial data alignment, 319, 320
   specific statement of need, 318–319
Information infrastructures, 353
   enterprise, 353–354
   ESB failure, 355
   evolution, 353
   GIG-E microservices, 354
   KIPS, 354
   multilevel defence, 355
   sharing resources, 353
Information management, 298, 307, 331, 336, 349, 350, 383–386
   adaptability, 339–340, 341, 342–344, 352
   authentication, 347
   capabilities, 342
   clearance, 346
   COI, 341, 343
   complexity, 345–346, 356

   control information, 347
   dynamic information, 340, 355
   geophysical time frames, 345
   identity, 346, 347
   information assurance, 348
   information infrastructures, 353–355
   information providers, 342
   information sharing, 340, 341, 348, 350
   loose coupling, 338
   metadata, 347
   network-centric, 341, 345, 346–347
   peers coalition, 344
   PKI, 347–348
   planning and configuration values, 341
   "pull mode" CONOPS, 331
   redundancy mechanisms, 350–352
   right budget, 348
   right people, 346
   right time, 344
   risk–cost–benefit analysis, 349–350
   semantically enabled web services, 355–356
   semantic formal models, 338–339
   technologies, 342, 343
   tenets, 331–335, 336–338
   time sources, 344–345
   transparency, 347
   use cases, 355
Information management tenets, 331–332, 336
   bitter end, 334
   C4ISR systems, 332
   challenges, 334–335
   complexity, 332, 334
   composite systems, 332, 333
   coupling, 336–337
   decoupling, 334
   encoding, 337
   five W questions, 335
   loose-coupling, 334
   mathematical models, 332–333
   modeling, 337–338
   state spaces, 333
   tight-coupling, 334
   upshot, 334
Information technology (IT), 10, 32, 61, 206, 229, 234, 290, 298, 389, 397, 436
   CMDB in, 283
   Infrastructure Library (ITIL), 281, 282
   processes, 283–284
   service-level management, 282

Infrastructure, 234n
    considerations, 276
    decentralized *vs.* centralized, 281
    ESM, 278–279
    framework, 397
    ITIL, 281
    loose-coupling, 397–398
    management, 275
    monitoring border architecture, 281
    monitoring techniques, 279
    people, 277–278
    process and planning, 278
    service concepts, 397
    SOA, 397
    technology, 276–277
Innovation, 9, 44, 217, 218
    C2 enterprise, 217
    CBP, in, 102
    integration processes, and, 404
    network, 76
    opportunistic, 10
    sensor technology, 13
Input, 6n, 145, 146
    data flow, 120
    vice, 101, 108
Integrated
    master schedule (IMS), 77
    product team (IPT), 176
Integrated Capabilities Review and Risk Assessment (I-CRRA), 397, 409, 431
    C2 constellation, 398
    capability allocation analysis, 415
    task force-level CONOPS, 410
Integration, 12, 44
Intelligence Surveillance and Reconnaissance (ISR), 382
    assets, 383
    deck, 383
Interaction, 48, 51–53
    canonical enterprise functions, 305
    interacting variables, 275, 284
    questions, 56
Interdependence, 43–44
Internal Revenue Service (IRS), 325
International
    Council on Systems Engineering (INCOSE), 2, 60
    Organization for Standardization (ISO), 11
Internet
    Architecture Board (IAB), 18
    Corporation for Assigned Names and Numbers (ICANN), 18
    Engineering Task Force (IETF), 12, 353
    protocol (IP), 12, 13, 206, 258
Interpersonal trust, 367–368
IP. *See* Internet—protocol (IP)
IPT. *See* Integrated product team (IPT)
IRS. *See* Internal Revenue Service (IRS)
ISO. *See* International—Organization for Standardization (ISO)
ISR. *See* Intelligence Surveillance and Reconnaissance (ISR)
IT. *See* Information technology (IT)
ITIL. *See* Information Technology—Infrastructure Library (ITIL)
IT Investment Portfolio System (I-TIPS), 139
I-TIPS. *See* IT Investment Portfolio System (I-TIPS)

J

JBFSA. *See* Joint Blue Force Situational Awareness (JBFSA)
JC2. *See* Joint C2 (JC2)
JCDRP. *See* Joint Concept Development and Revision Plan (JCDRP)
JCIDS. *See* Joint Capabilities Integration and Development Systems (JCIDS)
JCS. *See* Joint Chiefs of Staff (JCS)
JDL. *See* Joint Directors of Laboratories (JDL)
JEFX. *See* Joint Expeditionary Force Experiment (JEFX)
JFC. *See* Joint Force Commanders (JFC); Joint Functional Concept (JFC)
JICs. *See* Joint Integrating Concepts (JIC)
JOCs. *See* Joint Operating Concepts (JOC)
Joint Blue Force Situational Awareness (JBFSA), 433, 436
Joint C2 (JC2), 408, 437
Joint Capabilities Integration and Development Systems (JCIDS), 101, 191, 144, 147
    analysis process, 144, 146
    analytical output, 147
    concepts-centric approach, 144, 145
    criticisms of, 147–148
    FAA, 145
    FNA, 146
    FSA, 146
    governing policies, 199n

implementation, 203
  top-down approach, 144, 145
Joint Chiefs of Staff (JCS), 144, 412, 437
Joint Concept Development and Revision Plan (JCDRP), 108
Joint Directors of Laboratories (JDL), 310
  data fusion model, 311, 323. *See also* Data fusion
  information generation, 312
Joint Expeditionary Force Experiment (JEFX), 367, 420
Joint Force Commanders (JFC), 414
Joint Functional Concept (JFC), 401, 407, 408
Joint Integrating Concepts (JIC), 113, 407
Joint Operating Concepts (JOC), 149, 401
Joint Operations Concepts (JopSCs), 401, 437
Joint Requirements Oversight Council (JROC), 144
Joint Surveillance Target Attack Radar System (JSTARS), 420
JopSCs. *See* Joint Operations Concepts (JopSCs)
JROC. *See* Joint Requirements Oversight Council (JROC)
JSTARS. *See* Joint Surveillance Target Attack Radar System (JSTARS)

K

Key Interface Profiles (KIPS), 353, 354
Key management infrastructure (KMI), 262
Key performance parameter (KPP), 424
KIPS. *See* Key Interface Profiles (KIPS)
KMI. *See* Key management infrastructure (KMI)
Knowledge, 39n, 57
  generate, 304
  genetics, 33
  workers, 249
KPP. *See* Key performance parameter (KPP)

L

LAN. *See* Local area network (LAN)
Latency, 318
LDAP. *See* Lightweight Directory Access Protocol (LDAP)
Lightweight Directory Access Protocol (LDAP), 339
Local area network (LAN), 234n, 262, 354
London School of Economics (LSE), 329
Loose coupling, 11, 332n, 333, 334, 427

complexity science, 15–16
DNS, 13
HTTP, 12–13
information loss, 337
loose couplers, 12
network science, 15
PR, 13
LSE. *See* London School of Economics (LSE)

M

M&S. *See* Modeling and simulation (M&S)
M2M. *See* Machine-to-machine (M2M)
Machine-to-machine (M2M), 304, 340, 369, 377
Management Initiative Decision (MID), 410
Margin of error. *See* overengineering
Mashup, 12n, 13
Master Capabilities Library (MCL), 149, 150, 408
  mission planning support, 424
  review version, 421
MC2A. *See* Multisensor Command and Control Aircraft (MC2A)
MCL. *See* Master Capabilities Library (MCL)
Measures
  of effectiveness (MOEs), 115, 118, 211
  of performance (MOPs), 211
Mediation services, adaptive information, 343
  distilling data, 343
  encoding data, 343
  run-time mediation, 343
  standards, 343
  useful technologies, 344
MID. *See* Management Initiative Decision (MID)
Military transformation, 300
  AWACS, 301
  challenges, 300–301
  complex and adaptive threat, 300
  distributed operations, 300
  dynamism, 301
  needed capabilities, 300
  notional enterprise, 301
  recursive nature, 301
MIME. *See* Multipurpose Internet Mail Extensions (MIME)
Mission, 234, 238. *See also* System's context
  elements required, 234–235
  enterprise role, 235

Mission threads, 397, 405, 419
  analyses types, 420
  difficulties, 420
  operational threads, 420
  (OV)-5 and OV-6c, 420
  (SV)-5, 420
  uses, 419–420
Mitigation strategy, 256
Modeling and simulation (M&S), 129, 207
MOEs. *See* Measures of effectiveness (MOEs)
MOPs. *See* Measures of performance (MOPs)
Multidimensionality. See Complementarity
Multipurpose Internet Mail Extensions (MIME), 343
Multisensor Command and Control Aircraft (MC2A), 314

N

National Military Strategy (NMS), 100
National Security Space Office (NSSO), 412
Naturalistic decision making, 377
  mental models, 377–378
  MITRE's CAASD, 378
  project goal, 377
  research activities, 378–379
  synthetic task environments, 378
NCCT. *See* Network-Centric—Collaborative Targeting (NCCT)
NCES. *See* Net-centric Enterprise—services (NCES)
NCOW. *See* Network-Centric—Operations and Warfare (NCOW)
NCW. *See* Network-Centric—warfare (NCW)
NESI. *See* Net-centric Enterprise—Solutions for Interoperability (NESI)
Net-centric Enterprise
  services (NCES), 260
  Solutions for Interoperability (NESI), 353, 424
Net-Ready Key Performance Parameter (NR-KPP), 206
Network-Centric
  Collaborative Targeting (NCCT), 314
  Operations and Warfare (NCOW), 297, 317, 341
  warfare (NCW), 261
Network-centricity, 303
  clusters, 303
  expose data, 304
  generate knowledge, 304
  get connected, 304
  tenets, 303
Network science, 15
  Boolean networks, 16–17
  loose couplers in, 15
  phase change, 17
  preferential attachment, 15
  small worlds in, 16
NMS. *See National Military Strategy* (NMS)
Nodes. *See* Hubs
NORAD. *See* North American Aerospace Defense Command (NORAD)
North American Aerospace Defense Command (NORAD), 5
NR-KPP. *See* Net-Ready Key Performance Parameter (NR-KPP)
NSSO. *See* National Security Space Office (NSSO)

O

OEMs. *See* Original equipment manufacturers (OEMs)
Office of Management and Budget (OMB), 261
Office of the Secretary of Defense (OSD), 145, 431, 432
Offices of primary responsibility (OPR), 153
OIF. *See* Operation Iraqi Freedom (OIF)
OITL. *See* Operator-in-the-loop (OITL)
OMB. *See* Office of Management and Budget (OMB)
Open System Interconnect (OSI), 12
Operational
  capability, 150
  Requirements Document (ORD), 421
  view (OV), 199, 420
Operation Anaconda, 380
  information exploitation, 381–382
  key criticisms, 381
Operation Iraqi Freedom (OIF), 364
Operations support (OS), 194
Operator-in-the-loop (OITL), 207, 219
Opportunity, 168–169, 178–179, 254
  assessment, 169
  averse system, 170
  characterize continuously, in, 174
  classification, 170
  components, 169
  engineering scales, 163
  enterprise, 24–25
  ESE *vs.* SoS, 178
  establish rewards, 174

Hillson view, 162
identification, 254, 255
management, 177, 178
multiscale hypotheses, 176–177
neutral, 171
probability/impact grid, 171
process impacts, 256
pursuit plans, 256
seeking system, 170, 171
SoS and SE, 177
OPR. *See* Offices of primary responsibility (OPR)
ORD. *See* Operational—Requirements Document (ORD)
Organizations, 44n
Organized
complexity, 6, 45
simplicity, 6, 45
Original equipment manufacturers (OEMs), 259
OS. *See* Operations support (OS)
OSD. *See* Office of the Secretary of Defense (OSD)
OSI. *See* Open System Interconnect (OSI)
Outcome, 108, 244
enterprise, 71, 72, 74, 77, 78, 80
spaces, 73, 109
OV. *See* Operational—view (OV)
Overengineering, 350

P

Pairing pictures, 372, 373, 378
PASS. *See* Publish and Subscribe Services (PASS)
PB. *See* Program—baseline (PB)
PC. *See* Personal computer (PC)
PDF. *See* Probability density function (PDF)
Period three, 333
Personal computer (PC), 51, 165, 192
PIA. *See* Post independent analysis (PIA)
PKI. *See* Public Key Infrastructure (PKI)
Planning, programming, and budgeting system (PPBS), 247
Planning, Programming, Budgeting, and Execution (PPBE), 148, 155, 241, 410
PMs. *See* Program—managers (PMs)
POC. *See* Point of contact (POC)
POET. *See* Political, operational, economical, and technical (POET)
Point of contact (POC), 281

Political, operational, economical, and technical (POET), 17
POM. *See* Program—Objective Memorandum (POM)
Population, 48n
Post independent analysis (PIA), 409
Power-law distributions. *See* Heavy-tailed distributions
Power laws, 15
PPBE. *See* Planning, Programming, Budgeting, and Execution (PPBE)
PPBS. *See* Planning, programming, and budgeting system (PPBS)
PR. *See* Purchase—request (PR); Purchase—requisition (PR)
Preferential attachment, 15
Prisoners Dilemma problem, 19
Probability density function (PDF), 318
Process Sequence Models (PSMs), 150
Program
baseline (PB), 248
managers (PMs), 7, 19, 21, 169, 248, 258
Objective Memorandum (POM), 153
PSCM. *See* Purchase and Supply Chain Management (PSCM)
PSMs. *See* Process Sequence Models (PSMs)
PSTN. *See* Public switched telephone network (PSTN)
Public Key infrastructure (PKI), 262, 347
Public switched telephone network (PSTN), 281
Publish and Subscribe Services (PASS), 432–433
Purchase
request (PR), 197
requisition (PR), 13
and Supply Chain Management (PSCM), 197

Q

QDR. *See* Quadrennial Defense Review (QDR)
QSM. *See* Quorum sensing molecule (QSM)
Quadrennial Defense Review (QDR), 100
Quorum sensing molecule (QSM), 329

R

R&D. *See* Research and Development (R&D)
R4. *See* "*right information* to the *right people* at the *right time* to make the *right decisions*" (R4)

454 ■ *Index*

RAND. *See* Research ANd Development (RAND)
RCS. *See* Revision control system (RCS)
RDF. *See* Resource Description Framework (RDF)
Reductionism, 104n
Redundancy. *See* overengineering
Redundancy mechanisms, 350–351
    cascade failure risk, 352
    failover mechanism, 351
    information assurance aspects, 351
    issues, 351
    robust distribution, 351
    robust storage, 351
Research and Development (R&D), 297, 300
Research ANd Development (RAND), 2, 100
Resource allocation tasks, 382
    algorithm, 434
    capability requirement, 434
    evolutionary computation, 434–435
Resource Description Framework (RDF), 343
Revision control system (RCS), 272
"*right information* to the *right people* at the *right time* to make the *right decisions*" (R4), 297, 305, 360
Risk, 165
    and opportunities, 252
    assessment, 166
    assessment, 255
    averse system, 168
    classification, 167
    cost–benefit analysis, 349–350
    engineering scales, 163
    establish rewards, 174
    identification, 254
    mitigation, 256
    multiscale hypotheses, 176–177
    neutral, 168
    probability/impact grid, 171
    program-level, 256
    quantification, 167
    seeking system, 168
    TSE and SoS, 172
Risk management, 251
    enterprise system complexity, 252
    plan preparation, 253–254
    process, 253
    tolerance for change, 257–259
Road map, 150, 153, 397, 422. *See also* Strategic Master Plan
    AF CONOPS, 148
    AFMC template, 423
    applying technology, 418
    architecture products, 418–419
    bottom–up approach, 413
    capability, 140, 141, 397, 405, 412, 416
    capability-based flight plan, 422
    CSI, 412
    cycle, 413
    emergent behavior feedback, 418
    enterprise vision, 416–417
    evolution description, 141, 142
    furnishing indicators, 412
    hierarchical approach, 142
    integrate plans, 140
    key factors, 412
    MCL, 421, 424
    mission planning systems group, 421
    monitor user's environment, 418
    net-centric migration plan, 424–425
    objective, 414
    operational issues, 418
    process description, 421
    refinement, 419
    summary creation, 421
    supporting information compilation, 421
    technology, 418
    top–down method, 396, 397
    transition plan products, 416
    user needs, 418
ROE. *See* Rules of Engagement (ROE)
ROM. *See* Rough order of magnitude (ROM)
Rough order of magnitude (ROM), 126
Royal Saudi Air Force (RSAF), 198
RSAF. *See* Royal Saudi Air Force (RSAF)
Rules of Engagement (ROE), 301

S

SA. *See* Situational awareness (SA)
SAB. *See* Scientific advisory board (SAB)
SADL. *See* Situation Awareness Data Link (SADL)
SAF/AQC. *See* Secretary of the Air Force for Acquisition (SAF/AQC)
SAGE. *See* Semi-Automatic Ground Environment (SAGE)
SAMP. *See* Single Acquisition Management Plan (SAMP)
Santa Fe Institute (SFI), 7
SCDR. *See* Shared cross-domain resource (SCDR)
Scientific advisory board (SAB), 424
SE. *See* Systems engineering (SE)

Secretary of the Air Force for Acquisition (SAF/AQC), 197
Security guard, 261
Security management, 260
    agility and flexibility, 261
    bilateral, 260
    coalition, 260
    enclave, 261–262
    robustness and availability, 261
    speed and performance, 261
Selection, 48, 53, 57
    adaptive system, 54–55
    agent, 54
    questions, 57
    requirements, 53
    strategy, 54
Semi-Automatic Ground Environment (SAGE), 2
Service-oriented architecture (SOA), 86, 206, 257, 331, 343, 397, 425–426. *See also* Architecture (s)
    artificial dependencies, 427
    characteristics of services, 426–427
    coupling, 427
    loose coupling, 427
    RCA features, 427–428
    real dependencies, 427
    SOA framework, 426, 427
    transformation via, 427
SFI. *See* Santa Fe Institute (SFI)
Shared cross-domain resource (SCDR), 261
SIAP. *See* Single Integrated Air Picture (SIAP)
Simple Mail Transfer Protocol (SMTP), 343
Simple network management protocol (SNMP), 280, 354
Simple Object Access Protocol (SOAP), 429
Single Acquisition Management Plan (SAMP), 233
Single Integrated Air Picture (SIAP), 314
Sinusoids, 6n
Situational awareness (SA), 36, 311–312
    comprehension, 311
    perception, 311
    projection, 311
Situation assessment, 310, 311, 324
Situation Awareness Data Link (SADL), 385
Skill, 39n
SM. *See* Strategic management (SM)
Small world, 16
SMEs. *See* Subject matter experts (SMEs)
SMIO. *See* Space Systems Management Integration Office (SMIO)

SMTP. *See* Simple Mail Transfer Protocol (SMTP)
SNMP. *See* Simple network management protocol (SNMP)
SOA. *See* Service-oriented architecture (SOA)
SOAP. *See* Simple Object Access Protocol (SOAP)
Social network theory, 302
SOF. *See* Special Operations Forces (SOF)
SoS. *See* System(s)—of systems (SoS)
Space Systems Management Integration Office (SMIO), 420
Special Operations Forces (SOF), 304, 384
SPOs. *See* System(s)—program offices (SPOs)
Stakeholder Analysis, 66, 83, 115–116
State, 190
State space, 333
STP. *See* Strategic Technical Plan (STP)
Strategic management (SM), 233, 238, 242–243
    goals, 239–242
    key components, 238
    need for assessment, 239
    planning approaches, 242
    process—10 step model, 244
    vision, 238
Strategic Master Plan, 412
Strategic planning, 243
    adjusting to environment, 247
    closing the gap, 245–246
    enterprise, 404–405
    implementation and assessment, 246–247
    outcomes, 243–245
    progress measurement, 245
Strategic Technical Plan (STP), 11, 13, 14, 23, 67, 69, 90, 398
    architectural patterns, 11
    architecture control, 14
    attributes, 11
    complexity science, 15, 16
    efficiency–innovation trade space, 14
    loose coupling, 12
    network science, 15, 16
    phase change in networks, 17
    sample command, 14
Strategy, 48n
Strengths, weaknesses, opportunities, and threats (SWOT), 245
Structure, 108
Structure mapping, 372
Subject matter experts (SMEs), 117, 151
Subversion (SVN), 271, 272

Supervised classification, 325
Supervised learning. *See* supervised classification
SV. *See* System(s)—view (SV)
SVN. *See* Subversion (SVN)
SWOT. *See* Strengths, weaknesses, opportunities, and threats (SWOT)
Synthesis, 36n, 40–41. *See also* Complementarity
System(s), 4, 162, 191n
    acquisition, 387–388
    analysis, 39, 40
    of Systems (SoS), 33, 162, 190, 234, 396
    perspective, 33n
    program offices (SPOs), 286
    thinking, 33n, 40, 243. *See also Synthesis*
    view (SV), 199, 420
System's context, 234
    high-level interface, 235
Systems engineering (SE), 2, 162, 164, 234
    adjectives used, 38
    classical, 38

T

T&E. *See* Test and evaluation (T&E)
Tactical Data Information Link (TADIL), 385
Tactics, Techniques, and Procedures (TTPs), 374, 419
TADIL. *See* Tactical Data Information Link (TADIL)
Tai-Chi symbol. *See* Yin Yang
TAP. *See* Telocator alphanumeric (input) protocol (TAP)
Targeting C2 Architecture (TC2A), 419
Targets-of-interest (TOI), 328
Task, post, process, use (TPPU), 431
Task, process, exploit, and disseminate (TPED), 431
TC2A. *See* Targeting C2 Architecture (TC2A)
Technical view (TV), 199
Technology readiness level (TRL), 276
Telocator alphanumeric (input) protocol (TAP), 280
TELs. *See* Transport Erectile Launchers (TELs)
Test and evaluation (T&E), 208
Thinking, 33n
Time-Sensitive
    Target (TST), 383
    team decision making, 379
TOI. *See* Targets-of-interest (TOI)

Top–down approach, 396, 397. *See* Capabilities-Based Engineering Analysis (CBEA)
Top–down method, 396, 397, 414–415
TPED. *See* Task, process, exploit, and disseminate (TPED)
TPPU. *See* Task, post, process, use (TPPU)
Traditional Systems Engineering (TSE), 2, 3, 66, 164, 173, 182, 396
    Blind Men and the Elephant, The, 182
    tools and models, 402–403
    understanding of interactions, 403
Training, tactics, and procedures (TTP), 419
Transportation Security Agency (TSA), 401
Transport Erectile Launchers (TELs), 319
TRL. *See* Technology readiness level (TRL)
TSA. *See* Transportation Security Agency (TSA)
TSE. *See* Traditional Systems Engineering (TSE)
TSE and ESE differences, 400
    BSAHLS, 401
    decomposition from top down, 402
    enterprise management, 402
    JOC level, 401
    JopSCs model, 401
    key element, 402
    systems engineers, 402
TSTs. *See* Time-Sensitive—Target (TST)
TTP. *See* Training, tactics, and procedures (TTP)
TTPs. *See* Tactics, Techniques, and Procedures (TTPs)
TV. *See* Technical view (TV)

U

UAV. *See* Unmanned Aerial Vehicles (UAV)
UDDI. *See* Universal Description, Discovery, and Integration (UDDI)
UML. *See* Unified modeling language (UML)
Unified modeling language (UML), 249
Uninterruptible power supply (UPS), 279
United States (U.S.), 235, 299
United States Air Force (USAF), 235, 397
    capabilities road map, 397
    DoDAF, 397
    mission thread analysis, 397
United States Department of Defense (U.S. DoD), 100
    CBP, 100, 102
Universal Description, Discovery, and Integration (UDDI), 429

Unknown unknowns, 25
Unmanned Aerial Vehicles (UAV), 332, 359, 415
Unorganized complexity, 6, 45
UPS. *See* Uninterruptible power supply (UPS)
U.S. *See* United States (U.S.)
USAF. *See* United States Air Force (USAF)
U.S. DoD. *See* United States Department of Defense (U.S. DoD)
U.S. Naval Reserve (USNR), 364
USNR. *See* U.S. Naval Reserve (USNR)

V

Value-focused thinking (VFT), 150
Variation, 48, 49, 435
  exploration–exploitation trade, 50
  questions, 56
Verification, validation, and accreditation (VV&A), 227
VFT. *See* Value-focused thinking (VFT)
Video teleconferences (VTC), 368
Virtual private network (VPN), 281
VPN. *See* Virtual private network (VPN)
VTC. *See* Video teleconferences (VTC)
VV&A. *See* Verification, validation, and accreditation (VV&A)

W

W3. *See* World Wide Web (W3)
W3C. *See* WWW Consortium (W3C)

WAN. *See* Wide area network (WAN)
Warfare, future, 299. *See also* Battlefield
  battlespace, 299
  key aspects, 299
  response, 300
Weapon–target pairing (WTP), 372
Web browser, 9, 13
Web Ontology Language (OWL), 343
Web Services Definition Language (WSDL), 355–356, 390
Web Services Description Language (WSDL), 291, 433
Wide area network (WAN), 354
WOL. *See* Web Ontology Language (OWL)
World Wide Web (W3; WWW), 3, 32, 399
WSDL. *See* Web Services Definition Language (WSDL); Web Services Description Language (WSDL)
WTP. *See* Weapon–target pairing (WTP)
WWW. *See* World Wide Web (WWW)
WWW Consortium (W3C), 18

X

XML. *See* eXtensible Markup Language (XML)

Y

Yin Yang, 60n, 61